THE QUALITY
TECHNICIAN'S HANDBOOK

ABOUT THE AUTHOR

Gary Griffith is presently employed as manager of Process Control Engineering at Allied-Signal Aerospace Company, AiResearch Los Angeles Division, Torrance, California. He is also president of Griffith Training, a quality control consulting firm in Huntington Beach, California. Gary conducts public seminars and in-plant courses in geometric tolerancing and in statistical process control methods and implementation. He is a senior member of the American Society for Quality Control, an ASQC Certified Quality Engineer, and a Fellow of the Institute for Advancement of Engineering.

During his 24 years of experience Gary has held a variety of quality control positions, including inspector (various), inspection supervisor, process control supervisor, quality engineer, quality manager, and process control engineering manager. He has also taught courses in quality control, drafting, and machine tool technology at El Camino College. He is a firm believer in quality practices and in the promotion of national quality awareness in order to improve the posture of American businesses.

THE QUALITY TECHNICIAN'S HANDBOOK

SECOND EDITION

Gary K. Griffith

PRENTICE HALL, Englewood Cliffs, New Jersey 07632

Library of Congress Cataloging-in-Publication Data

Griffith, Gary.
    The quality technician's handbook / Gary K. Griffith. -- 2nd ed.
        p.   cm.
    Includes indexes.
    ISBN 0-13-747452-0
    1. Engineering inspection.   2. Machine-shop practice.   I. Title.
TS156.2.G75   1992
    658.5'68--dc20   91-32468
                                                                CIP

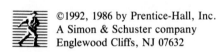

©1992, 1986 by Prentice-Hall, Inc.
A Simon & Schuster company
Englewood Cliffs, NJ 07632

Printed in the United States of America

10 9 8 7 6 5 4 3 2 1

ISBN 0-13-747452-0

PRENTICE-HALL INTERNATIONAL (UK) LIMITED, *London*
PRENTICE-HALL OF AUSTRALIA PTY. LIMITED, *Sydney*
PRENTICE-HALL CANADA INC., *Toronto*
PRENTICE-HALL HISPANOAMERICANA, S.A., *Mexico*
PRENTICE-HALL OF INDIA PRIVATE LIMITED, *New Delhi*
PRENTICE-HALL OF JAPAN, INC., *Tokyo*
SIMON & SCHUSTER ASIA PTE. LTD., *Singapore*
EDITORA PRENTICE-HALL DO BRASIL, LTDA., *Rio de Janeiro*

To my wife, Sharon, and my family
for their understanding while I worked on
this project at home and for their continuous support.

Contents

3 BLUEPRINTS 65

4 COMMON MEASURING TOOLS AND MEASUREMENTS   133

11 PROBLEM-SOLVING TECHNIQUES 444

GLOSSARY 452

APPENDIX TABLES AND CHARTS 464

Preface

This book acquaints its readers with a variety of the basic skills that contribute to outgoing quality in the mechanical trades. Most books cover only specific areas of study in depth and do not make many comparisons with other related areas. This book introduces a variety of areas and includes the "up-front," need-to-know basics of each area. Operators, apprentice machinists, inspectors, quality technicians, and junior quality engineers will find the contents useful as a quick reference that will aid them in their understanding of the subjects covered.

This book is designed specifically to reach operators, machinists, inspectors, and others who are in training, have had unidirectional experience, or who want to study the mechanical trades for the first time. It is an informative guide that aids them in their task of producing, inspecting, and controlling quality products. This text will also help other personnel in manufacturing, industrial engineering, and purchasing who contribute to quality.

Chapter discussions are arranged so that related subjects are covered with continuity, and wording is as simplified as possible without degrading the meaning. Numerous illustrations help the reader to understand the material. In many instances, step-by-step instructions are given to lead the user through some of the techniques.

Several changes have been made in this second edition: (1) an expanded introduction has been added to quality control in Chapter 1; (2) additions and clarification of text and problems have been made in Chapter 2, "Shop Mathematics"; (3) additional coverage on coordinate measuring machines and pneumatic gaging has been added to Chapter 4; (4) old Chapter 6, which covered basic casting layout techniques, has been deleted; (5) "Graphical Inspection Analysis" has replaced Chapter 6 in sequence; (6) Chapter 7, "Quality Costs," has been expanded and has more examples, (7) Chapter 9 has been retitled Lot-by-Lot Acceptance Sampling Plans; (8) old Chapter 11, "Quality Circles," has been retitled "Problem Solving Techniques" and the text has been expanded; (9) old Chapter 12, "Introduction to Reliability," has been deleted; (10) review questions have been added to each chapter; and (11) a glossary of terms and definitions has been added just before the Appendix.

The contents of this book have been very effective as a training aid within several divisions of one company, and they have also been helpful as a reference guide for persons taking ASQC Certified Quality Technician and Certified Mechanical Inspector examinations. It is one book that can be kept in a machinist's or inspector's toolbox as a reference when needed to provide a quick answer to some of the problems that come up in the shop.

Product quality is everyone's job, and everyone should be aware (at least basically) of the ways quality can be achieved. I sincerely hope that this book will be a contribution to that goal.

Gary K. Griffith

Acknowledgments

I thank my friends and co-workers for their assistance, particularly Arvind Shah who encouraged and helped me while I was working on this project.

Many others took part in the development of this book, offering editorial comments, and I also thank them.

A special thanks goes to George Pruitt of Technical Documentation Consultants, Inc., Ridgecrest, California, for Chapter 6 on graphical inspection analysis.

A special thanks also goes to Eugene Barker of Douglas Aircraft, Long Beach, California, for his input in Chapter 8, Inspection Planning.

I would appreciate any comments regarding the improvement of this book as it is my concern to provide a worthwhile body of knowledge for quality.

G.K.G.

THE QUALITY
TECHNICIAN'S HANDBOOK

1

Introduction to Quality

Quality is an important factor for the survival of any business. It directly impacts the other factors, such as cost, on-time delivery, and market share. Poor quality increases costs, delays delivery, and if there is competition, can decrease the company's market share.

The word *quality* has taken on a variety of definitions:

- Fitness for use
- Conformance to requirements
- Meeting customers' expectations
- Doing it right the first time
- Degree of excellence

In the broad sense, these definitions are all acceptable. The idea they try to convey is most important. You must first know what quality means to the customer, and then consistently ship that level of quality to them. Quality is not necessarily a tight tolerance, a shiny surface, or a perfect fit. Quality is what the customer wants, needs, and is willing to pay for.

Even though there may be a quality organization in the company, the responsibility for quality is everyone's. In the broad sense, the following are steps to quality.

1. Determine the customers' wants and needs (the goal).
2. Communicate those needs effectively to all concerned.

3. Assess your ability to meet the needs (or your ability to change to meet the needs).
4. Have an overall pro-quality attitude.

Once the customers' needs have been clearly defined, the product must next be designed accordingly. The quality of design is vital. The design activities are where quality begins. The phrase "Make it like the blueprint" loses meaning if the blueprint does not stem from a quality design or communicate design requirements effectively. Once you have a quality design, the next step is conformance.

Conformance to requirements has a lot to do with the following factors:

- The ability to meet design requirements
- A controlled process
- A well-trained and motivated work force
- Management support

COMMUNICATION

The largest factor in achieving quality is communication. Poor quality is usually the result of a lack of communication of some sort. A few examples of this are

1. Poor communication between:
   a. Consumer and producer (wants and needs are not well defined).
   b. Contracts and product engineering (an internal communication gap that can cause poor design quality).
   c. Engineering and manufacturing (drawings and specifications that do not clearly communicate design requirements).
2. A lack of communication between departments.
3. A lack of feedback from the field.
4. Poor (or no) training (the "how to" communication is open looped).
5. Poorly written procedures or work instructions.
6. No written procedures or work instructions.

THE QUALITY ORGANIZATION

The *quality organization* is actually a combination of all personnel and functions in the company. The traditional definition of the quality organization led many to believe that they had to have a separate department of skilled people who had direct responsibility for the various levels of quality in the company. All functional departments within a company have responsibilities for quality, not just one. Quality is not achieved just by having skilled monitors or a policing function. Quality is a multidisciplined team effort that begins with the identification of customers wants and needs and extends through all phases of production or service beyond delivery of the product or service to the customer. Various departments in the company play a key role in quality from the identification of customers requirements through field service. The following is a review of the various company functions and how they contribute to quality. It is important to note that most of these functions are best performed concurrently in order to get all departments involved as early as possible (or practical) in the process. Another aspect of quality is the realization that each entity within the company has internal suppliers and customers. All personnel involved with the product (or service) have a supplier and a customer in a "total quality" organization. In each step of the process of bringing a product or service to delivery, everyone involved has someone who supplies them with input and someone who receives their output. In manufacturing processes, for example, each operation deals with the output of a previous operation, and provides input to the next operation. People should be sensitive to their internal customers' needs, and they should identify their needs to their internal supplier.

Marketing (Sales)

Marketing (sales) personnel are responsible for identifying all customers' wants and needs. Considerable efforts should be made to understand the customer's application, special requirements, or particular performance criteria of the product. It is then the responsibility of marketing to communicate those requirements of the customer clearly and completely to their internal customers (such as product engineering) so that the design of the product meets the customer's requirement.

Product (or Design) Engineering

The quality of design is a direct responsibility of product engineering. The fit, form, function, and reliability of a product is dependent on the adequacy of the design and production process. Product engineering is responsible to design the product to meet the customer's requirements, identify the appropriate standards for the product, and prepare engineering drawings and specifications that communicate the requirements clearly and completely. Product engineering has been improved with recent methods in statistical experimental designs and other statistical methods that assist the product engineering department toward design improvements and appropriate tolerancing of the product. Another recent improvement in the design process has been called concurrent (or simultaneous) engineering. As early as practical in the design phases, other functional departments, such as quality assurance, manufacturing, manufacturing engineering, and process engineering should be involved so that the inspection, processing, process control, materials, producibility, and other requirements (or potential problems) can be identified. One of the direct internal customers of product engineering is manufacturing operations.

Process Engineering

This function exists in some companies where there are engineers who are responsible for the design and specifications for special processes such as heat treatment, brazing, welding, chemical processes, coating, and plating. Skills in chemical and metallurgical sciences and process design are required to ensure that the quality of these processes is controllable, capable, and will not have an adverse effect on previous operations, or the function or reliability of the product. With the advent of statistical experimental designs, many of these processes can be designed to provide a more uniform output (or with less variability). Process engineering supports their customer, manufacturing operations, by designing processes used by production to add that specific value to the product.

Manufacturing Engineering

Manufacturing engineering is responsible for supporting their customer (manufacturing) by identifying and selecting the most capable manufacturing processes, including machines, equipment, fixtures, cutting tools, processing sequences, speeds, feeds, and other factors in the production process. Work instructions and manufacturing sketches are prepared as necessary to ensure that the manufacturing sequence and proper machine tools or equipment are used as they were designed. Ongoing shop floor support is given to ensure that the production processes are controlled and capable of meeting specification requirements.

Purchasing

Purchasing is responsible for identifying and selecting qualified suppliers for all raw materials, semifinished products, and finished products, where the decision has been made to purchase these materials or products. Equal factors of price, delivery, and quality should be used in the supplier selection process, and suppliers should be monitored to verify that their processes are controlled and capable of meeting the company's specification requirements. Control over the quality of incoming raw materials, processing, and purchased products is an important aspect of quality. Purchasing's direct customer is manufacturing, the people who use the raw materials, purchased products, or processing to make the product.

Production

Quality control is a regulating activity that utilizes various methods during the production process in order to control the output variability of the process and produce quality products. The quality control function in the past was one that was confusing to many people, because it was presumed to be a responsibility of the quality assurance department. The control of quality (or quality control) is most effective when it is a clearly defined responsibility of production. Production supervision is responsible to ensure that operators are given the proper environment, machines, methods, materials, measuring equipment, training, and instructions in order to produce product and control the output of the process. Production operators are responsible for monitoring the quality of the output of the process, and take action to ensure that quality products are produced and sent on to the next operation (their customer).

Quality Assurance

Quality assurance is primarily a planning and analysis function that involves many people. The assurance of quality can be viewed as all of the necessary personnel and activities which are intended to provide assurance that the quality of the product delivered to the customer will meet the customer's requirements. The assurance of quality begins in the early stages of identifying the customer's requirements and extends, in stages, through shipment of the product to the customer. In the product design stage, for example, design reviews that are held at various levels of the design of a product are an assurance function. Several different people are, or should be, involved in design reviews in order to assess the adequacy of the design to meet requirements and to begin identifying needs for producing and evaluating the product in accordance with the design. The analysis functions of quality assurance are intended to provide analysis and feedback on both the design and production processes. These functions are also performed by many different people in the company. Some form of analysis is used in all phases of design and production of the product, such as experimental models, process yields analysis, statistical experimental designs, chemical and metallurgical analyses, quality data collection and reporting, and many other methods that provide feedback for design and process improvement.

Quality Assurance Department

The quality assurance department is responsible to provide support in terms of specific technical planning and analyses at various stages of the design and production process. Initial quality planning is a company-wide activity where the quality assurance department plays a key role. Initial quality planning functions performed by the quality assurance department include those specific functions which (1) identify quality measurement, corrective, and reporting requirements such as inspection sampling and corrective action requirements; (2) identify the appropriate inspection points in the process and levels (or amounts) of inspection; (3) select appropriate measuring and gaging methods for the product and training personnel to use these methods (also taking appropriate actions where required measurements exceed the state of the art), (4) establish a quality reporting system that will monitor process yields and quality costs, and reporting those yields and quality costs to those concerned for the purpose of improvement; and (5) plan and perform system and product audits for the purpose of improvement. Quality assurance also provides support during the process, such as maintaining calibration of measuring, gaging, and test equipment; establishing and controlling bonded areas for segregation of nonconforming materials and products; and performing applicable lot-by-lot inspections on the product. For lot-by-lot inspection points, quality assurance traditionally provides inspection functions (such as receiving inspection of raw materials, or purchased products from suppliers). With today's emphasis on product and process control, lot-by-lot inspections (by quality assurance departments) during production operations are being reduced and replaced with process control inspection by operators.

Storage

Stores are responsible to maintain products that are in inventory in such a manner that the quality of the product will not be degraded. Only products that are per specification requirements should be stored, and they should be stored in a way which will ensure that products are not damaged or altered from their as-built condition. Many things can go wrong with products in storage, depending on the type of product and those factors that will adversely affect quality. Storage records should be maintained and correlated with products in stores in order to maintain applicable traceability requirements. Issuance of correct materials requested is another quality responsibility of storage departments.

Shipping Department

The shipping department is responsible for (1) packaging the product in accordance with company and customer requirements for the purpose of protecting the quality of the product, (2) identifying and adhering to appropriate shipment locations and schedules, and (3) including with the product all required documentation.

Field Service

Another important aspect of quality is field service. Supporting the product in the field (after shipment to the customer) is a vital aspect of any business. Field service has responsibilities that include working with the customer on various problems, such as identifying and sometimes analyzing field failures, performing repairs, and identifying and communicating customer complaints or other needs to appropriate company personnel. There are several other functions that field service performs, depending on whether products delivered are consumables, repairable, replaceable, and so on, but one of the most important quality aspects of field service is the feedback of customer quality problems to the company. This important activity helps close the loop between customers wants and needs and the quality of the product delivered.

Summary

To summarize, it is obvious that quality assurance is not a department or a single group of people dedicated to a monitoring function. Quality assurance consists of all the related activities that occur from the initial review of the customer's wants and needs to use of the product in the field. Quality involves everyone in the company working through a team approach. The words *total quality* used in many communications, articles, and books are trying to convey this message.

PEOPLE IN QUALITY AND PRODUCTION

The Machine Operator

The operator plays a key role in quality, with the responsibility to

1. Follow operation instructions.
2. Make machine setups.
3. Attempt to meet production standards.
4. Check his or her own parts.
5. Make adjustments as necessary.
6. Understand his or her machine.
7. Be alert for problems.
8. Make sure of his or her material.
9. Report problems to his or her supervisor.
10. Plot statistical control charts (where used).

Any misunderstanding of the points listed above can be a cause for producing defective products. The operator has quite a job on his* hands. It is much more than simply putting the parts in the machine and pressing the *go* button.

There are causes of defects that are not necessarily controllable by the operator. To produce good parts, the operator must have good

Machines	Equipment	Tools	Material
Work standards	Specifications	Work instructions	

The operator needs to make a sound attempt to achieve production standards without letting quality fall. He should report quality problems to his super-

* The discussion here and throughout the book is intended to include both women and men.

visor as they occur, and try to help solve them. He should fully understand what is expected of him prior to pressing the *start* button.

The operator should continuously strive to strengthen his abilities, and the company should strive to support him with the "tools" to do the job. The operator should have an overview of quality knowledge in order to perform his duties. He has considerable influence on outgoing quality, because he is making the part.

The Machinist

The machinist is a person who can be handed raw materials and specifications, can be turned loose among a group of machines, and can come up with a finished product. The machinist is a "Jack of all trades" who generally gets along well with any machining situation. The machinist also needs to understand the various techniques of obtaining and measuring quality.

Machinists are generally the only inspectors of the products and must be aware of measurement techniques, drawing interpretation, and other areas related to quality in order to perform their function. They often have to "rig" setups on short-run components, as opposed to machine operators, who usually have a standard setup with instructions. The machinist has a wider scope of responsibility than an operator because he is *assumed* to be a person who has the skills and training to work alone, and if support is needed, he will ask for it.

The Inspector

Inspection is done by everyone in some form, whether it is inspecting your automobile for nicks in the doors, or shopping for groceries, or simply noticing that something is missing from your dresser top. In industry, the professional inspector plays a vital role. His main function is to *appraise* the product, report his findings, and assist in quality improvement wherever possible.

Knowledge and experience

An inspector needs to have a wide variety of knowledge and experience. The most important areas are

- Quality in general
- Quality standards
- Shop mathematics
- Sampling inspection
- Basic inspection planning
- Blueprint reading and interpretation
- Measuring tools, gages, and techniques

- Record keeping
- Basic machining methods

Other than these technical areas of knowledge, the inspector must also know the ways in which he should perform his function. There are some key words and phrases that express this.

Appraise. Inspection is simply a matter of comparing a part to its specifications. You must thoroughly understand the specification, accurately measure (evaluate) the part, and make the decision to accept or reject the part strictly per the specification and applicable standards.

Consistency. It is important that inspectors be consistent in what they do. It is most important that they consistently accept parts that are per specifications and reject parts that are not.

Bias. This is something that must be avoided at all times. Bias is basically the act of making something happen because you want it to happen. Bias must be avoided in measurement and sampling.

Accuracy. Accuracy (in measurement) means obtaining an unbiased true value. This is most nearly achieved by using calibrated tools and good measuring techniques, selecting the proper tool for the measurement, measuring clean parts, and avoiding other possible errors in measurement such as heat, burrs, dust, wear, geometry, and manipulation errors.

Inspector/shop relations

The inspector must try to maintain good relations with shop personnel at all times. People are very proud of their work, and when inspection rejects it, there is usually some discontent as a result.

One way an inspector can keep good relations is by advising the shop personnel that it is his job to reject defective products, and to assure them that it is not a personal matter. The inspector should not have an attitude of ''I got you'' at any time. He should feel good about stopping defective products, but at the same time, feel bad that he had to do so. The primary goal of all inspectors should be to inspect products as thoroughly as required, and find no defects.

If shop personnel understand that the inspection function is to help them improve quality (not to point fingers), a better relationship can be achieved.

Material review and the inspector

The material review function is to decide on what is to be done with defective products and to take corrective action. This review of the material comes after the inspector has rejected the product.

The decision can go one of four ways: scrap, rework, repair, or use as is. The latter decision ("use as is") is the one that inspectors generally have a hard time dealing with. Some inspectors view this decision as a waste of their time and efforts in finding the defects. This is simply not true.

Decisions on defective products are made based on several reasons such as: classification of defects, customer needs, quality, and others. The "use as is" decision is one that only the material review board can make, and should not be viewed by an inspector as a waste of his time. Parts that are *scrap* must be scrapped, parts that are *reworkable* can be reworked, and parts that can be *used as is* often are.

Hints for good inspection

1. Consistently accept products that are per specifications.
2. Consistently reject products that are not per specifications.
3. Understand the specifications used.
4. Use only the latest change drawings.
5. Never memorize drawings; review them each time (they frequently change).
6. Use calibrated tools and instruments.
7. Don't guess—*know*.
8. Maintain a positive attitude.
9. Keep accurate and complete reports on quality.
10. Attempt to help correct problems.
11. Stick strictly to inspection and sampling plans provided.
12. Clear up "gray areas" with your supervisor.
13. Avoid arguments with the shop; be open for discussion.
14. Seek training in your weak areas.
15. Be open-minded about what you are doing and willing to learn.
16. Ask questions at all times when you are not certain.
17. Be a part of the team.
18. When describing a defect, write rejections that are short, simple, and complete.

Types of inspection

Types of inspection include: source, receiving, in-process, final (components), and final (assembly).

Source inspection. When companies buy products from another company (source), there is sometimes the need for one of their own inspectors to inspect products at that source. The source inspector often works at the vendor's facility inspecting products just prior to shipping them to his company.

Receiving inspection. The receiving inspector plays a major inspection role. His effectiveness is often measured by the quality of raw materials on the shop floor, or purchased parts in assembly. If the receiving inspector is effective, defects can be caught early, before a company invests more time and money processing defective material or parts.

In-process inspection. In those companies with a need for in-process inspection, the process inspector also plays an important role. The best time to find defects is *as early as possible* in the process. Process inspectors are generally the ones who give that all-important first-piece setup approval. This is important because the permission to run many more of the same is either given or denied. If the first piece is bad, there is a strong possibility that the rest of the run will also be bad.

Final inspection (component). This is the last chance to find defective products prior to final assembly. At this point, final inspection can stop defects that have not been detected thus far in the process. It is more costly to find defects at this point because the company has already invested time and money into processing the parts.

Final inspection (assembly). This inspection is generally a visual inspection for obvious assembly problems such as whether or not the correct components have been used and assembled properly. This is the last opportunity to find defects that may have leaked through the process before the product is shipped to the customer.

The Auditor

The duties of an auditor are very important to outgoing quality and to quality improvement. There are two types of audits: *product* and *system*.

Product audits

A product audit requires the auditor to review the product dimensionally as an inspector would, comparing the product to the specification. Sometimes an auditor will be required to take the same sample an inspector took to audit for inspection effectiveness (methods, techniques, tools, sampling plans).

System audits

A system audit requires the auditor to review systems and procedures for both quality and manufacturing processes. He would make himself totally aware of the procedure and review what is supposed to be done against what is actually being done on the floor. Auditors have been referred to as inspectors of a slightly higher degree. They generally report directly to higher management and have the authority to audit virtually any product, or system (procedure), and report their findings to management.

The auditor shares about the same predicaments as those of inspectors. Auditors should not be biased and should not be unprofessional (e.g., argue with people). They should try to understand the reasons for conducting audits and be patient about improvement (knowing that things are not improved overnight in most cases). Auditors should get along well with people and not have an "I got you" attitude. They must earn credibility and keep that credibility strong. They do this by

1. Always stating facts.
2. Reporting accurately and clearly. Reports should be simple and complete.
3. Understanding the important areas.
4. Using their time wisely.
5. Showing findings that no one can dispute (or doubt).
6. Following audit plans except where obvious discrepancies (that are not in the plan) require attention.

Some key hints for auditors to be successful are the following:

1. Know the standards, or specifications.
2. Understand that you are trying to help people, not point fingers.
3. Know the measurement tools and techniques.
4. Report your findings as either deficient or acceptable on a consistent basis.
5. Work with people to help them correct their deficiencies.
6. Make your reports clear and concise so they are easily understood.
7. State the facts at all times.
8. Never act on opinions or guesses. Know the facts.
9. Follow audit plans to the extent practical.
10. Make suggestions for improvement.

Auditors who do all of the above on a consistent basis will earn credibility and keep it. Reports will be believed and acted upon, and they will get along very well with people. It often helps to explain thoroughly what you will be doing during an audit and invite the supervisor of that particular area to come along with you. Quality audits can mean a lot to improving quality in a company, and the auditor is the person who can "make it all happen."

Follow-up as a result of an audit

Once an audit has been performed, the task of correcting deficiencies is the responsibility of the function that was audited. Follow-up, to assure that deficiencies have been corrected, is usually done by management or, in some cases, the auditor.

REVIEW QUESTIONS

1. Name the largest single factor in achieving quality.
2. Name a key word that describes the function of inspection.
3. What are the four common material review decisions?
4. In a "quality organization," what function has the direct responsibility for supplier quality?
5. Give an example of an internal customer.
6. The identification of customer's wants and needs is a _____ responsibility.
7. What organization is engineering's internal customer?

8. Quality control is the responsibility of _____.
9. Name two activities that are a quality assurance department responsibility.

10. An important function, after the product has been shipped, is _____ _____.

11. Name the two types of audits.
12. Who is responsible for correcting deficiencies found during a quality audit?

2

Shop Mathematics

Inspectors, operators, and machinists are all involved in the use of shop math at one time or another. Sometimes the understanding of shop math can be the difference between getting the job done or not. Shop math is encountered in a variety of problems on the shop floor from tapers, to threads, drill sizes, tolerances, and hole patterns. The list of uses of shop math related to the mechanical industry is very long.

In the following pages we outline the necessary shop mathematics for most of the problems encountered on the shop floor. In these pages we discuss

Fractions	Triangles
Equations	Percentages
Circles	Volumes
Decimals	Angles
Perimeters	Areas
Ratios	Tapers

These subjects encompass a good basis for shop math as used by inspectors, operators, and machinists. There is much more to shop math than these subjects, but when "push comes to shove," most of them can be applied to solve the problem.

ADDING AND SUBTRACTING FRACTIONS

The Least Common Denominator

When adding or subtracting fractions it is necessary to be able to find the *least common denominator* (LCD). The LCD is the smallest denominator in the group of fractions to be added or subtracted in which all of the denominators will evenly divide.

Example

$$\frac{3}{8} + \frac{5}{16} + \frac{1}{4}$$

The LCD in the problem above is 16 because it is the least number that is common to the other denominators (4 and 8). The only step now is to change $\frac{3}{8}$ and $\frac{1}{4}$ to sixteenths, then add all three fractions with a common denominator of 16.

When working with fractions, you must remember to make them alike before attempting to add or subtract them. You cannot add apples to oranges and say that you have a box of apple–oranges. You either have apples, or oranges.

Finding the LCD

There are methods for finding the LCD in any given problem. One of them is simply inspection.

Example

$\frac{1}{4} + \frac{1}{2}$. You know that $\frac{1}{2}$ is $\frac{2}{4}$, so

$$\frac{1}{4} + \frac{2}{4} = \frac{3}{4}$$

Another method is calculation.

Example

$$\frac{1}{3} + \frac{1}{4} + \frac{3}{8}$$

See if the largest denominator (8) can be evenly divided by the other ones (3 and 4). No, it cannot.

Now multiply the largest (8) times the smallest (3) and see if 24 is common. In this case, 24 is the common denominator.

Converting the fractions to the LCD found

Once an LCD is found, convert the fraction to it in three steps.

Step 1. Divide the denominator into the LCD.

Step 2. Multiply this number times the numerator.

Step 3. Place the answer in step 2 over the LCD for an equivalent fraction.

Example

Converting $\frac{3}{16}$ into an LCD of 64.

Step 1. $16\overline{)64}$ giving 4

Step 2. $3 \times 4 = 12$ *Answer:* $\frac{12}{64}$

Example

$$\frac{3}{8} + \frac{5}{16} + \frac{3}{64} + \frac{1}{3}$$

Solution LCD is 192 (found simply by multiplying the lowest denominator, 3, by the highest denominator, 64).

Conversion: $\frac{3}{8}$ to 192nds is $192 \div 8$ then $\times 3 = \frac{72}{192}$

$\frac{5}{16}$ to 192nds is $192 \div 16$ then $\times 5 = \frac{60}{192}$

$\frac{3}{64}$ to 192nds is $192 \div 64$ then $\times 3 = \frac{9}{192}$

$\frac{1}{3}$ to 192nds is $192 \div 3$ then $\times 1 = \frac{64}{192}$

Next, add the numerators: $72 + 60 + 9 + 64 = 205$.

Now, you have an answer stated as an *improper fraction*, $\frac{205}{192}$. An improper fraction is one where the numerator is larger than the denominator. To convert this to the final answer (which will be a mixed number), divide the denominator, 192, into the numerator, 205, and put the remainder over the denominator.

$$\frac{205}{192} = 1\frac{13}{192} \quad Answer$$

Reducing Fractions

Sometimes you will add or subtract fractions and get an answer that is not in proper form (such as $\frac{2}{4}$). The fraction $\frac{2}{4}$ is not in proper form because it can be reduced to $\frac{1}{2}$. Reducing some fractions can get complicated unless you do it in a step-by-step manner.

Example

Reduce $\frac{20}{160}$ to lowest terms.

Solution   By inspection, find the largest number that will divide evenly into both 20 and 160. In this case it is 20. Now divide into both 20 and 160.

$$20 \div 20 = 1$$

$$160 \div 20 = 8 \qquad \text{The fraction } \frac{20}{160} \text{ reduces to } \frac{1}{8}.$$

Another way is to start dividing the number 2, 3, or 4 into both numerator and denominator, and continue to do this until the fraction is in lowest terms.

Example

$$\frac{16}{64} = \frac{8}{32} = \frac{4}{16} = \frac{2}{8} = \frac{1}{4}$$

As you get better, some of the steps above can be omitted.

Adding and Subtracting Fractions Using Cross Multiplication

Another way to add or subtract unlike fractions is by cross multiplying. The easiest method to add or subtract "unlike" fractions (where a LCD is necessary) is cross multiplication. By cross multiplying the two fractions and then reducing the "new fraction" to lowest terms, the answer is easily found. Other methods of finding the LCD are by inspection or deduction.

Example

$$\frac{3}{8} + \frac{1}{4} = ?$$

(The answer here is obviously $\frac{5}{8}$, since

$$\frac{1}{4} = \frac{2}{8} \qquad \left(\text{and } \frac{3}{8} + \frac{2}{8} = \frac{5}{8} \right)$$

To cross-multiply, simply follow these steps. Before long, with practice, you can perform these steps quickly in your head.

Step 1. Set up the problem as shown.

$$\frac{3}{8} + \frac{1}{4} = \frac{\text{(new fraction)}}{}$$

Step 2. Multiply the denominator (8) of the first fraction times the numerator (1) of the second fraction and put it in the numerator of the ''new fraction.''

$$\frac{3}{8} + \frac{1}{4} = \frac{8}{\rule{3cm}{0.4pt}}$$

Step 3. Multiply the numerator (3) of the first fraction times the denominator (4) of the second fraction and put it in the numerator of the new fraction.

$$\frac{3}{8} + \frac{1}{4} = \frac{8 \qquad\qquad 12}{\rule{3cm}{0.4pt}}$$

Step 4. Multiply both denominators (8) and (4) together and use that answer for the denominator of the new fraction.

$$\frac{3}{8} + \frac{1}{4} = \frac{8 \qquad\qquad 12}{32}$$

Step 5. Now, since you are adding, put a plus sign between the two numbers in the numerator. (If you were subtracting, this would be a minus sign, and you would simply subtract the smallest number from the largest number there.)

$$\frac{3}{8} + \frac{1}{4} = \frac{8 \quad + \quad 12}{32}$$

Step 6. Do the math in the new fraction.

$$\frac{3}{8} + \frac{1}{4} = \frac{20}{32}$$

Step 7. Reduce the new fraction (if necessary) by finding the largest single number that will divide evenly into both the numerator and the denominator (the number 4).

$$\frac{20}{32} = \frac{5}{8} \quad Answer$$

Some more difficult examples

1. $\dfrac{5}{16} + \dfrac{3}{64} = \dfrac{48 + 320}{1024} = \dfrac{368}{1024} = \dfrac{23}{64} \quad Answer$

2. $\dfrac{3}{8} - \dfrac{13}{64} = \dfrac{104 - 192}{512} = \dfrac{88}{512} = \dfrac{11}{64} \quad Answer$

MULTIPLYING AND DIVIDING FRACTIONS

Multiplying Fractions

Multiplying fractions is a three-step procedure: (1) multiply the numerators, (2) multiply the denominators, and (3) reduce it to lowest terms (if necessary).

Example

$$\frac{1}{8} \times \frac{3}{4} =$$

Step 1. Multiply the numerators (the top numbers).

$$\frac{1}{8} \times \frac{3}{4} = \frac{3}{}$$

Step 2. Multiply the denominators (the bottom numbers).

$$\frac{1}{8} \times \frac{3}{4} = \frac{3}{32}$$

Step 3. Reduce the answer (if necessary) to lowest terms.

$$\frac{3}{32} \quad \text{(this answer is in lowest terms)}$$

Dividing Fractions

When dividing fractions, the main thing to remember is to invert (or flip) the divisor and then multiply.

Example

$$\frac{\frac{3}{4}}{\frac{3}{8}} = \frac{3}{4} \times \frac{8}{3} = \frac{24}{12} = 2 \quad Answer$$

Note. The answer is often an improper fraction that needs to be changed back into a proper fraction.

Example

$$\frac{16}{12} = 12\overline{)16} = 1\frac{4}{12} = 1\frac{1}{3}$$

WORKING WITH DECIMALS

Decimals are widely used in industry. Some examples of these applications are

- Drawing and part tolerances (English or metric)
- Inspection and machining formulas
- Business percentages

There are a few things about decimals that need to be covered. These are

- Reading decimals
- Converting fractions to decimals
- Adding and subtracting
- Multiplying and dividing
- Percentages

Reading Decimal Places

All decimals are represented by numbers written to the right of a decimal point (.), and in some cases, this quantity of numbers could be infinite. For the purpose of this book we will only go as far as six decimal places.

Example

0 . A B C D E F   (Letters are used to identify the places of a decimal.)

0 place. The place for a whole number such as 1, 2, and so on.

A place. The "tenths" place.   *Example:* The fraction $\frac{1}{10}$ is shown as 0.1 (called one-tenth).

B place. The "hundredths" place.   *Example:* The fraction $\frac{48}{100}$ is shown as 0.48 (called forty-eight hundredths).

C place. The "thousandths" place.   *Example:* The fraction $\frac{1}{1000}$ is shown as 0.001 (called one-one thousandth).

D place. The "ten-thousandths" place.   *Example:* The fraction $\frac{14}{10,000}$ is shown as 0.0014 (called fourteen-ten-thousandths).

E place. The "one-hundred thousandths" place.   *Example:* The fraction $\frac{15}{100,000}$ is shown as 0.00015 (called fifteen-one-hundred thousandths).

F place. The "millionths" place.   *Example:* The fraction $\frac{2}{1,000,000}$ is shown as 0.000002 (called two-millionths).

Here are some examples of reading decimals.

Example 1

0.001 in. means one-thousandth of an inch.

Example 2

0.0015 in. means fifteen-ten-thousandths of an inch.

Example 3

0.2 in. means two-tenths of an inch.

Note. Many people in industry have gotten into the habit of calling the "ten thousandths" place *tenths;* this is a bad habit and should be avoided.

Converting a Fraction into a Decimal

This is done simply by dividing the numerator by the denominator of the fraction.

Example

Fraction		Decimal

$$\frac{5}{8} \begin{array}{l} \text{Numerator} \\ \text{Denominator} \end{array} = 8\overline{)5} = 0.625$$

Adding and Subtracting Decimals

Rule.   Keep all decimal points *in line*.

Example

Add 1.52 + 2.3 + 6.434.

Solution

$$\begin{array}{r} 1.52 \\ 2.3 \\ + 6.434 \\ \hline 10.254 \quad Answer \end{array}$$

Example

Subtract 3.84 from 5.393.

Solution

$$\begin{array}{r} 5.393 \\ - 3.84 \\ \hline 1.553 \quad Answer \end{array}$$

Multiplying Decimals

Rule.   You do not need to line up decimal points, but you must have the same number of decimal places in the answer as you have in the two numbers being combined.

Example

Multiply 4.536 × 3.12.

Solution

$$\begin{array}{r} 4.536 \quad \text{(there are three decimal places here)} \\ \times 3.12 \quad \text{(there are two decimal places here)} \\ \hline 9072 \\ 4536 \\ 13608 \\ \hline 1415232 \quad \end{array}$$ (since there are a total of five decimal places used above, the answer must also have five decimal places)

Answer: 14.15232   (five places)

Dividing Decimals

Rule.   Make the divisor a whole number by moving the decimal point as many places to the right as necessary, then move the decimal point in the dividend the same amount of places. Next, bring the decimal point straight up in the answer.

Example

Divide 3.0 by 0.24:

$$\overset{\text{quotient}\quad Answer}{\text{divisor}\overline{)\text{dividend}}}$$

Solution   First, set up the problem. $0.24\overline{)3.0}$. Next, make the divisor a whole number: $24\overline{)3.0}$ (had to move two places). Next, move the dividend point the same number of places: $24\overline{)300}$. Then the decimal point in the dividend comes straight up into quotient.

$$24\overset{\cdot}{\overline{)300.}}$$

Now, divide the numbers.

$$\overset{\qquad\qquad Answer}{24\overline{)300.}}$$

Converting Decimals to Percentages

A decimal can be converted into a percentage simply by multiplying it by 100. (An easier method is simply moving the decimal point exactly two places to the right.)

Example

0.12   (convert it to percent)

Solution

$$100 \times 0.12 = 12 \text{ percent}$$

or

     0.12   (move decimal two places to the right) = 12.0 or 12 percent

If one knows and follows the rules for decimals, they are very easy to work with. A good beginning in the study of the metric system, in fact, is a thorough knowledge of working with decimals.

Converting Percentages to Decimals

A percentage can be converted into a decimal by dividing the percentage by 100 or by moving the decimal point two places to the left. Keep in mind that whenever a whole number is written, it can be viewed as having a decimal point immediately after the whole number.

Example

When writing a whole number such as 10, the decimal point after the 10 is understood (such as 10.). If the whole number is stated as a percentage (e.g., 10%), it means 10.%.

To change the 10.% to a decimal fraction, divide the value by 100.

$$10.\% \div 100 = 0.10$$

or move the decimal point two places to the left.

$$10.\% = 0.10 \text{ decimal fraction}$$

ADDING AND SUBTRACTING ANGLES

There are occasions where addition and subtraction of angles that are in degrees (°), minutes ('), and seconds (") become necessary to make a job easier, or simply to get the job done. Performing these two operations on angles of this kind is very simple, once you know the rules.

$$\text{Conversion:} \quad 1' = 60 \text{ seconds } (")$$

$$1° = 60 \text{ minutes } (')$$

The key conversion number is 60 for either one.

Addition of Angles

This is simply a matter of working with the problem from right to left and remembering the conversions.

Example

Add:

$$3° \ 12' \ 40''$$
$$+ \ 2° \ 51' \ 30''$$

Solution   First: Add the seconds: 40″ + 30″ = 70″. Since there are 60 seconds in 1 minute, you can leave 10 seconds in the "seconds" column, and carry over 1 minute, to the "minutes" column.

$$1'$$
$$3° \ 12' \ 40''$$
$$+ \ 2° \ 51' \ 30''$$
$$\overline{\qquad\qquad 10''}$$

Next, Add the minutes: 1′ + 12′ + 51′ = 64′. Since there are 60 minutes in 1 degree, leave 4′ in the "minutes" column, and carry over 1 degree to the "degree" column.

$$
\begin{array}{r}
1° \ 1' \\
3° \ 12' \ 40'' \\
+ \ 2° \ 51' \ 30'' \\
\hline
4' \ 10''
\end{array}
$$

Last step: Add the degrees: $1° + 3° + 2° = 6°$.

$$
\begin{array}{r}
1° \ 1' \\
3° \ 12' \ 40'' \\
+ \ 2° \ 51' \ 30'' \\
\hline
6° \ 4' \ 10''
\end{array}
$$

Answer: 6° 4′ 10″

Subtraction of Angles

In subtraction, you still work from right to left in the same manner, but must often "borrow" from the next area to solve the problem.

Example

Subtract:

$$
\begin{array}{r}
12° \ 13' \ 10'' \\
- \ 9° \ 35' \ 40'' \\
\hline
\end{array}
$$

Solution   First: You can see that you cannot subtract 40″ from 10″, so you borrow 1 minute (which is 60 seconds). Now you have 70″ minus 40″ = 30″.

$$
\begin{array}{r}
12' \ 70'' \\
12° \ \cancel{13'} \ \cancel{10''} \\
- \ 9° \ 35' \ 40'' \\
\hline
30''
\end{array}
$$

Next: You see that you cannot subtract 35′ from 12′, so you must borrow 1 degree (which is 60 minutes). Now you have 72′ minus 35′ = 37′.

$$
\begin{array}{r}
72' \\
11° \ \cancel{12'} \ 70'' \\
\cancel{12°} \ \cancel{13'} \ \cancel{10''} \\
- \ 9° \ 35' \ 40'' \\
\hline
37' \ 30''
\end{array}
$$

Last step: Subtract 9 degrees from the remaining 11 degrees.

$$
\begin{array}{r}
72' \\
11° \ \cancel{12'} \ 70'' \\
\cancel{12°} \ \cancel{13'} \ \cancel{10''} \\
- \ 9° \ 35' \ 40'' \\
\hline
2° \ 37' \ 30''
\end{array}
$$

Answer: 2° 37′ 30″

WORKING WITH SIMPLE EQUATIONS

Simple equations are frequently used in production and inspection to make or inspect products. Knowing how to work with these equations correctly is often a prerequisite to getting the job done.

Simple equations are defined as equations with only one unknown (or one variable). Examples of these equations are seen every day.

Examples

$$\text{sine} = \frac{1}{2} \qquad C = \pi(12 \text{ in.}) \qquad A = 0.785(10^2)$$

Now consider an even simpler equation.

Example

$$X = \frac{20}{2}$$

Note. In *all* equations, everything on one side of the "equal" sign is always equal to what is on the other side. You can see here that X (unknown value) equals 20 divided by 2.

All single-variable equations are equations that have only one unknown quantity.

Example

$$50(2) = \frac{2(X)}{4}$$

Note. ( ) means multiply, and the line means divide.

As you see here, the left side is simply numbers: (50 times 2), or 100. The right side cannot be worked out directly because you don't know what X is. The simplest way to find the number X is to get it all alone on one side of the equals sign. Then you can solve the other side, (and since they are always equal) you will know what X is.

Important Rule. You can move any number from one side of the equal sign to the other side by changing its *sign*. For example, if a number is multiplied ($\times$) on one side, move it and divide ($\div$) with it. Or, if a number is added ($+$) on one side, move it and subtract ($-$) it on the other side.

The object is to get X all alone on one side of the equals ($=$) sign, and then do the math on the other side, to solve the problem.

Solution of Preceding Example

$$50(2) = \frac{2(X)}{4}$$

First, move the 4 and multiply: $50(2)(4) = 2(X)$.

    Remember. When you move a number, never write it down again in the same area from which you moved it.

Next, move the multiplied 2 and divide with it on the other side:

$$\frac{50(2)(4)}{2} = X$$

Now you have X all by itself. Finally, you simply do the math.

50 times 2 is 100

100 times 4 is 400

400 divided by 2 is 200

Now you have $200 = X$, and since it is an equation, X is 200.

Proving That Your Answer Is Correct

If you want to prove that your answer of $X = 200$ is correct, you simply try 200 in place of X in the original problem.

$$\text{Original problem:} \quad 50(2) = \frac{2(X)}{4}$$

$$\text{Now replace } X \text{ with your answer:} \quad 50(2) = \frac{2(200)}{4}$$

Now do the math on both sides, and if the numbers on both sides are the same, your answer is correct.

$50(2) = 100$ $2(200) = 400$ divided by $4 = 100$

$100 = 100$ Your answer is correct.

Another Example

$$\text{sine} = \frac{10}{20} \qquad \begin{array}{l} 10 \text{ divided by 20 is 0.5; therefore, the} \\ \text{sine is 0.5.} \end{array}$$

Constants

In many formulas there are constants. These are simply letters or figures that always mean one particular number.

Example

π(pi) is always 3.1416. Anywhere you see this sign, just replace it with the number 3.1416.

Example Formula

Area = 0.785 (D^2)
D means the diameter of the circle. 0.785 is a constant.

So, knowing the diameter of a circle, and using the constant number 0.785, you can find the surface area of that circle.

THE CIRCLE

These are situations that arise during production and inspection that require the solution of parts of the circle. Those who have a working knowledge of circles can solve many of these problems. The circle has several different parts. Some of these are shown in Figure 2.1.

Angular Relationships

All circles have 360 degrees, as shown in Figure 2.2. Circles may also be divided into four "quadrants" of 90 degrees each, as shown in Figure 2.3. It is also a good idea to picture the circle on the $X-Y$ coordinate axes (called the *Cartesian coordinate system*) as shown in Figure 2.3.

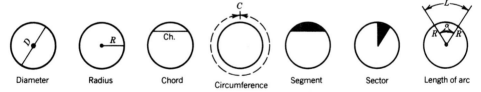

| Diameter | Radius | Chord | Circumference | Segment | Sector | Length of arc |

Figure 2.1 Parts of a circle.

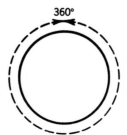

Figure 2.2 Circle

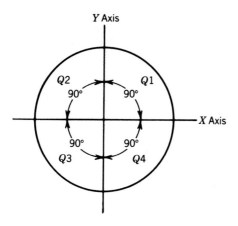

Figure 2.3 Cartesian coordinate system.   **Figure 2.4** Inscribed angle.

All angular relationships start at quandrant 1 on the X axis and go counter-clockwise from there. There are also inscribed angles (Figure 2.4) and central angles (Figure 2.5). A central angle forms the sector.

Linear Distances

Examples of linear relationships of a circle are

1. Length of circumference (or the distance around the circle)
2. Diameter (the distance from one point on the circle, going through the center to its opposite point)
3. Length of chord (a straight line drawn from one point to another that does not go through the center)
4. Length of arc (a specific distance around the circumference of a circle that is not the circumference) (see Figure 2.6).
5. Radius (exactly one-half of the diameter)

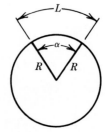

Figure 2.5 Central angle.   **Figure 2.6** Length of an arc.

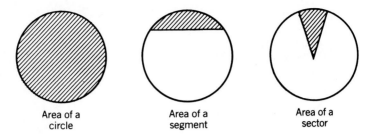

Figure 2.7   Areas of circle parts.

Areas

Surface area is always in square inches, square feet, square yards, and so on. There are three main surface area problems that are related to the circle. Examples of these are shown in Figure 2.7.

Solving for Lengths and Areas

Abbreviations used in formulas:

D = diameter
L = length of arc
α = angle
Ch. = chord
A = area
R = radius

θ = one-half of an included angle (see Figure 2.10)
C = circumference
π = 3.1416 (constant called pi)
[ ], ( ) = multiply

Some Formulas Related to Circles

Note. Symbols or numbers next to each other must be multiplied with or without ( ), as shown in these formulas in Figures 2.8 to 2.11.

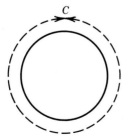

Figure 2.8   Circumference of a circle.

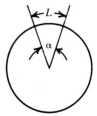

Figure 2.9   Length of an arc.

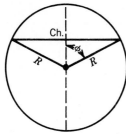

Figure 2.10   Length of a chord.

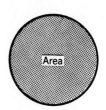

Figure 2.11   Area of a circle.

a. $C = (\pi)D$ or $C = 2(\pi)(R)$

b. $L = 0.017453(R)(\alpha)$ or $L = \pi(D)\dfrac{\alpha^\circ}{360^\circ}$

c. $\mathrm{CH.} = 2(R \sin \theta)$

d. $A = \dfrac{\pi(D^2)}{4}$ or $A = \pi(R^2)$

     There are many uses for the formulas above and others that apply to the circle. One example of this is the bolt circle shown in Figure 2.12.

Circle Practice Problems and Solutions

1. Find the circumference of a circle that has a 3-in. diameter.
   Solution: The formula is $C = (\pi)D$.
   Answer: $C = 3.1416 \times 3 = 9.4248$ in.
2. Find the circumference of a circle with a 1.5-in. radius.
   Solution: The formula is $C = 2(\pi)R$.
   Answer: $C = 9.4248$ in.

    Note. This is the same answer as above because a 1.5-in. radius is equal to a 3-in. diameter.

3. Find the length of an arc produced by a 30-degree included angle on a circle with a diameter of 6 in.

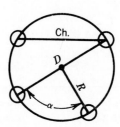

Figure 2.12   Bolt (hole) circle.

Solution: The formula is

$$L = \pi(D) \left(\frac{\text{included angle}}{360 \text{ degrees}} \right)$$

$$= 3.1416 \times 6 \times \frac{30}{360}$$

$$= 3.1416 \times 6 \times 0.0833$$

Answer: $L = 1.57$ in.

4. Find the area of a circle that has a 10-in. diameter.
   Solution: The formula is

$$A = \frac{\pi(D^2)}{4}$$

$$= \frac{3.1416 \times 10^2}{4}$$

$$= \frac{314.16}{4}$$

Answer: $A = 78.54$ sq in.

5. Find the length of a chord where the included angle is 30 degrees, and the radius of the circle is 3 in.
   Solution: The formula is

$$\text{Ch.} = 2[R(\sin \theta)]$$

$$= 2[3(0.26)]$$

$$= 2 \times 0.78$$

Answer: Ch. = 1.56 in.

6. If I begin at the starting point of all angles in the Cartesian coordinate system, and revolve counterclockwise by 185 degrees, what quadrant would I be in?
   Solution: See Figure 2.3. All angles begin on the right side of the X axis, and each quadrant has 90 degrees; I would be 5 degrees within the third quadrant.
   Answer: Quadrant 3.

7. If I have a circle with a circumference of 6.2832 in., what is the diameter?
   Solution: The basic formula is

$$C = \pi(D)$$

$$D = \frac{C}{\pi}$$

changed to

$$D = \frac{6.2832}{3.1416}$$

Answer: 2 in. diameter.

8. What is the radius of the same circle?
Solution:

$$\frac{2}{2}$$

Answer: 1 in. radius.

PERIMETERS

The perimeter is simply the distance "around" any object. Perimeter is a length that is expressed in linear units such as inches, feet, yards, and meters. The only exception to the word *perimeter* when discussing the distance around something is a circle. The distance around a circle is called the *circumference.*

For the purpose of this chapter, we discuss only the perimeters of the four basic shapes: square, rectangle, triangle, and circle.

Perimeter of a Square

The formula is $P = 4s$ (Figure 2.13).

Example

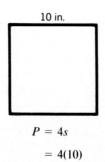

10 in.

$$P = 4s$$
$$= 4(10)$$

Answer: $P = 40$ in.

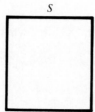

S

Figure 2.13 Square.

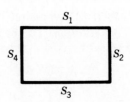

S_1

S_4 S_2

S_3

Figure 2.14 Rectangle.

Perimeter of a Rectangle

The formula is $P = S_1 + S_2 + S_3 + S_4$ (Figure 2.14).

Example

10 ft

3 ft

$$P = S_1 + S_2 + S_3 + S_4$$
$$= 10 + 3 + 10 + 3$$

Answer: $P = 26$ ft.

Perimeter of a Right* Triangle

The formula is $P = H + B + A$ (see Figure 2.15). You must use the Pythagorean theorem to find any side not given.

Example

Not given

3 in.

4 in.

$$P = H + B + \sqrt{B^2 + H^2}$$
$$= 3 + 4 + \sqrt{4^2 + 3^2}$$
$$= 3 + 4 + \sqrt{16 + 9}$$
$$= 3 + 4 + 5$$

Answer: $P = 12$ in.

Using the Pythagorean theorem ($A^2 = B^2 + H^2$) and inserting values for the two given sides, the third (unknown) side may be calculated. If it is not a right triangle, the formula $P = H + B + A$ will give you the perimeter (see pages 40–41 on the Pythagorean theorem).

Circumference of a Circle

The formula is $C = \pi(D)$, where π is 3.1416 (Figure 2.16).

---

* One angle is a right angle (90 degrees).

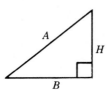

Figure 2.15 Right triangle.

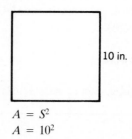

Figure 2.16 Circle.

Example

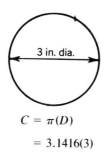

$$C = \pi(D)$$

$$= 3.1416(3)$$

Answer: $C = 9.4248$ in.

AREAS

In some industrial applications, the surface area of an object must be calculated. The surface area has no thickness; it is simply the area of a surface, such as the top of a table or the floor of a building. In this chapter, only the four basic geometrical shapes will be discussed, but they can lead you to the solution of other shapes.

When you are concerned with the area of something, your answer is in *square* units such as square inches (sq in.) and square feet (sq ft). For example, the surface of a box that is 1 foot long and 1 foot wide is equal to 1 square foot.

Area of a Square

The formula is $A = S^2$ (Figure 2.17).

Example

10 in.

$$A = S^2$$
$$A = 10^2$$

Answer: $A = 100$ sq in.

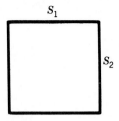

Figure 2.17 Square.

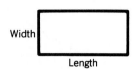

Figure 2.18   Rectangle.

Area of a Rectangle

The formula is $A = L \times W$ (Figure 2.18).

Example

3 in.

12 in.

$$A = L \times W$$
$$= 12 \text{ in.} \times 3 \text{ in.}$$

Answer: $A = 36$ sq in.

Area of a Right Triangle

The formula is $A = \frac{1}{2}(B)(H)$ (Figure 2.19).

Note. A right triangle is one that contains one angle of 90 degrees.

Example

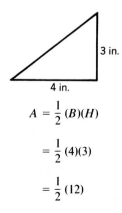

3 in.

4 in.

$$A = \frac{1}{2}(B)(H)$$

$$= \frac{1}{2}(4)(3)$$

$$= \frac{1}{2}(12)$$

Answer: $A = 6$ sq in.

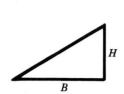

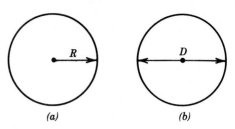

Figure 2.19 Right triangle.   **Figure 2.20** Circle: (*a*) radius; (*b*) diameter.

Area of a Circle

Formula 1 is $A = \pi R^2$, where π is 3.1416 (Figure 2.20*a*).

Example

$$R = 4 \text{ in.}$$

$$A = \pi(R^2)$$

$$= 3.1416 \times 4^2$$

$$= 3.1416 \times 16$$

Answer: $A = 50.2656$ sq in.

The alternate formula 2 is $A = \dfrac{\pi(D^2)}{4}$ (Figure 2.20*b*).

Example

$$R = 4 \text{ in. or } D = 8 \text{ in.}$$

$$A = \frac{\pi(D^2)}{4}$$

$$= \frac{3.1416 \times 8^2}{4}$$

$$= \frac{3.1416 \times 64}{4}$$

$$= \frac{201.0624}{4}$$

Answer: $A = 50.2656$ sq in.

VOLUMES

Volume is a *cubic* measure. It is simply the area times thickness (or height), and expressed in cubic units such as cubic inches (cu in.) and cubic feet (cu ft). Volume is the total space within a solid geometrical shape.

The following examples are only of simple geometrical shapes, but are the foundation for the more complex examples found in industry.

Volume of a Cube

The formula is volume = $L \times W \times Th$ (Figure 2.21).

Example

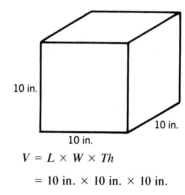

10 in.

10 in.

10 in.

$$V = L \times W \times Th$$
$$= 10 \text{ in.} \times 10 \text{ in.} \times 10 \text{ in.}$$

Answer: $V = 1000$ cu in.

Note. For a cube, $Th = W = L$.

Volume of a Rectangular Solid

The formula is volume = $L \times W \times Th$ (Figure 2.22).

Example

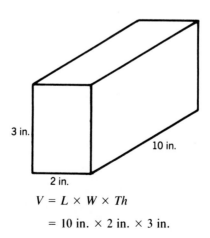

3 in.

10 in.

2 in.

$$V = L \times W \times Th$$
$$= 10 \text{ in.} \times 2 \text{ in.} \times 3 \text{ in.}$$

Answer: $V = 60$ cu in.

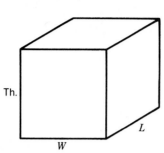

Figure 2.21 Cube.

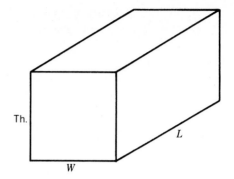

Figure 2.22 Rectangular solid.

Volume of a Right Triangular Solid

The formula is volume $= \frac{1}{2}(B \times H)(Th)$ (Figure 2.23).

Example

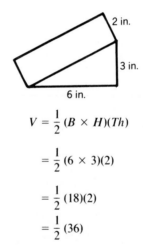

$$V = \frac{1}{2}(B \times H)(Th)$$

$$= \frac{1}{2}(6 \times 3)(2)$$

$$= \frac{1}{2}(18)(2)$$

$$= \frac{1}{2}(36)$$

Answer: $V = 18$ cu in.

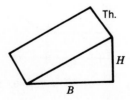

Figure 2.23 Right triangle solid.

Volume of a Cylinder

Formula 1 is (if radius is known): volume $= \pi(R^2)(L)$, where π is 3.1416 (Figure 2.24).

Example

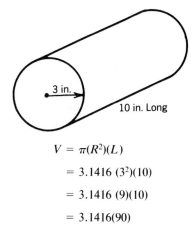

$$V = \pi(R^2)(L)$$
$$= 3.1416 \, (3^2)(10)$$
$$= 3.1416 \, (9)(10)$$
$$= 3.1416(90)$$

Answer: $V = 282.744$ cu. in.

Formula 2 is (if diameter is known): volume $= \dfrac{\pi(D^2)}{4}(L)$.

Volume of a Cone

The formula is $V = \frac{1}{3}\pi r^2 h$ (Figure 2.25).

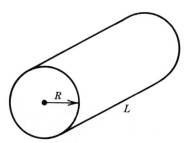

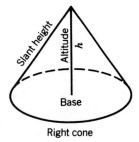

Figure 2.24 Cylinder.

Figure 2.25 Cone.

SOLVING THE RIGHT TRIANGLE FOR SIDES ONLY: THE
PYTHAGOREAN THEOREM

In many manufacturing and inspection applications, it becomes necessary to be able to find the length of one side of a right triangle when you know the length of the other two sides. It is always an asset to have the ability to see a triangle within certain geometrical shapes. Then, once you see the triangle, you can solve it for an unknown side by using the *Pythagorean theorem*. The Pythagorean theorem is simply stated by Figure 2.26 and the formula $c^2 = a^2 + b^2$.

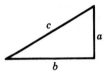

Figure 2.26   Sides of a right triangle.

To avoid the lengthy solving of the one formula each time to calculate a side, it is better simply to write out each of the three possible formulas and have them handy to use when needed. These are shown in Figure 2.27. Now you can quickly solve for the unknown side (a, b, or c).

$$a = \sqrt{c^2 - b^2}$$
$$b = \sqrt{c^2 - a^2}$$
$$c = \sqrt{a^2 + b^2}$$

Figure 2.27   Formulas for sides of a right triangle.

Remember. The three formulas above work only when you call the sides a, b, and c, as shown in the figure.

An example of how to use the formulas is best described by the 3–4–5 triangle, shown in Figure 2.28. This triangle says that side $c = 5$ in., side $a = 3$ in., and side $b = 4$ in.

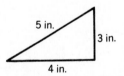

5 in.

3 in.

4 in.

Figure 2.28   A 3–4–5 triangle.

To prove that the formula works (and for practice), let's say that we had the same triangle but did not know the length of the longest side (the 5-in. side). So when looking at the three choices of formulas, we see that the side we are looking for is called side c. Now use the formula for side c, which is

$$c = \sqrt{a^2 + b^2}$$

The steps in using the formula are shown below.

Step 1. Enter the sides you know into the formula

$$c = \sqrt{3^2 + 4^2}$$

Step 2. Square the numbers. (Squaring is simply multiplying the number times itself.)

$$c = \sqrt{9 + 16}$$

Step 3. Add the two numbers together.

$$c = \sqrt{25}$$

Step 4. Find the square root.

$$\sqrt{25} = 5$$

Answer: $c = 5$ in.

As you can see, the answer is 5, as shown in the original triangle. Be sure to follow the steps above at all times.

Practice Problems and Answers

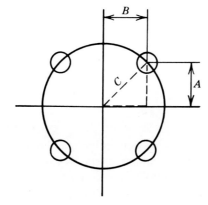

ANSWER. 3.5 in.   ANSWER. 6 ft

Figure 2.29   Right triangles with one unknown side.

Examples of Seeing the Triangle in a Problem

If you know the length of *A* and *B* (in Figure 2.30), you could solve the triangle for side *C*, which is the radius of the bolt circle, then double it to find the Bolt circle

Figure 2.30   Finding the diameter of a bolt circle.

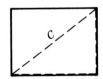

Figure 2.31 Finding the diagonal distance, *C*, of a rectangle.

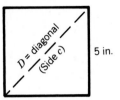

Figure 2.32 Diagonal of a square.

diameter. Or find the corner-to-corner (or diagonal) distance of a rectangle if sides *a* and *b* are given (Figure 2.31).

Diagonal of a Square

The Pythagorean theorem can also be used to calculate the diagonal (or corner-to-corner distance) of a square. In this case sides *a* and *b* are the same, and the diagonal is equal to side *c*.

Example

A 5-in. square (Figure 2.32).

$$\text{Diagonal } (c) = \sqrt{a^2 + b^2}$$
$$= \sqrt{25 + 25}$$
$$= \sqrt{50}$$
$$= 7.07 \text{ in.} \quad \textit{Answer}$$

Another method is with a direct formula.

$$D = S \times 1.4142$$

where *D* is the diagonal
 S is the side
 1.4142 is a constant number

$$D = S \times 1.4142$$
$$= 5 \times 1.4142$$
$$= 7.07 \text{ in.} \quad \textit{Answer}$$

SOLVING THE RIGHT TRIANGLE

The right triangle is used in many ways to solve many problems in the production and inspection of machined parts and castings. Knowing how to solve the right triangle for unknown sides and angles is an asset to everyone involved.

A right triangle has three sides and three angles as shown in Figure 2.33. The three angles are *a*, *b*, and *c*. The three sides are *A*, *B*, and *C*. Angle *c* is always 90 degrees. The reason you need to solve for the rest of the angles and sides is that in most cases, you are already given part of the information and you need to know the rest.

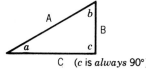

C   (*c* is *always* 90°)

Figure 2.33   The three sides and three angles of a right triangle.

The steps to solving the right triangle are simple once you practice with them. There are six trigonometry functions used to solve these triangles. These functions are simply formulas to use and should not be regarded as too challenging.

The functions are

sine   cosine   tangent   cotangent   secant   cosecant

(or abbreviated):

sin   cos   tan   cot   sec   csc

The formulas are

$$\sin = \frac{\text{opp.}}{\text{hyp.}} \qquad \cos = \frac{\text{adj.}}{\text{hyp.}} \qquad \tan = \frac{\text{opp.}}{\text{adj.}}$$

$$\cot = \frac{\text{adj.}}{\text{opp.}} \qquad \sec = \frac{\text{hyp.}}{\text{adj.}} \qquad \csc = \frac{\text{hyp.}}{\text{opp.}}$$

Usually, the only ones used are the first three (the other three are reciprocals of the first three).

Example:

Reading the Formulas

$$\sin = \frac{\text{opp.}}{\text{hyp.}}$$

This says that the sine function is equal to the length of the opposite side, divided by the length of the hypotenuse side.

There are three sides to a right triangle:

hypotenuse, adjacent, and opposite

The hypotenuse is *always* the *longest* side of any right triangle.

One thing to remember is that the adjacent side and the opposite side change, depending on which of the angles you are dealing with. The thing to remember in knowing which is which is that the opposite side is the side that is directly *opposite* the angle you are working with at the time, and the adjacent side is always the side next to the angle you are working with at the time.

Figure 2.34   Naming the opposite sides of a right triangle: *(a)* for angle *a*; *(b)* for angle *b*.

The drawings in Figure 2.34 show how to name the sides of the triangle. For example, if you are working with angle *a* above, the opposite side is as shown, but if you are working with angle *b* above, the other side is the opposite side. *It is important to understand this.*

The next step (after knowing the formulas, and how to name the sides) is to decide which formula to use for the problem. A good rule is to take what you know about a problem, plus what you want to find, and look for the only formula that has both.

Example

The right triangle in Figure 2.35 has a 5-in. hypotenuse and a 1-in. opposite side. You can see in the formulas that the only formula having these things in it is the sine formula. This is the one you need to solve the problem. You will also need a table or calculator to convert your sine of 0.2 into the angle because the formula only gives you the sine function of the angle, which is simply a number related to the angle.

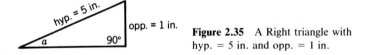

Figure 2.35   A Right triangle with hyp. = 5 in. and opp. = 1 in.

This sine 0.2 can be changed into an angle by using a calculator.

$$\sin = \frac{\text{opp.}}{\text{hyp.}} = \frac{1}{5} = 0.2$$

Right Triangle Practice Problems and Solutions

Below are some practice problems with step-by-step solutions to assist you in learning to solve the right triangle.

Example 1

Solve the triangle in Figure 2.36 for the length of side *A*.

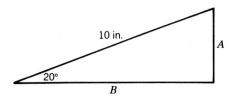

Figure 2.36 Right triangle with unknown sides *A* and *B*.

Solution

Step 1. Name the sides you are working with. (They are hyp. and opp.)

Step 2. Since the sides are hyp. and opp. use the formula

$$\sin = \frac{\text{opp.}}{\text{hyp.}}$$

Step 3. Change the formula to solve for side *A*.

$$\sin(\text{hyp.}) = \text{opp.}$$

Step 4. Enter the known values into the formula.

$$\sin 20°(10) = \text{opp.}$$

Step 5. Find the sine of 20 degrees and enter that number in the proper place.

$$0.3420(10) = \text{opp.}$$

Step 6. Work the simple math for the answer.

$$0.3420(10) = 3.420 \text{ in.}$$

Answer: 3.420 in.

Example 2

Solve the triangle in Figure 2.36 for the length of side *B*.

Solution

Step 1. Name the sides you are working with. (They are hyp. and adj.)

Step 2. Since the sides are hyp. and adj., use the only formula with them in it.

$$\cos = \frac{\text{adj.}}{\text{hyp.}}$$

Step 3. Change the formula to solve for the adj. side.

$$\cos(\text{hyp.}) = \text{adj.}$$

Step 4. Enter the known values where they belong in the formula.

$$\cos 20°(10) = \text{adj.}$$

Step 5. Find the cosine of 20 degrees and enter it into the formula. (The cosine of 20 degrees is 0.9397.)

Step 6. 0.9397(10) = 9.397 in.

Answer: 9.397 in.

Example 3

Find the angle α in Figure 2.37.

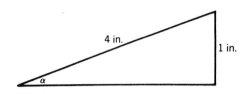

Figure 2.37   Right triangle with unknown angle α.

Solution

Step 1. Name the sides you are working with. (They are the 4-in. hypotenuse and the 1-in. opposite side.)

Step 2. Choose the only formula (of the basic three) that has both of these sides in it.

$$\text{Formula:}\quad \sin = \frac{\text{opp.}}{\text{hyp.}}$$

Step 3. The formula above says that the sine function of the angle is equal to the opposite side length divided by the hypotenuse side length. *This formula solves only for the sine function of the angle, not the angle itself.*

Step 4. Enter all known values in their proper place.

$$\sin_? = \frac{\text{opp.}}{\text{hyp.}} = \frac{1}{4}$$

Step 5. Do the simple math.

$$\sin_\alpha = \frac{1}{4} = 0.25$$

Step 6. Convert the sine function of 0.25 to degrees.
If $\sin_\alpha = 0.25$, then arcsin = 14.48 degrees (rounded off).

Answer: 14.48 degrees.

Example 4

What is angle α in Figure 2.38?

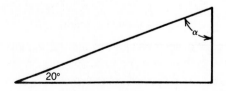

Figure 2.38   Right triangle with unknown angle α.

Solution   The sum of all angles in a triangle is 180 degrees; therefore, the unknown angle must be

$$180° - 20° - 90° = 70°$$

Answer: 70 degrees.

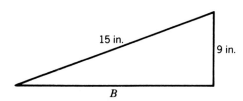

15 in.

9 in.

B

Figure 2.39   Right triangle with unknown side *B*.

Example 5

Solve the triangle in Figure 2.39 for side *B* using the Pythagorean theorem.

Solution $B = \sqrt{15^2 - 9^2} = 12$ in.

Answer: 12 in.

Example 6: Word Problem

A ladder is 8 feet long (Figure 2.40) and is leaning against a building. The point where the top of the ladder is touching is 6 feet high. How far is the base of the ladder from the base of the building?

Step 1. Draw the problem so that there is a triangle (as in Figure 2.40).

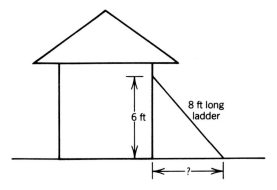

6 ft

8 ft long
ladder

?

Figure 2.40   Right triangle formed by a ladder leaning against a building.

Step 2. Solve this problem with the Pythagorean theorem.

Step 3. ? $= \sqrt{8^2 - 6^2}$.

Step 4. ? $= 5.292$ ft (rounded off).

Answer: 5.292 ft.

Triangle Practice Problems and Solutions

Note. The total of the internal angles of a triangle always equals 180 degrees (Figure 2.41).

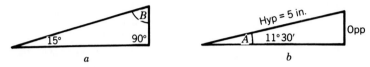

Figure 2.41 *(a)* Right triangle with unknown angle *B*. *(b)* Right triangle with unknown opposite side.

1. Find angle *B*.
   Solution: 180 − (90 + 15) = 75°
   Answer: 75 degrees.
2. Find the opposite side.
   Solution:

$$\sin \text{ angle } A = \frac{O}{H}$$

$$O = \sin \times H$$

$$= 0.199 \times 5 \text{ in.}$$

$$= 0.995 \text{ in.}$$

   Answer: 0.995 in.
3. Find the sine of 11° 0′.
   Answer: 0.19081.
4. Find the tangent of 78° 30′.
   Answer: 4.9151.

MEMORIZING THE TRIGONOMETRY FORMULAS FOR THE RIGHT TRIANGLE

The trigonometry formulas discussed earlier are necessary in solving problems in manufacturing and inspection. Some people choose to memorize the formulas as they become comfortable with solving a right triangle. This is not easily done without some kind of system.

The system discussed here is very effective for accurately memorizing the formulas. It is simply a short cut to memorizing the formulas without having to look them up each time. If used correctly, it can be very helpful.

$$\text{sine} = \frac{\text{opposite}}{\text{hypotenuse}} \qquad \text{cosine} = \frac{\text{adjacent}}{\text{hypotenuse}} \qquad \text{tangent} = \frac{\text{opposite}}{\text{adjacent}}$$

$$\text{cotangent} = \frac{\text{adjacent}}{\text{opposite}} \qquad \text{secant} = \frac{\text{hypotenuse}}{\text{adjacent}} \qquad \text{cosecant} = \frac{\text{hypotenuse}}{\text{opposite}}$$

Next are five steps to memorizing the formulas.

Step 1. Memorize these letters *in order.* O   A   O   A   H   H

Step 2. Next, underline them. <u>O   A   O   A   H   H</u>

Step 3. Then write them backward under the first set.

O	A	O	A	H	H
H	H	A	O	A	O

Step 4. Now draw vertical lines separating each set of letters.

O	A	O	A	H	H
H	H	A	O	A	O

Step 5. Now (in the proper order as shown) put the abbreviated names of the functions over the sets of letters.
The proper order is sin, cos, tan, cot, sec, csc:

sin	cos	tan	cot	sec	csc
O	A	O	A	H	H
H	H	A	O	A	O

Now you can see that by remembering *OAOAHH* and the order of the functions, you can effectively memorize the six trigonometry functions of the right triangle for quick use in the shop or anywhere.

SOLUTION OF OBLIQUE TRIANGLES

Triangles that have no 90-degree angles are called *oblique triangles.* It is necessary to know three things to solve oblique angles. When three parts are known (e.g., two sides and an angle), the formulas can be used to solve for the other parts.

There are still three sides and three angles (the same as the right triangle). The formulas below match this picture (Figure 2.42) of the oblique triangle. When

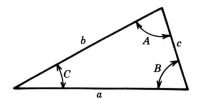

Figure 2.42 Oblique triangle.

solving a problem you simply take the parts you know and match them to this sample triangle; then find what you want to know from the formula.

TO FIND THIS	KNOWING THIS	USE THIS FORMULA
A	B and C	$180° - (B + C)$
B	A and C	$180° - (A + C)$
C	A and B	$180° - (A + B)$
b	a, A, B	$b = \dfrac{a \times \sin B}{\sin A}$
c	a, A, C	$c = \dfrac{a \times \sin C}{\sin A}$
area of	a, b, C	$\text{area} = \dfrac{a \times b \sin C}{2}$
c	a, b, C	$c = \sqrt{a^2 + b^2 - (2ab \times \cos C)}$
$\cos A$	a, b, c	$\cos A = \dfrac{b^2 \times c^2 - a^2}{2bc}$
$\sin B$	a, A, b	$\sin B = \dfrac{b \times \sin A}{a}$
$\tan A$	a, b, C	$\tan A = \dfrac{a \times \sin C}{b - (a \times \cos C)}$

Note above that the abbreviations sin, cos, and tan are used. In formulas where they are used, remember that they mean the sine, cosine, and tangent function of the angle.

Example

The sin 30 = 0.5 (or 0.5 is the sine function of 30 degrees).

Oblique Triangle Practice Problems and Solutions

For the following problems, refer to the previous table of formulas to find an unknown side or angle.

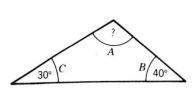

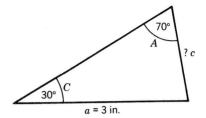

Figure 2.43 Oblique triangle with unknown angle A.

Figure 2.44 Oblique triangle with unknown side c.

1. Find the unknown angle in Figure 2.43.
   Solution: The unknown angle is angle A. The formula is

$$A = 180 - (B + C)$$
$$= 180 - (40 + 30)$$
$$= 180 - 70$$

Answer: $A = 110$ degrees.

2. Find the unknown side in Figure 2.44.
   Solution: The unknown side is the side c in the reference triangle. The formula is

$$c = \frac{a \times \sin C}{\sin A}$$
$$= \frac{3 \times 0.5}{0.9397}$$
$$= \frac{1.5}{0.9397}$$

Answer: $c = 1.596$ in.

3. Find the area of the triangle in Figure 2.45.

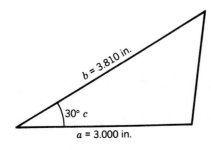

Figure 2.45 Finding the area of an oblique triangle.

Solution: The formula is

$$\text{area} = \frac{a \times b \times \sin C}{2}$$

$$= \frac{3 \times 3.810 \times 0.5}{2}$$

$$= \frac{5.715}{2}$$

Answer: Area = 2.8575 sq in.

4. Find the unknown angle in Figure 2.46.

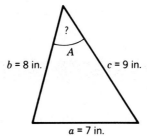

$b = 8$ in. $c = 9$ in.

$a = 7$ in.

Figure 2.46 Triangle with unknown angle A.

Solution: Sides a, b, and c are known. The unknown angle is angle A. The formula is

$$\cos A = \frac{b^2 + c^2 - a^2}{2bc}$$

$$= \frac{8^2 + 9^2 - 7^2}{2 \times 8 \times 9}$$

$$= \frac{64 + 81 - 49}{144}$$

$$= \frac{96}{144}$$

Answer: $\cos A = 0.66667$.

Now, if the cosine of angle A is 0.66667, then convert the cosine of A to angle A using a calculator, or the trigonometry tables.

Answer: arccos 0.66667 = 48.2 degrees.

Miscellaneous Geometric Problems

Area of a trapezoid

EXAMPLE   EQUATION

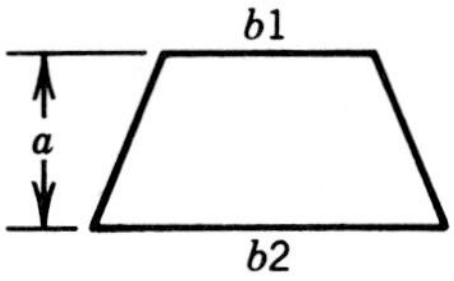

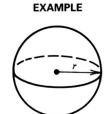

$$\text{Area} = \frac{1}{2} a(b_1 + b_2)$$

Figure 2.47 Trapezoid.

Area and volume of a sphere

EXAMPLE   EQUATION

$$\text{Area} = 4\pi r^2$$

$$\text{Volume} = \frac{4}{3}\pi r^3$$

Figure 2.48 Sphere.

Diameter of a circle inscribed in an equilateral triangle

EXAMPLE   EQUATION

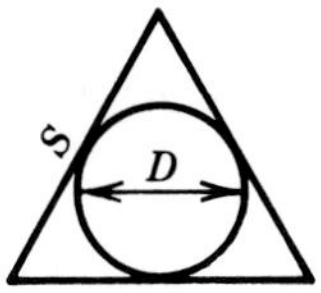

$$D = S(0.57735)$$

Figure 2.49 Equilateral triangle with an inscribed circle.

Side of an equilateral triangle inscribed inside a circle

EXAMPLE   EQUATION

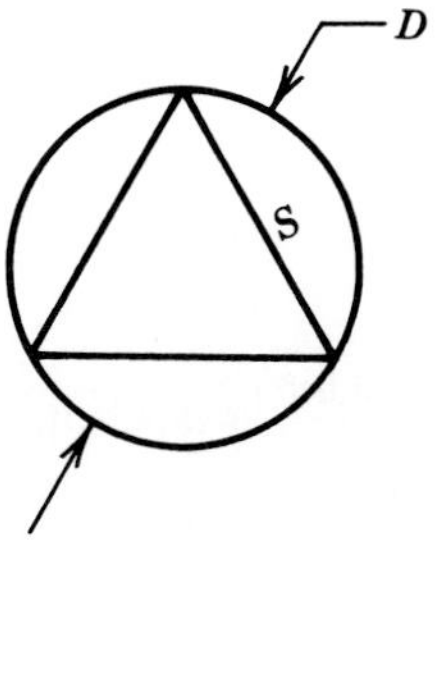

$$S = D(0.866)$$

Figure 2.50 Equilateral triangle with a circumscribed circle.

TAPERS

There are a variety of applications on engineering drawings that require tapers. Taper on a single surface can be explained by referring to it as the *rise versus the run* (See Figure 2.51).

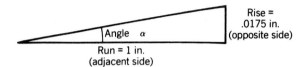

Figure 2.51   Triangular representation of a tapered surface.

Surface Taper

For 1 degree of taper on a single surface, the rise will be approximately 0.0175 in. per 1 in. of run. The computation necessary for this is

$$\text{tangent}_\alpha \times \text{adjacent side} = \text{opposite side}$$

where "tangent" is the tangent function of the angle
  α is the angle
  "adjacent side" is the side nearest the angle
  "opposite side" is the side opposite the angle

Diametral Taper

With respect to a diameter (or thickness) where two surfaces are involved, there are three parts to consider. There are

1. Large diameter (or thickness)
2. Small diameter (or thickness)
3. Length between diameters (or thicknesses)

The taper is usually referred to as *taper per inch* (TPI) or *taper per foot* (TPF). In many cases, the inspector or machinist is given only parts of the information for a taper and must calculate the remaining parts.

Example: Three parts Are Given, but Not the Taper (Figure 2.52)

To find the taper in Figure 2.52, subtract the small diameter from the large diameter and divide the answer by the length.

$$\frac{\text{large diameter} - \text{small diameter}}{\text{length}} = \text{taper (inch per inch)}$$

$$\frac{1.000 \text{ in.} - 0.500 \text{ in.}}{4.000 \text{ in.}} = 0.125 \text{ in. per in.}$$

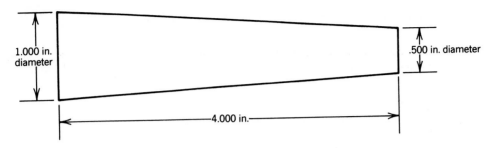

Figure 2.52 Tapered shaft.

Example: The Amount of Taper Is Given, but One of the Diameters Is Not Given (Figure 2.53)

Diameter (d) is found by first multiplying the taper per inch (TPI) times the length (4.000 in.), then subtracting that amount from the large diameter.

$$\text{Small diameter } (d) = \text{large diameter} - (\text{TPI} \times \text{length})$$

$$= 1.000 \text{ in.} - (0.125 \text{ in.} \times 4.000 \text{ in.})$$

$$= 1.000 \text{ in.} - 0.500 \text{ in.}$$

$$= 0.500 \text{ in.}$$

If the missing diameter were the large diameter, the equation would be

$$\text{large diameter } (D) = \text{small diameter} + (\text{TPI} \times \text{length})$$

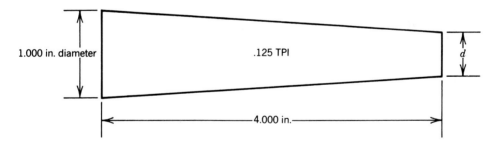

Figure 2.53 Tapered shaft.

Converting TPI to TPF

Taper per inch (TPI) can be converted to taper per foot (TPF), or vice versa, using the following equations:

$$\frac{\text{TPF}}{12} = \text{TPI}$$

$$\text{TPI} \times 12 = \text{TPF}$$

Converting Taper into the Corresponding Angle

The angle (per side) can be found by dividing the taper per foot (TPF) by 24, then finding the angle whose tangent function corresponds to that quotient.

Example

$$\text{Tangent angle} = \frac{\text{TPF}}{24}$$

$$= \frac{1.5}{24}$$

$$= 0.0625$$

The angle can then be found by referring to the table of natural trigonometric functions in the Appendix. The angle is 3° 35″. To find the included angle, multiply the taper angle by 2.

Practice Problems and Solutions

1. If the rise were 0.035 in. for a run of 1 in., what would be the corresponding angle?
   Solution: There is approximately 0.0175 in. rise per inch per degree; therefore, 0.035 in. is twice that amount, or 2 degrees.

2. If a diameter were tapered and the large diameter were 0.500 in., the small diameter were 0.100 in., and the length were 1.000 in., what would the taper be?
   Solution:

$$\frac{0.500 - 0.100 \text{ in.}}{1.000 \text{ in.}} = 0.400 \text{ in. per in.}$$

3. A shaft that is 4.000 in. long is tapered. The small diameter is 1.000 in. and the TPI is 0.125 in. What is the large diameter?
   Solution: The larger diameter (D) = the small diameter + (TPI × length).

$$D = 1.000 \text{ in.} + (0.125 \times 4.000 \text{ in.})$$

$$= 1.000 \text{ in.} + 0.500 \text{ in.}$$

$$= 1.500 \text{ in.}$$

4. If there were a shaft that had a taper per inch (TPI) of 0.100 in., what would be the corresponding TPF?
   Solution:

$$\text{TPI} \times 12 = \text{TPF}$$

$$0.100 \text{ in.} \times 12 = \text{TPF}$$

$$\text{TPF} = 1.200 \text{ in.}$$

5. If a shaft had a taper per foot (TPF) of 0.144 in., what is the corresponding taper per inch (TPI)?
Solution:

$$TPI = \frac{TPF}{12}$$

$$= \frac{0.144 \text{ in.}}{12}$$

$$= 0.012 \text{ in.}$$

METRIC OR ENGLISH: A COMPARISON

In comparing the English and metric systems, there are several factors involved. The conversion from English to metric units is a subject that has been under study for a long time. The reason for the length of time is the impact the conversion would have on U.S. industry. One major consideration is the large amounts of money that would have to be spent for

Drawings   Tooling   Training   and so on

The bottom line is to answer the question: Is it something we should do, or not?

The English System: Conversion

In the English system, conversions (such as feet to miles, yards to inches, gallons to ounces, etc.) usually cause confusion because nothing is straightforward. For example, there are many different constants.

1 mile = 5280 feet = 1760 yards
1 yard = 3 feet = 36 inches.
1 foot = 12 inches.
1 inch can be divided into a wide variety of fractions and decimal fractions (usually, halves, fourths, eighths, sixteenths, thirty-seconds, and sixty-fourths).

All of these constants (5280 or 1760, 3 or 36, 12, etc.) cause confusion in conversion. For example, there are many calculations involved in converting 1 mile to inches.

1 mile = 5280 feet = 5280 × 12 inches/feet = 63,360 inches

or   1 mile = 1760 yards = 1760 × 36 inches/yard = 63,360 inches

Our system (English) is convenient simply because we have been trained to use it from childhood, and have been using it ever since. Many people are shaking with the thought of changing over to metric units because they do not understand the system. The metric system is much more logical than the English system and much easier to use.

The Metric System: Conversion

The metric system is easier to use than the English system because it uses a base of 10.

Example

Any number multiplied by 10 simply moves the decimal point one place to the right.

$$6.0 \times 10 = 60$$

Any number divided by 10 simply moves the decimal point one place to the left.

$$5.7 \div 10 = 0.57$$

When converting a metric measurement to another, the decimal point is simply moved to the right or left. The following example shows how easy conversion is in the metric system.

0	0	0	1.	0	0	0
↑	↑	↑	↑	↑	↑	↑
kilometer	hectometer	dekameter	meter	decimeter	centimeter	millimeter
1000 meters	100 meters	10 meters	1 meter	0.1 meter	0.01 meter	0.001 meter

You can see above that 1 centimeter equals 10 millimeters, 1 meter equals 100 centimeters, 1 dekameter equals 10 meters, and so on. All of these conversions have just one multiplier to use in converting from one unit to another, for example, from meters to centimeters.

Remember the terms

kilo   means 1000

hecto means 100

deka   means 10

deci   means $\dfrac{1}{10\text{th}}$

centi means $\dfrac{1}{100\text{th}}$

milli means $\dfrac{1}{1000\text{th}}$

These terms (and the conversion) are used in all measurements, whether it be grams, liters, or meters. Further study of the metric system reveals how easy the system is to use. In fact, you will wonder why we did not convert long ago. Refer to the Appendix for English-to-metric conversion tables.

RATIOS

Most people use ratios every day without realizing it. When working with fractions, you are actually working with ratios.

Example

The fraction $\frac{1}{2}$ is simply a ratio of 1 to 2, written as $1:2$.

Ratios are simply a statement that compares two numbers. When a pie is cut into eight equal pieces and you get one of those pieces, your piece is a ratio of $1:8$ (because there are a total of eight pieces, and you got one of them).

Another example is the sprocket on a bicycle. The size of the rear sprocket related to the size of the front sprocket is important. If the front sprocket turns in a ratio of $4:1$ compared to the back sprocket, the rider will have to pedal faster. If the front sprocket were a ratio of $2:1$, the rider will pedal slower but hold the same speed. Why, you ask?

A $4:1$ ratio between the front and rear sprocket means that the front sprocket must make four full turns to turn the rear sprocket just one full turn. Therefore, the rider pedals faster. A $2:1$ ratio means that the front sprocket has to turn only two full turns for the rear sprocket to make a full turn. Therefore, the rider can pedal slower than a rider on the same type of bike with a $4:1$ ratio.

Ratios are used throughout industry in several applications. Some of these are machines and others are measuring equipment. Regardless of use, if you understand ratios, you will be in a better position to work with them.

TEMPERATURE CONVERSION

At times it becomes necessary to convert temperature from one scale to another. Today, the main scales to be considered are Fahrenheit and Celsius (sometimes called centigrade).

Converting from Fahrenheit to Celsius, or from Celsius to Fahrenheit, is very simple. Formulas can be used for each conversion, or tables are available to read the conversions directly.

The formulas are

$$T_c = \frac{5}{9}(T_f - 32) \qquad T_c \text{ stands for temperature in Celsius}$$

$$T_f = \frac{9}{5}(T_c) + 32 \qquad T_f \text{ stands for temperature in Fahrenheit}$$

Example 1

Convert 68 degrees Fahrenheit to Celsius. The Formula is $T_c = \frac{5}{9}(T_f - 32)$.

Solution $T_c = \dfrac{5}{9}(68 - 32)$

$$= \frac{5}{9}(36)$$

$$= \frac{180}{9}$$

Answer: T_c = 20 degrees Celsius (20°C).

Example 2

Convert 38 degrees Celsius to Fahrenheit. The formulas is $T_f = \frac{9}{5}(T_c) + 32$.

Solution $T_f = \dfrac{9}{5}(38) + 32$

$$= 68.4 + 32$$

Answer: T_f = 100.4 degrees Fahrenheit (100.4°F).

The ability to convert temperatures (specifically Celsius and Fahrenheit) will assist you in many ways. An example of this is in temperature measurement. If the specification calls for degrees Celsius and you have equipment that measures Fahrenheit, you can measure it and convert your value to Celsius.

FACTORIALS

There are various applications where one needs to find the factorial (!) of a number. In quality control, most of these applications have to do with statistical sampling and probability.

The factorial of any number is easily calculated. It is found by starting with 1 and multiplying the successive numbers until you get to the number in question. This is represented by the following equation:

$$n! = 1 \times 2 \times 3 \times 4 \times \cdots \times n$$

Example

$$3! = 1 \times 2 \times 3 = 6$$

$$5! = 1 \times 2 \times 3 \times 4 \times 5 = 120$$

$$2! = 1 \times 2 = 2$$

$$1! = 1$$

THE REAL NUMBER LINE

There are many mathematical situations where an understanding of the real number line can be an asset to the person trying to solve the problem. Most frequent are those problems that contain positive and negative numbers. The real number line is helpful in gaining an understanding of positive and negative numbers and looking at them in terms of magnitude (or distance). The line itself is very simple. It begins with a value of 0 in the middle and shows that all numbers to the right are plus and all numbers to the left are minus, as shown in Figure 2.54.

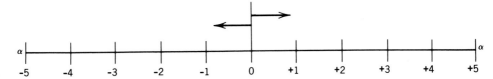

Figure 2.54 Real number line.

If one simply looks at the real number line in terms of distance one can clearly see that the distance between +3 and −2 is 5 since the distance is 5 divisions, regardless of whether the values are positive or negative. Everyone understands that the distance from 0 to +5 is 5, and the distance from 0 to −5 is also 5. When working with positive and negative numbers, remember the real number line and distance and they will not be such a problem.

Practice Problems

1. What is the magnitude from −10 to +2?
2. What is the magnitude from +2 to +8?
3. What is the magnitude from −3 to −12?
4. What is the magnitude from 0 to +7?

Solution to practice problems

1. 12
2. 6
3. 9
4. 7

THE CARTESIAN COORDINATE SYSTEM

An understanding of the Cartesian coordinate system is essential to solving some geometrical relationships and many inspection problems. The system begins with

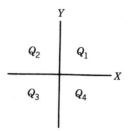

Figure 2.55 Cartesian coordinate system.

two axes, X and Y. The X axis is horizontal and the Y axis is vertical. These axes separate the space into four *quadrants*. The quadrants begin in the upper right space called quadrant 1 and proceed counterclockwise to quadrant 4. An example of the system is shown in Figure 2.55.

Plotting Points

When a scale is placed on the X and Y axis with certain divisions, points can be plotted. From the center of the system where the axes converge, values are assigned a positive or negative number, depending on which direction they are plotted. For example, on the X axis, a value to the right is positive and a value to the left is negative. On the Y axis, a value upward is positive and a value downward is negative.

Now the points plotted in any given quadrant can be assigned positive or negative signs. Points in quadrant 1 are positive, since it takes an X value to the right and a Y value upward. Points in quadrant 2 have a positive and negative sign. Points in quadrant 3 are both negative, and points in quadrant 4 are positive and negative. Examples of this relationship are shown in Figure 2.56.

When plotting points on the system the first value spoken is the X value. For example, if one is referring to the point at $(+2, +3)$ they are talking about an X value of $+2$ and a Y of $+3$. The point for these values would fall in the first quadrant, two units to the right on the X axis, and three units upward on the Y axis.

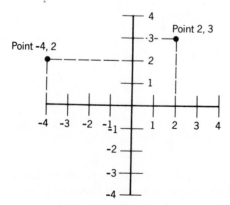

Figure 2.56 Points plotted on the X axis.

Graphs

When graphs (such as a line graph) are used, they usually appear in the first quadrant of the system. In this application, the horizontal X axis is usually referred to as the *abscissa* and the vertical Y axis is called the *ordinate*. In this relationship, all values plotted on the graph are plus unless a specific scale is assigned to the ordinate to dictate otherwise.

Angles

Angles on the Cartesian coordinate system begin on the first quadrant X axis at 0 degrees and rotate counterclockwise. This angular relationship is shown in Figure 2.57. The relationship is clearer when a circle is superimposed on the system (or vice versa).

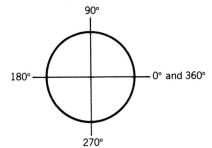

Figure 2.57 Angular relationship on the Cartesian coordinate system.

Practice Problems

1. Two values are plotted. The X value is $+3$ and the Y value is $+5$. In which quadrant will the plotted point for these values fall?
2. Two other values are plotted. The X value is -2 and the Y value is $+3$. In which quadrant will the point be plotted?
3. Which quadrant of the system contains the line for a 75-degree angle?
4. Two values are plotted. They are $X = +3$ and $Y = +3$. When you plot this point and draw a line from the center of the system through this point, what will the angle be?

Note. Draw the problem before answering.

Solutions to practice problems

1. Quadrant 1 (since both values are plus).
2. Quadrant 2.
3. Quadrant 1 (since it contains everything from 0 to 90 degrees).

4. 45 degrees (since the line goes through the middle of quadrant 1 and quadrant 1 contains everything from 0 to 90 degrees).

REVIEW QUESTIONS

1. Add the following fractions: $\frac{3}{8} + \frac{3}{16}$.

2. Reduce $\frac{12}{32}$ to lowest terms.

3. Multiply the following fractions: $\frac{5}{8} \times \frac{3}{4}$.

4. Write the decimal forty-three hundredths inch.

5. Multiply the following decimals: 0.502×0.003 (round the answer to four places).

6. Divide the following decimals: $0.502 \div 0.2$.

7. If $646 = X \div 12$, what is X?

8. Find the circumference of a 5-in.-diameter circle (round off to one decimal place).

9. Find the area of a circle that has a 7-ft diameter (round off to one decimal place).

10. What is the perimeter of a 12-in. square?

11. What is the area of the above 12-in. square?

12. Find the volume of a cylinder with a 5-in. radius and a height of 10 in.?

13. The hypotenuse of a right triangle is 10 in. long, and the shortest side is 6 in. long. What is the length of the other side?

14. Find the diagonal distance of a 0.010-in square (round the answer to three decimal places).

15. The sum of two angles in a right triangle is 140°. What is the other angle?

16. Find the largest square that will fit inside a 0.010-in.-diameter circle (round the answer to three decimal places).

17. Convert 98°F to Celsius (round answer to one decimal place).

3

Blueprints

To those working in industry, the engineering drawings used must be understood. These drawings relate design requirements to people such as operators, machinists, inspectors, assemblers, and many others who use engineering drawings to make, inspect, or assemble products.

Therefore, drawings are the industrial language. This language must be learned by all who have the need. There is a phrase that expresses the importance of reading blueprints:

as designed = as built

It is the intent of an engineering drawing (blueprint) to describe clearly all of the requirements that the product has to meet to be a quality product. Many quality problems occur solely because someone did not understand how to read or interpret a blueprint. To avoid getting involved in these problems (or causing them), it is necessary to be able to understand how to read and interpret blueprints.

Blueprints are made up of various "lines" and "views" that are easily learned. Once you learn these, you will be halfway to learning blueprint reading. The other half is understanding the title block, footnotes, change block, tabulations, and in some cases symbols. The title block gives information about the part including name, part number, engineers, draftsman, and tolerances that are applicable. The change block tells the latest revision letter, date, and what was revised. Tabulation is simply a way of making several similar parts using only one drawing. Footnotes are simply notes that are applicable and necessary for the user to build or inspect the part. Symbols are used on some drawings to save time and space. In the following pages we review some highlights of blueprint reading and interpretation that may be helpful.

LINES ON A BLUEPRINT

A basic requirement for reading blueprints is understanding the types of lines used and what they are used for. The following shows the type of lines and gives examples of how they are used on engineering drawings.

Object line (visible line). ━━━━━━━━ (heavy–solid)   Used to show the outer and inner boundaries of surfaces and features in a particular view.

Example: a washer (Figure 3.1).

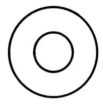

Figure 3.1   Washer.

Centerline. ━━━ ─ ━━━ (long and short dashes)   Used to show the geometrical center of an object.

Example: a washer (Figure 3.2).

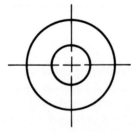

Figure 3.2   Washer.

Hidden line. ─ ─ ─ ─ ─ ─ (short dashes)   Used to show surfaces and features that exist, but cannot be seen in that particular view.

Example: a washer (Figure 3.3).

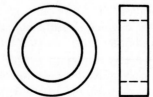

Figure 3.3   Washer.

Leader line. ⟶ (medium–solid arrow)   Used to point (or lead) the reader from a dimension (or note) to the feature or surface where the dimension (or note) applies.

Example: a washer (Figure 3.4).

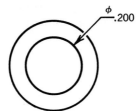

φ
.200

Figure 3.4   Washer.

Extension line. ———————— (light–solid)   Used to extend a feature (or surface) from the view so that it can be dimensioned.

Example: a washer (Figure 3.5).

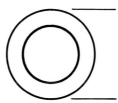

Figure 3.5   Washer.

Dimension line. ⟵——————⟶ (light–solid arrows)   Used to dimension the part or to show the distance between the extension lines.

Example: a washer (Figure 3.6).

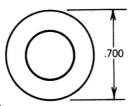

.700

Figure 3.6   Washer.

Section lines. ⧄ (light–solid) 45 degrees   Used to show the surfaces of a part that has been sectioned (or cut).

Example: a washer (Figure 3.7).

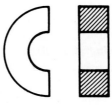

Figure 3.7   Washer.

Phantom lines. ——— —— ——— (one long and two short dashes)   Used in four main ways:

1. To show an existing structure that needs modification.

Example: A box must be put on an existing pole in the ground, and the pole is drawn in phantom lines to show that it is already there (Figure 3.8).

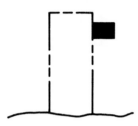

Figure 3.8   Structure needing modification.

2. To show alternate positions of an object.

Example: A lever that moves on an assembly would be drawn solid in one position and drawn in phantom lines to show its other possible positions (Figure 3.9).

Figure 3.9   Alternate positions of an object.

3. To avoid unnecessary detail.

Example: Avoid drawing the coils of a spring when it is not necessary.

4. To show the direction of a unilateral profile tolerance zone in geometric tolerancing. [See profile of a surface (unilateral zones).]

Cutting plane (or viewing plane lines). ⬆︎__ __·__ __⬆︎ (thick–broken) Used to show where a part is to be cut, or viewed by the reader.

Example: a washer viewing plane (Figure 3.10).

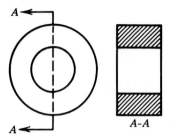

Figure 3.10   Washer.

Short break line.  (medium–jagged)   Used often to break up a particular view and show some of it in section (Figure 3.11).

Example:

Figure 3.11   Short break line.

Long break line. 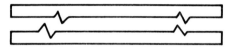 (thin)   Used to break up a long object along its length (Figure 3.12).

Example:

Figure 3.12   Long break line.

Cylindrical break line. (S-shaped)   Used to show a break in a cylindrical object (such as a drill rod) (Figure 3.13).

Example:

Figure 3.13   Cylindrical break line.

Chain line. (heavy and shaped like a centerline)   See chain line applications on page 123.

ORTHOGRAPHIC PROJECTION VIEWS

On engineering drawings, various views are used to pictorially represent the object. There are many different views that are used for this, but the main ones are *front, right side, left side, top,* and *bottom.* The most important view of all is the *front* view. This is the view that is selected to show (in visible lines) the most information about the shape of the part.

Example 1

As you see in Figure 3.14, the front view tells a lot about the actual part configuration because it shows clearly the hole, the slot, and the step.

After the all-important front view is chosen, using it as a reference, the right side, left side, top, and bottom views may be used. I say *may be used* because it is important that only the necessary views be used.

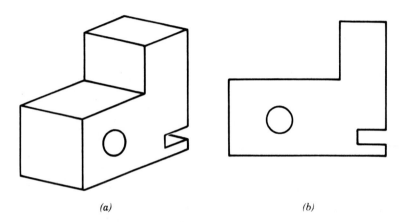

(a)   *(b)*

Figure 3.14   *(a)* The part. *(b)* The front view.

Example 2

A washer needs to be drawn with the necessary views.

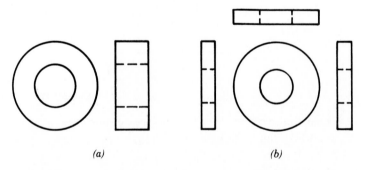

(a)   *(b)*

Figure 3.15   Washer drawing with *(a)* correct and *(b)* incorrect views.

You can see in Figure 3.15 that the washer needs only to have the front and right-side view (because all other views look exactly like the right-side view).

The blueprint reader must understand the views on a drawing and be able to "fold" them together in his or her mind, and imagine what the real object looks like.

Example 3

Figure 3.16 is a picture of a cardboard box that has been cut, and the sides have been folded out in the same way that drawing views are folded from the front view.

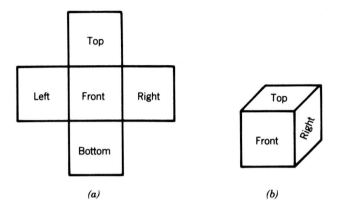

(a)

(b)

Figure 3.16 (*a*) A box with sides folded out. (*b*) The box itself.

Look at the views drawn in parts *a* and *b* of Figure 3.17 and see if you can form the solid object in your mind. The object is in part *b* of the figure. Only with practice can you learn this; a way to learn is to sketch the views of simple objects.

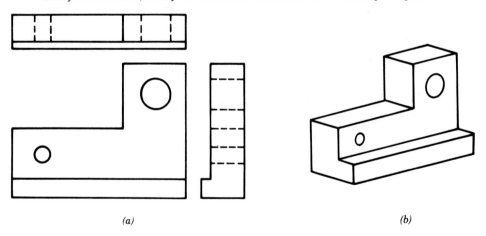

(a)

(b)

Figure 3.17 (*a*) View of an object. (*b*) The object.

CHANGE BLOCK, NOTES, AND TABULAR DIMENSIONS

Change Block

The change block is usually shown in the upper right-hand corner of a drawing. It is the block that keeps records on changes that are made. Drawings have *revision* letters.

Example

The first revision (or change) on a drawing would be given letter "A," and further revisions would be made using the alphabet. When the alphabet is used up, changes would start with "AA" and continue.

Generally, the information given in the change (or revision) block will be as shown in Figure 3.18.

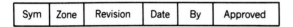

Figure 3.18   Revision block.

Sym. This is the revision letter (*A*, *B*, *C*, and so on).

Zone. This is the drawing area in which the change was made (Figure 3.19).

Example:

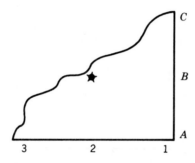

Figure 3.19   Drawing zone.

The asterisk is in zone *B*2 of this drawing.

Revision. This will describe the revision that was made. It usually describes what it was and what it has been changed to.

Example: 0.510 ± 0.010 was 0.510 ± 0.005.

Date. This is simply the date the revision was made.

Example: 5-4-82.

By. The person who made the revision on the drawing.

Approved. The person who approved the revision.

In the revision block other bits of information may be found, such as the *DCN* (drawing change notice) *number*. This is the document that is written to cause a change in the drawing. Drawings are not just changed. They must be approved by many people in most cases, and drawing change notices are part of the paperwork required to get this approval, after the change is reviewed and found to be acceptable.

Notes

There are two types of notes used on a drawing: *general notes* and *local notes*.

General notes

General notes are usually located at the bottom of a drawing. They are used when the note applies to several areas of the drawing and when it would clutter the drawing to put the note in all of those areas.

General Note Example

1. Finish all over.

General notes are numbered, as shown, and are used to save time and space, and to keep a drawing from becoming cluttered and hard to read.

The above is a good example of a general note. It would be very time consuming if all surfaces had a finish mark on them. This note says that all surfaces are to be finished (or machined).

Local notes

Local notes are notes that apply locally to one or more particular characteristics on the drawing. They are usually used with a leader line pointing directly to the characteristic; or they may be flagged, in which case a triangle is used with the note number within.

Local Note Example

This surface to be medium knurled.

Regardless of the notes used, there is one thing to remember: **Read the notes thoroughly before doing anything.**

Tabular Dimensions

There are times when several similar parts are made with only slight differences. An example of this is when several flat washers are produced, the only difference being the inside diameter.

Part Number	Inside Diameter	Part Number	Inside Diameter
9165–1	.250	9165–6	.500
9165–2	.300	9165–7	.550
9165–3	.340	9165–8	.580
9165–4	.420	9165–9	.630
9165–5	.475	9165–10	.750

Figure 3.20 Tabulation block.

These washers will usually have the same basic part number but will have different dash numbers to identify them. When something like this is done, tabular dimensioning can be effectively used to save time in drawing a different print for each washer (as shown in Figure 3.20).

Now with one drawing and all other dimensions the same, several washers can be made. The only difference between them is the hole size.

Example

A 9165-4 washer must have a 0.420 diameter hole in it.

THE TITLE BLOCK

The title block of a drawing is always in the lower right-hand corner of the page. It contains information that identifies the part and other particulars about that part. There is no standard arrangement for the information in the title block, but there are standard pieces of information found in it. Some of these are: part name, part number, material to use, next assembly, title block tolerances, draftsman, checker, company name, and scale.

1. *Part name*. The name of the part is always shown in the title block. Sometimes it is shown with a *descriptive modifier*, which is simply a word that follows the name to describe it further.

Example of Descriptive Modifier

CAP, HUB.

In the example above, the name of the part is *CAP* and the descriptive modifier is *HUB*. It is usually printed as shown, but when someone in the shop says the part name, he or she would say "HUB CAP."

2. *Part number*. All parts, assemblies, and subassemblies have a part number that should not be exactly the same as any other in the company. This is shown in the title block also.
3. *Material to use*. The title block may also list what material the part is to be made of, such as aluminum, steel, bar stock, and so on.
4. *Next assembly*. Some title blocks may tell you where (or in what assembly part number) the part will be used.
5. *Title block tolerances*. On almost all drawings, there are title block tolerances. These are usually in the form of a note, as shown in Figure 3.21.

Unless Otherwise Specified

.XXX = ±.010
.XX = ±.03
Angles = ±1 degree   **Figure 3.21** Title block tolerances.

This means that on all dimensions shown on the face of the drawing that do not have tolerances already specified, use the tolerances stated here. The exception is any BASIC dimension (which is always used with geometric tolerances).

There are also dimensions without tolerances shown.

Example

There are four different examples shown here: 0.50; 0.500; 0.510 BSC.; and 10°.

If you have the 0.50 dimension above, the title block must be used to establish the tolerance. Since the title block says that 0.XX = ±.03, the 0.50 is supposed to be 0.50 ± 0.03. Therefore, the limits are 0.53 and 0.47. The next one above is 0.500 with no tolerance shown. Since the title block says that 0.XXX = ±0.010, the 0.500 dimension is to be 0.500 ± 0.010, so the limits would be 0.510 and 0.490. The next one is 0.510 BSC. (which means BASIC). This is one dimension for which you cannot use the title block tolerance; to establish its limits, you must use the position tolerance.

The last one above is 10°. The title block is used here, and it says that angles are ±1 degree, so that the tolerance band on the 10-degree angle is 9 to 11°.

Remember. If dimensions have a tolerance shown, use it. If not, use the title block tolerance (unless it is BASIC).

Three ways a BASIC dimension can be shown:

1. 0.500 BSC.
2. 0.500 BASIC
3. [0.500] (preferred method)

BASIC Dimensions Always Locate a Geometric Tolerance Zone.

6. *Drafter.* The title block also shows the name of the drafter who drew the print.
7. *Checker.* The checker checks the drafter's work to make sure that there are no errors.
8. *Company name.* The name of the company that makes the part is usually shown in the title block.
9. *Scale.* The title block also shows the scale of the drawing. This is where the actual drawing size and the actual part size are compared to each other.

Example

Full scale (1 : 1). The drawing is the same size as the actual part.
Half scale (1 : 2). The drawing is one-half the size of the actual part.
Not to scale. The drawing is not to any particular scale.

Note. One should never take measurements from a print, but work from the dimensions given.

10. *Size.* Drawing sizes are indicated by letters. The following table indicates letters and the drawing size:

> *A.* $8\frac{1}{2}$ in. × 11 in. (or 9 in. × 12 in.)
> *B.* 11 in. × 17 in. (or 12 in. × 18 in.)
> *C.* 17 in. × 22 in. (or 18 in. × 24 in.)
> *D.* 22 in. × 34 in. (or 24 in. × 36 in.)
> *E.* 34 in. × 44 in. (or 36 in. × 48 in.)
> *J.* 36 in. × any length
> *R.* 48 in. × any length

Depending on the company, type of product, and other variables, there could be more information in the title block, but the information above is usually there.

RULES TO FOLLOW WHEN USING BLUEPRINTS

Wherever engineering drawings are used, there are some basic rules that should be followed.

1. Never get into the habit of memorizing a drawing regardless of how long you have used it. Always pull the drawing from the file on each job. Engineering drawings do change from time to time, and if you practice memorization, you are likely to make an error due to the change.

2. Unless otherwise specified, always keep only the latest change drawing in the file. When a new drawing is received, tear up the old one or dispose of it in the prescribed manner.

3. To avoid damage to drawings, handle them carefully and fold them properly.

4. Whenever there is a question about anything on the drawing, clear it up with your supervisor. Do not guess at anything.

5. When you suspect something is wrong with a drawing (unusually wide tolerances, unclear statements, missing dimensions, etc.), report it to your supervisor.

6. Drawings must always be followed to the letter.

7. Never mark on a drawing unless you are authorized to do so. If you are authorized, it is sound practice to sign and date your markings.

8. Make sure that you read and understand all notes on a drawing before you begin working.

TOLERANCES

There is variation in every process. No two parts can be made exactly alike. Therefore, all dimensions must have a tolerance. "Tolerancing" simply shows how much variation is allowable in a dimension; it is the area between the maximum and minimum limits. There are various types of tolerances on drawings related to various dimensions, or "characteristics." Some of these are shown in the following examples.

Example 1: Bilateral Tolerance

Variation is permitted in both directions from a specified dimension (see Figure 3.22).

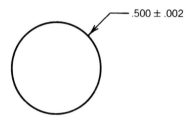

Figure 3.22 Bilateral tolerance.

Example 2: Unilateral Tolerance

Variation is permitted in only one direction from a specified dimension (see Figure 3.23).

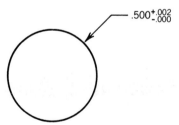

Figure 3.23 Unilateral tolerance.

Example 3: Limit Tolerance

Tolerances that show the high and low limits of a dimension (see Figure 3.24).

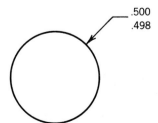

Figure 3.24 Limit tolerance.

Example 4: Single Limit Tolerance

A tolerance specifying a maximum or minimum limit only (see Figure 3.25).

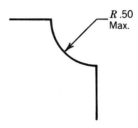

Figure 3.25   Example of a single limit tolerance.

Example 5: Title Block Tolerance

Many dimensions are shown without any tolerance. These are controlled by the title block tolerances at the bottom of the drawing. For example (see Figure 3.26):

.xxx = ±0.010   (a three-place decimal is ±0.010)

.xx   = ±0.03   (a two-place decimal is ±0.03)

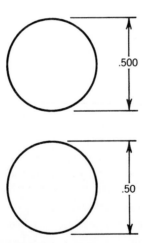

Figure 3.26   These dimensions are controlled by title block tolerances.

Example 6: Positional Tolerancing

Position of a feature is the theoretical exact location of that feature. Then the given position tolerance controls the amount of allowable variation from that exact location (Figure 3.27). All dimensions locating position are BASIC (sometimes shown as BSC, or with a box around it). BASIC dimensions have no tolerance. They simply locate a geometric tolerance zone.

Position is a complete course of study in itself and cannot be covered in one page. We will cover position and what it means later.

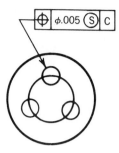

Figure 3.27 Bolt circle with its position tolerance. A 0.400–0.390 diameter with three holes equally spaced and located within 0.005 of position.

ALLOWANCES AND FITS

The following is a small collection of terms and examples that are necessary in the mechanical trades. The list is not all-inclusive but does stress some major points about allowances and fits, and relationships between mating parts.

Clearance fit (positive allowance). When two mating parts can be assembled easily (with airspace between them after assembly), as in Figure 3.28.

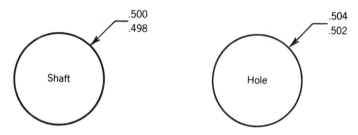

Figure 3.28 Mating parts with clearance fit.

No matter what size the shaft or the hole is, they will easily go together if they are both within their tolerance limits.

Interference fit (negative allowance). When two mating parts must be forced together (because the male part is larger than the female part), as in Figure 3.29.

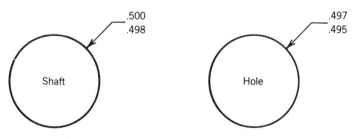

Figure 3.29 Interference fit of two mating parts.

No matter what their sizes (if they are both within their tolerance limits), there will be an interference fit.

Allowance (positive or negative). Allowance is defined as the *intended difference* between the virtual sizes of mating parts (whether it is clearance or interference).

Transition fit. This is a form of allowance that can be clearance *or* interference, depending solely on the *actual* sizes of each part (see Figure 3.30).

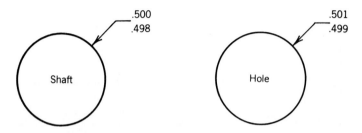

Figure 3.30   Transition fit of two mating parts.

There is interference if the shaft were actually 0.500 and the hole were actually 0.499. There is clearance if the shaft were actually 0.498 and the hole were actually 0.500.

Line to line fit. This is an interference fit where both parts are the same size (see Figure 3.31).

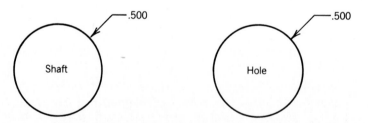

Figure 3.31   Line fit where both parts are the same size.

Virtual size. This is the *effective* size of a feature. For example, if a short shaft is supposed to be 0.500–0.498 diameter, and it is bent by 0.004, its virtual size is equal to the shaft's MMC (maximum material condition) plus the 0.004 crookedness. The MMC of the shaft above is its largest size (0.500), so that the virtual size is

$$0.500 + 0.004 = 0.504$$

TIR, FIR, AND FIM

Many dimensions are shown on drawings as TIR, FIR, or FIM. These abbreviations stand for

TIR   total indicator reading
FIR   full indicator reading
FIM   full indicator movement

Although according to the American National Standards Institute (ANSI) Y14.5, the preferred term is FIM, TIR and FIR are still used in industry and will probably continue to be used for years to come. The important thing to know is that all three of these mean the *same thing*. They mean the total distance that the indicator point travels from the start of the measurement to the finish.

Example

As shown on the print callout (Figure 3.32).

Figure 3.32   Feature control frame.

B surface is to be parallel to *A* surface within 0.010 TIR (Figure 3.33).

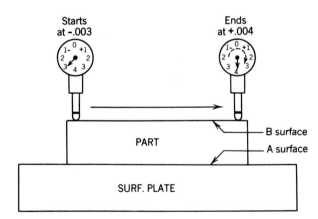

Figure 3.33   TIR: Parallelism.

Note. In this case, the part must still be within its thickness tolerance and parallel within 0.010 TIR at the same time.

Fully indicator movement (FIM) is −0.003 in. to +0.004 in., a total of 0.007 in.

One thing to remember when making a measurement calling for TIR, FIR, or FIM is to watch the extreme minus and the extreme plus amount on the indicator during the measurement.

Example

Indicator reading begins at −0.002 and ends at +0.002 (TIR is 0.004).

Indicator reading begins at 0 and ends at +0.002 (TIR is 0.002).

TIR is not a plus or minus value. It is simply a total value.

An easier way is to set your indicator purposely at 0 in the beginning so you can observe the movement in one direction more conveniently.

MAXIMUM MATERIAL CONDITION–LEAST MATERIAL CONDITION

The maximum material condition (MMC) or least material condition (LMC) of mating parts is very important because these conditions can cause problems between mating parts. The ANSI symbols for them are Ⓜ (for MMC), and Ⓛ (for LMC).

MMC. That condition of a dimension where the *most material* allowed (by the tolerance) is still there.

LMC. That condition of a dimension where the most *material allowed to be cut off* (by the tolerance) has *been cut off*.

Example 1

A shaft 0.502. Here the MMC is the largest size (0.502).

0.498. The LMC is the smallest size (0.498).

Example 2

A hole 0.600. Here the MMC is the smallest size (0.590).

0.590. The LMC is the largest size (0.600).

A good way to remember this is:

• MMC is the largest shaft and the smallest hole (per the size tolerance).
• LMC is the smallest shaft and the largest hole (per the size tolerance).

Example 3: Actual Problem Example

A shaft is supposed to fit inside a hole with only 0.001 clearance between them at MMC as shown in Figure 3.34. So

$$
\begin{array}{rr}
\text{MMC of hole} = & 0.503 \\
\text{minus MMC of shaft} = & -\,0.502 \\
\hline
\text{clearance} = & 0.001
\end{array}
$$

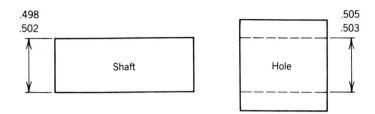

Figure 3.34 Example of fitting a shaft inside a hole.

With only 0.001 clearance, if the shaft happens to be oversize to 0.503, it will not fit and will require rework. Or, if the hole happens to be undersize to 0.502, it will not accept the shaft and will also require rework.

Most products are designed with clearances based on MMC and LMC. As the MMC increases, the clearance decreases. The MMC is very important for assembly because what will not go together cannot be shipped. The LMC is important also because clearances greater than LMC cause parts to be too loose in the assembly.

THE VIRTUAL CONDITION

The virtual condition (or virtual size) is a very important aspect of any size feature, particularly of mating parts. The collective effects of feature size and geometric error all contribute to a feature's virtual condition.

Example 1

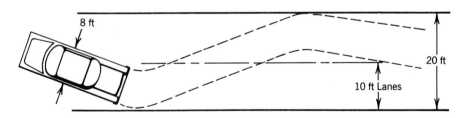

Figure 3.35 Virtual size.

Let's say that a truck is 8 feet wide and the street it is on is 10 feet wide in each lane (see Figure 3.35). Let's also say that the driver has been drinking. Therefore, he is going to swerve from side to side a little. As you can see, the outer edges of where the driver was swerving measure 20 feet and will not fit in the 10-foot lane. So an accident may occur. This example shows how the virtual size of something is different from its actual size.

Example 2

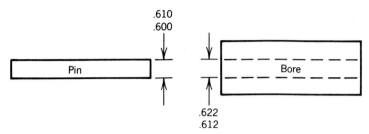

Figure 3.36   Pin and bore.

In Figure 3.36, a pin is toleranced at 0.610–0.600 diameter and must pass through a bore that is toleranced 0.622–0.612 diameter. At MMC this is only 0.002 clearance between the two parts. As shown in Figure 3.37, if the pin was tapered on one end and oversize to 0.615, the virtual size of the pin becomes 0.615 also, and there would be problems in assembly of the two parts.

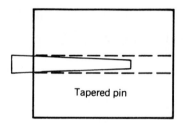

Figure 3.37   Effective size due to taper.

Example 3

Now let's say that the same pin in Example 2 was not tapered, but crooked. The pin measured 0.600 diameter all along its length. As shown in Figure 3.38, the distance to the highest area of the pin is 0.623. This pin would never go through any of the mating parts unless they were oversize or a different size, because the virtual size of the pin is now 0.623 and the largest diameter that the hole can be is 0.622.

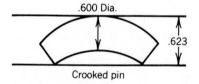

Figure 3.38   Virtual size of a crooked pin.

Remember. The *virtual size* of any dimension is dependent on its *size, form,* and *position* (or location).

- Size examples: (1) oversize shaft; (2) undersized hole.
- Form examples: (1) tapered; (2) barrel-shaped; (3) crooked.
- Position example: The feature is out of location (position).

COORDINATE METHOD OF POSITIONING FEATURES

One of the older methods of defining where a feature must be in position (related to datums, or reference surfaces) is the coordinate method. It is still used in some areas and therefore should be discussed. Coordinate dimensioning usually involves two dimensions locating the position of a feature (in this case, a hole), as shown in Figures 3.39 and 3.40.

Note. This type of tolerancing is not recommended because of its tolerance restrictions.

The drawing shown in Figure 3.39 means what is shown in Figure 3.40.

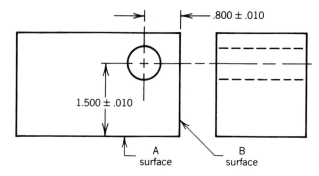

Figure 3.39   Coordinate ± tolerancing.

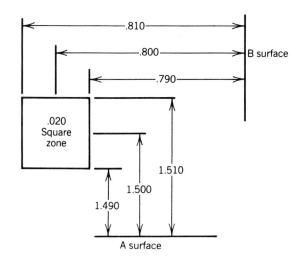

Figure 3.40   Square tolerance zone.

To inspect this part for hole location, you can use the following steps.

Step 1. Measure the size of the hole and record it. For example, we will say that the hold is 0.402 diameter.

Step 2. Get a pin that will fit the hole with a snug fit and put it through the hole.

Step 3. Set the part (with surface A down, as shown in Figure 3.41*a*) on the surface plate and measure the distance from the plate to the *top* of the *pin*. Then subtract one-half of the hole size from that distance. If the answer is between 1.490 and 1.510, the part is acceptable in this direction.

Step 4. Now rest surface B on the surface plate (as shown in Figure 3.41*b*) and again, measure the distance from the plate to the *top* of the *pin*. Again, subtract one-half of the hole size (which in this case is 0.201) from that distance. If the answer is between 0.790 and 0.810, the part is acceptable.

Remember. The part is good only if *both* of the coordinate dimensions are in tolerance at both ends of the hole.

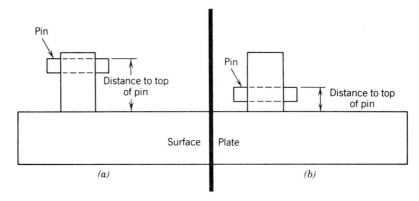

Figure 3.41   Measuring hole locations: *(a)* surface A located; *(b)* surface B located.

CONVERTING POSITION TOLERANCES TO COORDINATE TOLERANCES

On those tolerancing methods that use diametral zones, such as ANSI Y14.5 tolerances related to hole patterns, it becomes necessary to convert these diametral zones to coordinate ± tolerances for measurement. This is especially true when the hole pattern must be measured on a surface plate.

The coordinate tolerances on the hole positions with a diametral zone can be found using the standard chart in Figure 3.42. Reading this chart is easy once the user is familiar with it. The chart shows diametral tolerance zones in 0.001 increments up to 0.010 diameter. These diameters are shown along the left side of the chart. If your drawing says that the holes are to be within .xxx diameter of position, you first look for this diameter along the left side of the chart.

Now, note the diagonal line going from corner to corner of the box. This is the point where the coordinate tolerance you need to know meets the diametral tolerance zone you are working with. Once you find the point where the diametral tolerance meets the diagonal line, you simply follow that point to the right and straight up to get the tolerances you are allowed in the vertical and horizontal direction.

Remember. Follow the line from the diametral zone you are dealing with. Stop at the diagonal line. From that point, look right, and straight up for the plus and minus tolerance.

Example 1

A 0.021 diametral zone has coordinate tolerances of ± 0.0075.

Example 2

A 0.044 diametral zone has coordinate tolerances of ± 0.0155.

Example 3

A 0.010 diametral zone has coordinate tolerances of ± 0.0035.

You can also use the chart to convert coordinate tolerances to the appropriate diametral zone.

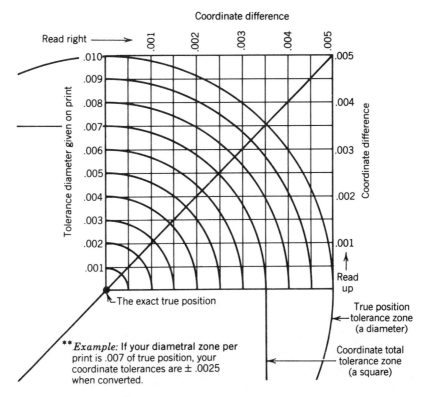

Figure 3.42 Sample conversion chart.

Example 4

The coordinate tolerances are ± 0.0035 *in both directions*. Find 0.0035 in the right column and the top row, and follow them to the diagonal line. Then follow the diametral line to find the diametral zone. The *answer* is the 0.010 diametral zone.

It is good practice for the user to become well versed in the conversions in either direction using the chart.

Note. This chart shows a maximum capability of only 0.010 diameter, but you can easily increase its capacity simply by moving the decimal points equally in all of the numbers.

Example 5

Change 0.010 to 0.10; then the ± 0.0035 becomes ± 0.035.

GENERAL RULES TO ANSI Y14.5M–1982 TOLERANCING

American National Standards Institute (ANSI) specification Y14.5M–1982 covers geometric tolerancing. This tolerancing method is widely used on drawings to maintain control over geometrical problems that are inherent in machined parts. When using the ANSI specification, there are five rules to consider. These rules govern the interpretation of the drawing symbols, datums, and tolerances.

The five rules are:

Rule 1. Regarding size tolerances.

Rule 2. Regarding position tolerance modifiers.

Rule 3. Regarding modifiers for tolerances other than position.

Pitch diameter rule. Regarding geometric tolerances that are applied to screw threads. This rule was formerly called Rule 4.

Datum/virtual size rule. Regarding datum features of size that are controlled by geometric tolerances to another datum. This rule was formerly called Rule 5.

When using geometric tolerancing, these rules must be kept in mind, so that the requirements of the drawing are clear. The following is a breakdown of the rules to simplify their meaning.

Rule 1.   This rule simply means that unless otherwise specified, the size tolerance of a feature controls form as well as size. No part of a feature shall be undersize or oversize. Size tolerances apply all over the feature. Per this rule, there is a maximum boundary of perfect form at MMC that cannot be violated. This rule controls geometric error within the size boundaries.

Rule 2.   This rule is very simple. When you use tolerances of position, you must put Ⓜ, Ⓛ, or Ⓢ if it applies (symbols are defined further in this chapter).

Rule 3.   On all tolerances (except position), Ⓢ is understood (does not have to be in the feature control frame). If Ⓜ applies, you must put it in the feature control frame.

Rule 4:   Pitch Diameter Rule. All geometric tolerances applied to screw threads apply directly to the axis of the pitch diameter of the thread unless otherwise specified. Figure 3.43 shows an example of *otherwise specified*.

Feature control frame

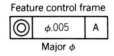

Major φ   **Figure 3.43**   Feature control frame.

Note. Geometric tolerances applied to gears and splines must designate the specific feature to which they apply (including the pitch diameter if applicable).

Rule 5:   Datum/Virtual Size Rule. A datum feature of size that is being geometrically controlled to another datum has a virtual size. Using this controlled datum as a reference must be at its virtual size (even when MMC is shown next to the datum letter). If it is not intended for the virtual condition to apply, the designer should apply a zero tolerance at MMC.

ANSI Y14.5M–1982 SYMBOLS

Some industrial drawings use ANSI Y14.5 tolerancing methods. This method uses symbols for geometric control instead of worded notes. The symbols are very easily learned because they generally take the shape of the geometric characteristic under consideration.

Example

The symbol for circularity is a circle.

The charts that follow represent a direct comparison of each geometric symbol and some particular rules and guidelines for them (see Figures 3.57 to 3.59).

Things to Remember

Maximum material condition (MMC) Ⓜ. Is that condition of a feature where the maximum amount of material that is allowed to be there (per the size tolerance) is still there, for example, the largest allowable shaft, or the smallest allowable hole.

Least material condition (LMC) Ⓛ. Is that condition of a feature of size where the most material that is allowed to be removed (per the size tolerance) has been removed, for example, the smallest allowable shaft, or the largest allowable hole.

Bonus tolerances and functional gages. Are directly applicable to any geometric characteristic that is modified by Ⓜ or Ⓛ.

Functional gages. Measure the collective effects of size and geometric tolerances simultaneously.

Regardless of feature size (RFS) Ⓢ. Means that the stated geometric tolerance applies no matter what size the feature is. No bonus tolerance is allowed, and no functional gage can be used.

Bonus tolerancing. Means that when MMC is shown modifying a particular tolerance, the stated tolerance applies only when the feature being controlled is at MMC size. *As the feature size departs from MMC size, you may add the amount of departure directly to the original tolerance.*

Example

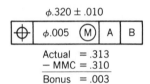

Actual = .313
− MMC = .310
───────────────
Bonus = .003

Figure 3.44   Example of bonus tolerancing.

In Figure 3.44 the hole must be in position within a cylindrical tolerance zone of 0.005 diameter when at MMC. *If the hole size were larger than MMC by 0.003,* you would get 0.003 bonus tolerance. You can add this 0.003 bonus tolerance to the original 0.005 tolerance and get a total of 0.008 zone. Using MMC gives you more tolerance.

Datum symbol. This symbol represents physical features or surfaces that must be used for location in inspection (see Figure 3.45). Any letter except I, O, or Q can be used. Use "AA," "AB," and so on, if the alphabet runs out. Datums must be physical features that can be contacted.

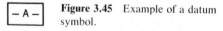

Figure 3.45   Example of a datum symbol.

Diameter symbol. This symbol replaces the word *diameter* (Figure 3.46). It should be used anywhere there is a diameter on the drawing and when the tolerance zone is diametral.

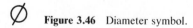

Figure 3.46   Diameter symbol.

Datum target symbol. This is the symbol for a datum on irregular parts such as castings and forgings (Figure 3.47). It can be a target point (see Figure

Figure 3.47 Datum target symbol.   **Figure 3.48** Datum targets.

3.48*a*) which must be located by a rounded pin, or a target area (see Figure 3.48*b*) which must be located by a flat pin of correct size. Datum targets must be located during machining.

Dimension not to scale. This symbol is used to state when a drawing dimension is not to the scale on the drawing (Figure 3.49).

.500   **Figure 3.49** Symbol meaning "dimension not to scale."

MMC modifier. Means that the maximum material condition applies (Figure 3.50).

Ⓜ   **Figure 3.50** Maximum material condition applies.

LMC modifier. Means that the least material condition applies (Figure 3.51).

Ⓛ   **Figure 3.51** Least material condition applies.

RFS modifier. Means regardless of feature size (Figure 3.52).

Ⓢ   **Figure 3.52** Symbol meaning regardless of feature size.

Projected tolerance zone modifier. When this symbol is shown (Figure 3.53), it means that the stated tolerance zone is projected above the feature, extending out in space. You must measure the feature compared to this extended zone and in the proper direction.

Ⓟ   **Figure 3.53** Symbol meaning that the stated tolerance zone projects above the feature extending into space.

Basic dimension. Basic dimensions (Figure 3.54) locate geometric tolerance zones. They could also be shown like this: 0.500 BSC, 0.500 BASIC. A basic dimension *has no tolerance;* it simply locates a tolerance zone.

.500   **Figure 3.54** Symbol meaning basic dimension.

Feature control frame (Figure 3.55)

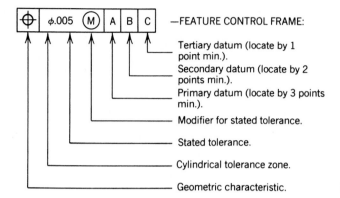

—FEATURE CONTROL FRAME:

Tertiary datum (locate by 1 point min.).

Secondary datum (locate by 2 points min.).

Primary datum (locate by 3 points min.).

Modifier for stated tolerance.

Stated tolerance.

Cylindrical tolerance zone.

Geometric characteristic.

Figure 3.55 Feature control frame.

Reference dimension. For computation only, not to be inspected (Figure 3.56).

(.500)

Figure 3.56 Symbol for reference dimension.

Feature. A feature is a physical portion of a part such as a hole, surface, or slot.

Tolerance zones. All tolerance zones shown in the feature control frame are total. For example, the position within a 0.005 cylindrical tolerance zone means that the tolerance zone is a 0.005 cylinder where the actual centerline of the feature must lie within. The exact position lies in the center of the 0.005 zone.

COMPARISON OF ANSI Y14.5 SYMBOLS

The geometric tolerancing symbols have undergone several changes over the years since their conception in the late 1950s. The most recent specifications involved are the 1973 and 1982 revisions. Figures 3.57 to 3.59 show the 1982 revisions in detail. Figure 3.60 is a comparison of the symbols in these two revisions.

PROJECTED TOLERANCE ZONES

There are certain cases where a particular geometric tolerance will be shown on a drawing with a Ⓟ next to it. This Ⓟ is the symbol for *projected tolerance zone*. To introduce projected tolerance zones, we must first understand how tolerance zones work. A tolerance zone for any feature is the same length as that feature.

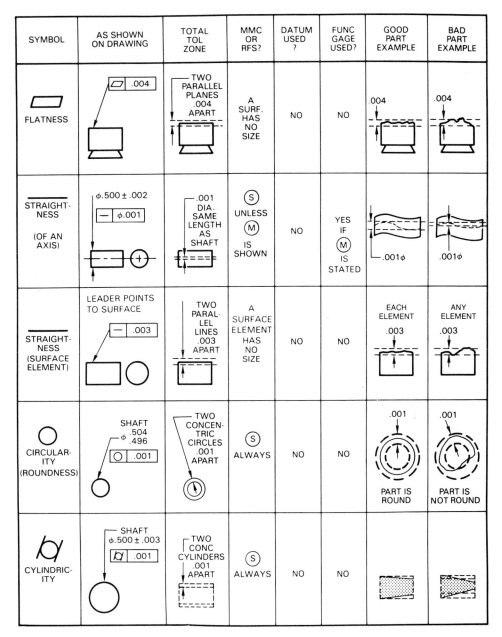

SYMBOL	AS SHOWN ON DRAWING	TOTAL TOL ZONE	MMC OR RFS?	DATUM USED ?	FUNC GAGE USED?	GOOD PART EXAMPLE	BAD PART EXAMPLE
FLATNESS	.004	TWO PARALLEL PLANES .004 APART	A SURF. HAS NO SIZE	NO	NO	.004	.004
STRAIGHT-NESS (OF AN AXIS)	φ.500 ± .002 — φ.001	.001 DIA. SAME LENGTH AS SHAFT	Ⓢ UNLESS Ⓜ IS SHOWN	NO	YES IF Ⓜ IS STATED	.001φ	.001φ
STRAIGHT-NESS (SURFACE ELEMENT)	LEADER POINTS TO SURFACE — .003	TWO PARALLEL LINES .003 APART	A SURFACE ELEMENT HAS NO SIZE	NO	NO	EACH ELEMENT .003	ANY ELEMENT .003
CIRCULAR-ITY (ROUNDNESS)	SHAFT φ .504 .496 ◯ .001	TWO CONCENTRIC CIRCLES .001 APART	Ⓢ ALWAYS	NO	NO	.001 PART IS ROUND	.001 PART IS NOT ROUND
CYLINDRIC-ITY	SHAFT φ.500 ± .003 ⌭ .001	TWO CONC CYLINDERS .001 APART	Ⓢ ALWAYS	NO	NO		

Figure 3.57 ANSI Y14.5 symbols.

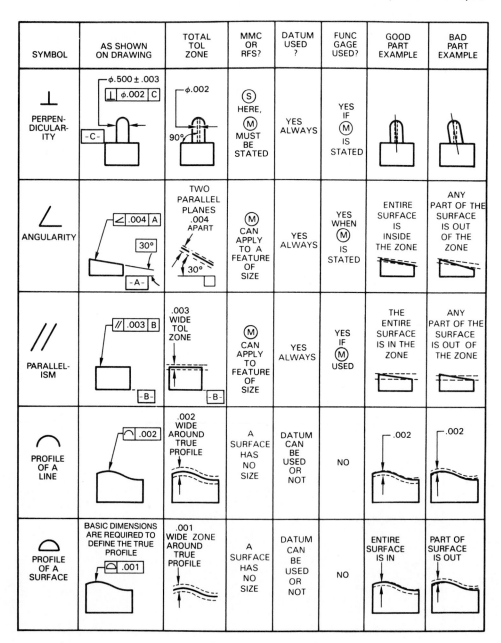

Figure 3.58 ANSI Y14.5 symbols.

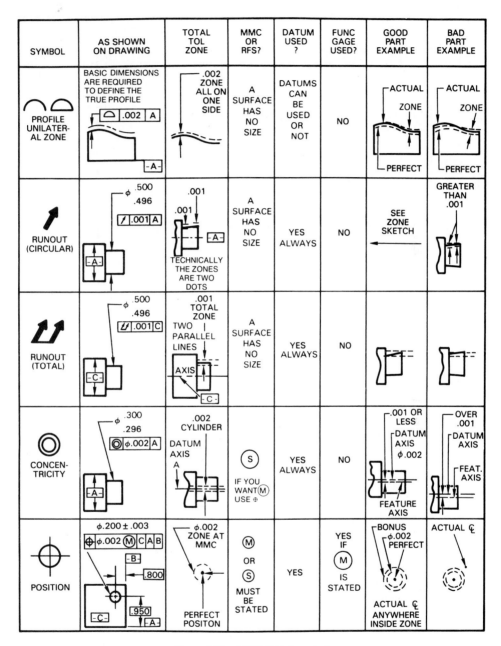

SYMBOL	AS SHOWN ON DRAWING	TOTAL TOL ZONE	MMC OR RFS?	DATUM USED ?	FUNC GAGE USED?	GOOD PART EXAMPLE	BAD PART EXAMPLE
PROFILE UNILATER-AL ZONE	BASIC DIMENSIONS ARE REQUIRED TO DEFINE THE TRUE PROFILE △ .002 A -A-	.002 ZONE ALL ON ONE SIDE	A SURFACE HAS NO SIZE	DATUMS CAN BE USED OR NOT	NO	ACTUAL ZONE PERFECT	ACTUAL ZONE PERFECT
RUNOUT (CIRCULAR)	φ .500 .496 / .001 A -A-	.001 .001 -A- TECHNICALLY THE ZONES ARE TWO DOTS	A SURFACE HAS NO SIZE	YES ALWAYS	NO	SEE ZONE SKETCH	GREATER THAN .001
RUNOUT (TOTAL)	φ .500 .496 ⟋⟋ .001 C -C-	.001 TOTAL ZONE TWO PARALLEL LINES AXIS -C-	A SURFACE HAS NO SIZE	YES ALWAYS	NO		
CONCEN-TRICITY	φ .300 .296 ◎ φ.002 A -A-	.002 CYLINDER DATUM AXIS A	Ⓢ IF YOU WANT Ⓜ USE ⊕	YES ALWAYS	NO	.001 OR LESS DATUM AXIS φ.002 FEATURE AXIS	OVER .001 DATUM AXIS FEAT. AXIS
POSITION	φ.200 ± .003 ⊕ φ.002 Ⓜ C A B -B- .800 -C- .950 -A-	φ.002 ZONE AT MMC PERFECT POSITON	Ⓜ OR Ⓢ MUST BE STATED	YES	YES IF Ⓜ IS STATED	BONUS φ.002 PERFECT ACTUAL ₵ ANYWHERE INSIDE ZONE	ACTUAL ₵

Figure 3.59   ANSI Y14.5 symbols.

Tolerance type	Name	1973 Standard Symbol	1982 Standard Symbol
Form	Flatness	▱	▱
	Straightness	—	—
	Roundness	○	○
	Cylindricity	⌭	⌭
Orientation	Parallelism	//	//
	Angularity	∠	∠
	Perpendicularity	⊥	⊥
Runout	Circular runout	↗	↗
	Total runout	↗ Total	↗↗
Profile	Profile of a line	⌒	⌒
	Profile of a surface	⌓	⌓
Location	Position	⊕	⊕
	Concentricity	◎	◎
	Symmetry*	≡	Use ⊕
Other symbols	Datum target	(A / 1)	(A1) or (φ.5 / A1)
	Dia. symbol	∅	∅
	Datum placement	[//	A	.002] or [//	.002	A]	[//	.002	A]

*Symmetry symbol replaced by position symbol in the 1982 standard.

Figure 3.60   Comparison of symbols.

As shown in Figure 3.61, the tolerance zone for this hole is always the same length as the hole itself (unless Ⓟ is used).

There are times, however, when the tolerance zone for a feature must be extended past the feature into space for very good reasons. The best reason in these cases is "fit." Sometimes it is not enough to control the feature alone, but to control the component that goes into it also.

For example, part 1 (shown in Figure 3.62) must have a hole put into it that is the correct size, and positioned to datum surfaces within a 0.002 cylindrical tolerance zone. According to this tolerance, you could have a hole that looks like Figure 3.63.

If part 1 must have a press-fit pin installed, the pin will follow the centerline of the hole and look like the drawing in Figure 3.64. The next assembly for part 1 requires that it fit into part 2 and that the mating surfaces close together completely, as shown in Figure 3.65. To ensure that the two parts fit together properly, we must make certain that the press-fit pin is also positioned to its datums. Since the pin extends above datum A by 1.00 in., we will simply project the tolerance zone for the hole above datum A by 1.00 in. This is shown in Figure 3.66.

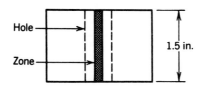

Figure 3.61 Tolerance zone and length of hole are the same. This hole is 1.5 in. long, so the tolerance zone is also 1.5 in. long.

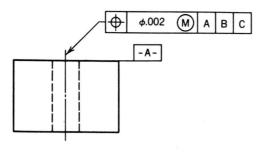

Figure 3.62 Part with a 0.002 cylindrical tolerance zone.

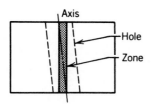

Figure 3.63 Tolerance zone of part 1.

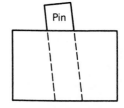

Figure 3.64 Part 1 with centerline pin.

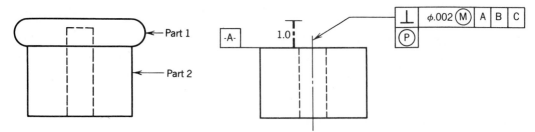

Figure 3.65 Assembly of parts 1 and 2.

Figure 3.66 Projected tolerance zone. The chain line (a thick long dash, short dash) shows the direction of the projected zone. If it were a blind hole, the chain line would not be necessary, and the 1.0 would go next to the circled P.

Now the position tolerance that is applied to the hole is also applied to the pin, and it will easily fit the mating part time after time (as long as both parts are within blueprint tolerance). Projected tolerance zone, Ⓟ, is a method of controlling a feature in such a way that you also control components that mate with that feature.

Projected tolerance zones are more restrictive than the original tolerance because the pin sticking up above the surface is being controlled as shown in Figure 3.67. The most important purpose for projected tolerance zones is to make sure that parts will fit together at assembly.

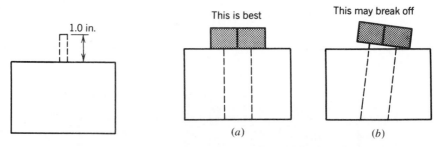

Figure 3.67 Press-fit pin also controlled.

Figure 3.68 (a) Proper seat for a bolt head. (b) Improper seat for a bolt head.

Another example for projected tolerance zones is to apply them to a part where a bolt must be torqued down with the bolt head meeting the surface flat all around. In this case, if the bolt head does not meet flat with the surface, it could break off when torque is applied. This example is shown in Figures 3.68.

Projected tolerance zones should be used only where they are absolutely necessary because they increase the cost of making the part.

Measurement Technique

The measurement technique for tolerances with projected tolerance zones is slightly different in that if you use a pin in the hole, the pin must extend out from the hole in the proper direction and your measurements will be taken on the pin at the distance the projection is indicated. If it is a 1-in. projection, you will measure at the pin and distance from the end of the hole to 1 in. above it. The measurement takes place only above the hole (in the projected tolerance zone) and not in the hole.

OPTIONAL DRAWING SYMBOLS

Some drawing symbols are optional and can be used to describe certain characteristics that usually require a word to explain. These optional symbols are listed in Figure 3.69.

Word	Symbol	Example
Counterbore or spotface		$6X\ \phi.224 \pm .005$ $\phi.375 \pm .010$
Depth (Deep)		.250 Min.
Quantity	X	$6X\ \phi.500 \pm .002$
Countersink		$\phi.320 \pm .010$
Square		$5X$
All around		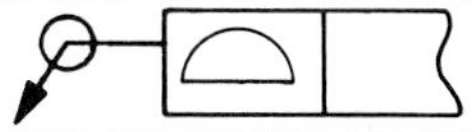
Conical taper		.002:1
Flat taper slope		$.010 \pm .003{:}1$
Radius	R	$R.250 \pm .005$
Spherical radius	SR	$SR.500 \pm .005$
Spherical diameter	$S\phi$	$S\phi.500 \pm .005$
Arc length		2.350

Figure 3.69   Optional drawing symbols.

UNDERSTANDING ANSI Y14.5 BONUS TOLERANCING

Geometric tolerancing has one major advantage built in—the *bonus tolerance application*. Bonus tolerances can be obtained on geometric controls where the maximum material condition or least material condition is used.

Before going further, a few definitions need to be repeated.

MMC. Simply defined, it is the smallest allowable Ⓜ hole *or* the largest allowable shaft (per the size tolerance).

Original tolerance. This is the size of the tolerance zone when (and only when) the feature being controlled is at its MMC.

Bonus tolerance. As the actual feature size departs from MMC, you may add this amount to the original geometric tolerance.

Total possible tolerance. This is the sum of the original tolerance (at MMC) *plus* the bonus tolerance (if any).

Example of How the Bonus Tolerancing Concept Works

Below are two mating parts. Part 1 has a hole in it in a certain position, and part 2 has a pin in it in a certain position. The only important thing in this assembly is that the two mating parts fit together while all four sides are matched evenly (as shown in Figure 3.70).

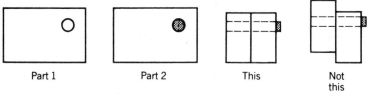

Figure 3.70   Two mating parts.

If the hole position and the pin position are not the same (and they seldom are), a virtual size is created as shown in Figure 3.71. However, if either the hole were larger, or the pin were smaller by a significant amount, the parts would still fit together as shown in Figure 3.72. This is how bonus tolerances work. They effectively use all of the possible clearances between mating parts to allow each individual part to be slightly farther away from its location tolerance and still work.

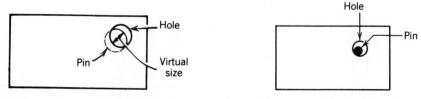

Figure 3.71   Virtual size with position tolerance.   **Figure 3.72**   Parts fit together.

How to calculate bonus tolerances

In one statement, you simply measure the feature, find out how far it is from its MMC, and add that much to the stated (original) geometric tolerance. This is your new tolerance (including bonus).

In Figure 3.73, the *perfect* position would be if the centerline of the hole was exactly 0.500 from datum B and exactly 1.250 from datum A.

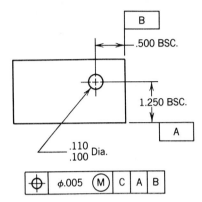

Figure 3.73 Bonus tolerance.

The original tolerance here is 0.005 MMC if the hole is 0.100. Let's say that we measured the hole and the actual size was 0.103. The actual size is 0.003 larger than its MMC, so you get 0.003 bonus tolerance on its position.

$$
\begin{array}{r}
0.005 \text{ original tolerance} \\
+\ 0.003 \text{ bonus tolerance} \\
\hline
0.008 \text{ total tolerance}
\end{array}
$$

Now you see that due to the maximum material condition concept, you have more tolerance to work with. What started out to be 0.005 position tolerance became 0.008 position tolerance simply because of the *actual* clearance between pin and hole.

Restrictions to bonus tolerancing

1. You can never go beyond print size dimensions to get more bonus tolerance.
2. Bonus tolerances can only be used in cases where if all the possible bonus tolerance were used, it would not affect the function of the part assembly.
3. Bonus tolerancing applies only on geometric controls where MMC or LMC applies.

ADDITIONAL TOLERANCE FROM SIZE DATUMS

In applications where a datum feature of size is shown in a feature control frame at MMC, a "bonus" tolerance is allowed. This is not the usual type of bonus tolerance, however. In this case a feature or pattern of features is allowed to shift or rotate as the datum feature of size departs from is MMC size. An example of this is shown in Figure 3.74.

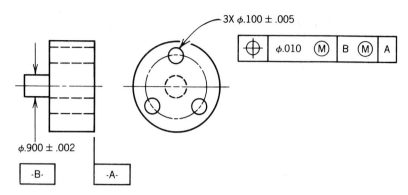

Figure 3.74 "Double" bonus.

In Figure 3.74 datum B is applied at MMC. The 0.010-in. position tolerance zone on the holes can increase in size as the 0.100-in. holes depart from their MMC size (0.095 in.), and the pattern of holes can shift off centerline or rotate as datum B departs from its MMC size (0.902 in.)

Therefore, bonus tolerance from a datum feature of size is not added directly to the position tolerance diameter, but allows the pattern to shift or rotate as the datum feature of size departs from its MMC size. Bonus tolerances do increase the diametral tolerance zone size as the controlled holes depart from their MMC size.

POSITION (PER ANSI Y14.5)

Positional tolerancing has been widely used since its introduction to manufacturing. It is designed to aid manufacturing in production, which it does in many ways. It allows you a wider tolerance zone to work with than the old coordinate methods. You can have bonus tolerances, depending on feature size and other conditions. Datums are specified for the purpose of location for measurement. There are many more advantages not stated here.

Things to remember

1. Dimensions used with position will always be BASIC.
2. Position tolerances apply with the feature at MMC, RFS, or LMC—whichever one is stated.
3. Position means "the theoretical exact location of a feature at its center axis or plane."

4. Position gives you a *circular* tolerance zone instead of a square zone. This circular (diametral) zone gives you 57 percent more tolerance area than the square zone (see Figure 3.75).

5. All the holes together in a bolt circle are called the *pattern*.

6. Some position tolerances apply only to the position of the pattern as a whole, and some apply to the position of the holes to each other.

7. When using position, (M), (S), or (L) must be shown in the frame.

The example given in Figure 3.76 is a positional tolerance that applies to each hole. There is also a 5.000 and a 4.000 dimension that locate the entire pattern to the edges of the part. This is called *combined position tolerance* (not recommended in the 1982 ANSI specification).

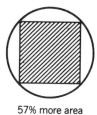

57% more area
in the circle

Figure 3.75   Circular and square zones. It is best to use all BASIC dimensions and a composite position tolerance.

Combined position and ± coordinates

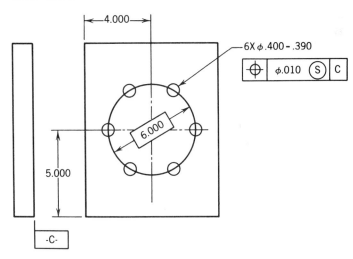

Figure 3.76   Combined position and coordinate ± tolerances.

The drawing in Figure 3.76 means that the centerlines of each of the six holes must be in their proper location within a 0.010 diameter tolerance zone. Note, however, that the pattern itself must also be within the coordinate tolerances.

This applies to the centerlines of the pattern. Before you can inspect this part, you must find a way to convert the hole positions from a bolt circle diameter to linear coordinate dimensions. There are basically two ways to do this. Dimensions and tolerances are shown in Figure 3.76.

One Way to Convert Is Shown in Figure 3.77

Here you must solve it for boths sides, knowing that the angle is 60 degrees because there are six holes divided into 360 degrees and knowing that the hypotenuse of the triangle is one-half of the bolt circle diameter. Then the opposite side length is typical to hole numbers 1, 6, 4, and 3 vertically, and the adjacent side is typical to numbers 1, 6, 4, and 3 horizontally. The horizontal positions of hole numbers 2 and 5 are exactly one-half of the bolt circle diameter (in this case, 3.000 in.).

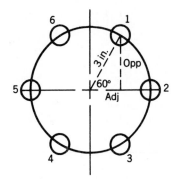

Figure 3.77 Conversion using a right triangle.

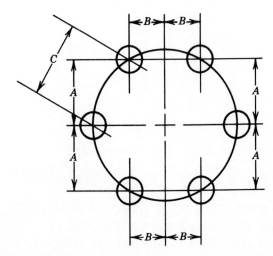

Figure 3.78 Conversion using the bolt circle chart.

Another (Easier) Way: The Bolt Circle Chart (Figure 3.78*)

Bolt circle charts are available in most manufacturing plants. You simply use the chart by multiplying the actual bolt circle *diameter* times the constant numbers in the table.

CONSTANTS FOR A SIX-HOLE CIRCLE

A dimensions = 0.43302×6 in.
B dimensions = 0.250×6 in.
C dimensions = 0.500×6 in.

To inspect

After you know what the dimensions should be, you simply clamp the part on an angle plate as shown in Figure 3.79 and measure each hole position vertically. Then you rotate the angle plate 90 degrees and measure each hole position that was horizontal. If A and B dimensions all fall within the 0.010 diameter zone, the part is acceptable.

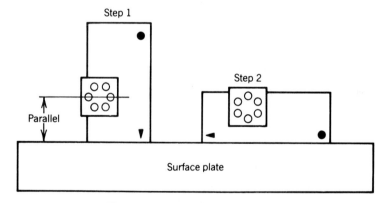

Figure 3.79 Inspecting hole locations.

WHEN FEATURES ARE OUT OF POSITION TOLERANCE

Position tolerances ($\oplus$) are used to control the location of features with respect to datum planes or features. One matter of controversy when using position tolerances is how to report those features that are not in the proper position when they are inspected.

---

* For a complete chart, refer to the Appendix.

The answer to the problem is very simple. All you need to do is report the deficiency according to the drawing requirements. For tolerances of position you have to know two things:

1. The basic coordinate(s) that locate the feature
2. The tolerance zone

Example 1

This example is concerned with a simple part where one hole is drilled that is located by two BASIC coordinate dimensions (see Figure 3.80). The 0.100–0.090 hole must be within a positional tolerance (cylindrical zone) of 0.005 RFS located by the 0.500 and 2.000 BASIC dimensions.

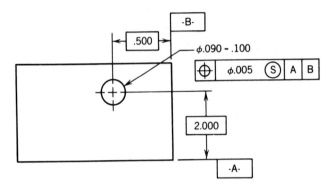

Figure 3.80 Position—RFS.

Actual Part Measurements

Let's say that during the inspection, the coordinate dimensions measured as follows.

The .500 BASIC measured 0.508.
The 2.000 BASIC measured 2.007.

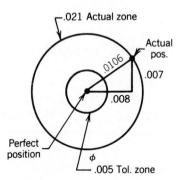

Figure 3.81 Actual diametral zone.

The differences in the actual (measured) coordinates and the BASIC drawing coordinates is 0.008 and 0.007, respectively. Figure 3.81 shows what the true diametral zone of the actual part is (Refer to the Appendix table on page 485 for the formula, or direct reading.)

Reporting the Actual

The rejection notice should say: "The 0.100–0.090 diameter is out of position tolerance to 0.021* diameter."

0.500 BSC IS 0.508
2.000 BSC IS 2.007

The report above tells everything that is necessary to describe the deficiency (the hole size would have come into play if the position tolerance was at MMC and the part exceeded the bonus tolerance zone).

Hole Patterns on a Bolt Circle

Another example is when there are a multiple of holes located on a bolt circle, as shown in Example 2, Figure 3.82. When holes are located on a bolt circle diameter, each hole has its own specific coordinates. Since the bolt circle diameter is BASIC, so are the coordinate dimensions.

Example 2

The four holes in Figure 3.82 are to be in position on a 6.000 BASIC bolt circle diameter within 0.010 cylindrical tolerance zone (when they are at MMC). As the hole size departs from MMC size (or gets bigger), the 0.010 tolerance zone also gets bigger by the same amount.

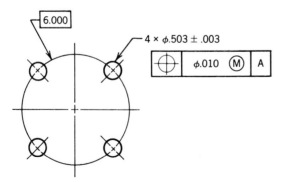

Figure 3.82 Bolt circle—position tolerance.

* Determined from the table (or formula) in the Appendix.

The coordinates in Example 2 for each hole are 2.121 in. when you calculate them. Since the angle is 45 degrees, both coordinates are 2.121 in. Let's say that we measured all four hole diameters and coordinates and got the following

Hole 1. Actual size is 0.503
X coordinate is 2.122
Y coordinate is 2.120

Hole 2. Actual size is 0.502
X coordinate is 2.120
Y is 2.121

Hole 3. Actual size is 0.505
X coordinate is 2.129 ⟩—— out of position
Y coordinate is 2.127

Hole 4. Actual size is 0.502
X coordinate is 2.120
Y coordinate is 2.121

Using your conversion chart, you find that the coordinates for each hole are ±0.0035 in each direction. This means that hole 3 is well out of position even when considering the bonus tolerance. Since 0.500 is MMC size, you can get 0.005 bonus tolerance to add to the 0.010 original zone for a total of 0.015 possible tolerance zone. The actual zone of the hole above is 0.020, which is much larger than the 0.015 total zone. This part cannot even be reworked because opening the hole size out to LMC size (0.506) will only give you a total allowable tolerance of 0.016. On this part, the rejection notice should read: "The 0.503 ± 0.003 diameter (at hole 3 only) is out of positional tolerance to a 0.020 diameter (exceeding the total possible tolerance). The 2.121 BASIC coordinates are 2.129 and 2.127."

Now you have reported the problem, and it is understood by those reading the report what is wrong with the four-hole pattern. Again, accurate reporting of parts that are out of position is simply a matter of comparing the drawing requirements with your actual measurements and describing the difference.

DATUMS

There are certain geometric tolerances that specify a datum surface or feature of size. The feature control frame will contain these datums in the last portion. When three datums are specified, this establishes a *datum reference frame.* The datum reference frame consists of the *primary, secondary,* and *tertiary* datums.

Datum Planes

When datum surfaces are used to establish datum planes, the following rules apply. With respect to functional datum planes:

1. *Primary datum.* This is the supporting datum that must be contacted at the three highest points on the surface. This is usually accomplished by a flat datum locating surface such as surface plates (see Figure 3.83).

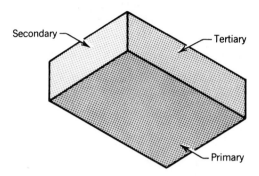

Figure 3.83 Three-plane concept.

2. *Secondary datum.* This is the aligning datum that must be contacted at the two highest points on the surface.
3. *Tertiary datum.* This is the stopping datum that must be contacted at the highest point on the surface.

These three planes are mutually perpendicular to each other and establish the datum reference frame.

Note. The same rules for the number of points contacted apply to datum targets except that the points contacted are specified points that are not necessarily the highest ones.

With respect to a cylindrical part the three datum planes are established as follows.

1. *Primary datum (axis).* This is established by two intersecting planes through the parts that are perpendicular to each other (see Figure 3.84).

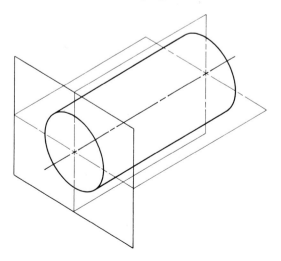

Figure 3.84 Three planes on a cylindrical part.

2. *Primary datum (end surface)*. This is established by the three highest points on the end surface.

3. *Secondary datum (axis)*. This is established by contacting the extremes of the diameter after the primary (end surface) has been contacted.

4. *Secondary datum (end surface)*. This is established by the extremes of the end surface after the primary (axis) has been contacted.

5. *Tertiary datum*. The tertiary datum for a cylindrical part is usually a feature that stops rotation (such as a slot) or one that is used for clocking.

The Selection of Functional Datums

During design of the product, functional datums are selected according to their functional importance, especially when they are interfaces in the next assembly. In some applications, only the primary datum is necessary. How many and what features are designated as datums depend on functional need.

Typically, the primary datum for a hole is that face which is perpendicular to that hole and is the mating surface (or interface). When positioning a hole from the outer edges of a part, the primary, secondary, and tertiary datums are needed. When positioning a pattern of holes to each other, only the primary datum (face) need be specified. This means that the position tolerance applies to the holes from each other and that they must be perpendicular to the datum plane in the same tolerance zone.

Datum axes are typically primary to certain geometric controls such as runout and concentricity. These datum axes are established by *datum features*. They are primary because they establish the functional axis of rotation for the unit. There are also times on a cylindrical part when a functional datum axis may be defined by contacting the diameter (primary) and an end face (secondary).

Who Should Contact Datums?

Inspection

Functional datums must always be contacted by inspection. A datum plane or axis on a finished part is a functional datum. Nonfunctional datums must also be contacted by inspection. Nonfunctional datums (such as datum targets) are important to the relationships on a part (e.g., casting) for verification of casting relationships but are not functional to the assembly.

Manufacturing

Functional datums *do not* have to be contacted in manufacturing (although they are allowed to). Manufacturing can use any means of locating the part that is practical. The important thing for manufacturing to consider is that the end result must be verified with respect to the functional datums. Manufacturing must locate nonfunctional datum targets, however. These datum targets are established for consistency in the manufacturing process. Datum targets are used on parts such as castings, forgings, weldments, and sheet metal parts.

Datums play an important role in manufacturing and design. If they are properly classified (functional/nonfunctional) and located by those concerned, they make measurements on the finished product more functional (therefore, meaningful) and the manufacturing process more consistent part after part.

DATUM TARGETS (PER ANSI Y14.5M–1982)

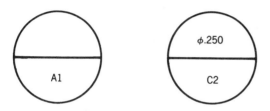

Figure 3.85   Datum target symbols.

Datum targets (symbols shown in Figure 3.85) are used to establish the location of parts that have irregular surfaces to make the initial cut. Castings and forgings are examples of parts where datum targets are applicable. Generally, you locate on datum targets to machine the primary datum plane, then locate on the machined datum plane for subsequent machining.

Example

In Figure 3.86 is a casting. As shown, the first machining pass will establish multiple datum A–B. Target symbols applied to the casting drawing for this part may look like the drawing in Figure 3.87. Now, according to the targets, the location of the part on the machine that will make the first cut may resemble the drawing in Figure 3.88.

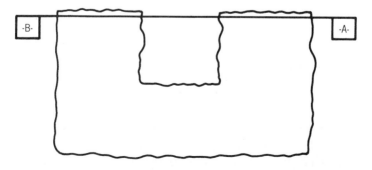

Figure 3.86   Casting.

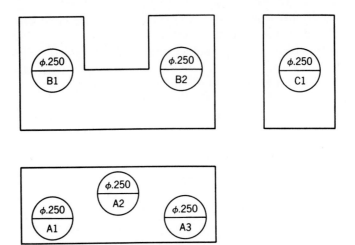

Figure 3.87 Targets shown.

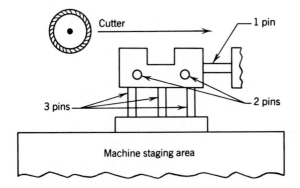

Figure 3.88 Path of machining cutter.

All pins are 0.250 diameter and flat at the contact area since the target symbol specifies a target area.

Using Datum Targets

When you use datum targets effectively, the initial location of irregular parts is more consistent part after part. This consistency in location of irregular parts for machining can be an asset to producing quality parts in many ways. Some of these ways are listed below.

1. Consistency in location for machining operations. Where to locate on each part is clearly defined.
2. Proper selection of tooling for production.
3. Types of locators to use are defined (flat or rounded).

4. Casting (or forging) dimensional problems (relative to the first cut) are readily discovered.

5. Casting or forging is securely located with respect to subsequent areas that are to be machined.

6. The casting can be inspected in the same way that it will be machined.

Types of Datum Targets

There are three basic forms a datum target may take.

Target point   Target line   Target area

The symbol describes which one is to be used on the part (see Figure 3.89).

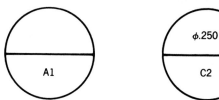

Figure 3.89   Target symbols.

Target points (and lines)

When a datum target symbol has nothing specified in the upper portion, as shown in Figure 3.90, it means that it is a target *point* or *line*. Target points must be located by a spherical (rounded) locator pin such as shown in Figure 3.91.

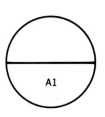

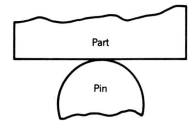

Figure 3.90   Target point symbol.   Figure 3.91   Target point located.

Target lines are contacted by the outside edge of a round pin, as shown in Figure 3.92.

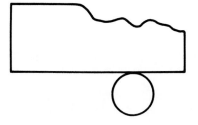

Figure 3.92   Target line located.

Target areas

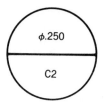

Figure 3.93   Target area symbol.

When a target area is specified in the upper portion of a symbol, as shown in Figure 3.93, it means that the specified diameter is the datum target area. Target areas are located by locator pins that are the diameter specified in the symbol and flat on top. An example of this is shown in Figure 3.94.

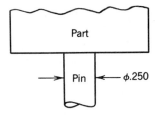

Figure 3.94   Target area located.

Where to locate

Datum targets (points, lines, or areas) are usually shown on a drawing with coordinate dimensions to describe exactly where the targets are located. Figure 3.95 shows an example of how this is done.

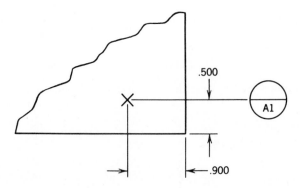

Figure 3.95   Targets on the drawing.

Using datum targets correctly can give you a greater chance of transforming an irregular mass of material into a finished part that is cleaned up on all machined areas.

BIDIRECTIONAL POSITION TOLERANCES

There are situations where the position of a feature in one direction is more important than its position in another direction. The example in Figure 3.96 shows how this used to be handled by old coordinate (plus and minus) tolerancing methods.

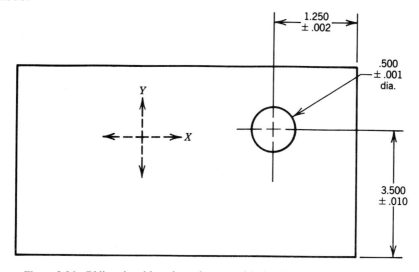

Figure 3.96   Bidirectional location tolerance with the old ± coordinate system.

Notice in Figure 3.96 that the hole has a larger tolerance in Y direction (±0.010) than in X direction (±0.002). This tolerancing method forms a rectangular tolerance zone that is 0.020 high and 0.004 wide for the location of the centerline of the hole. Figure 3.97 shows how this rectangular tolerance zone is constructed.

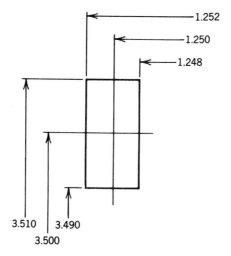

Figure 3.97   Rectangular tolerance zone.

As with all old coordinate methods of tolerancing, this rectangular zone is impractical and restrictive because a round hole is being located with a square-shaped zone. The ANSI tolerancing used on this part would show two different feature control frames. The example in Figure 3.98 shows how the part would be dimensioned per ANSI Y14.5M methods.

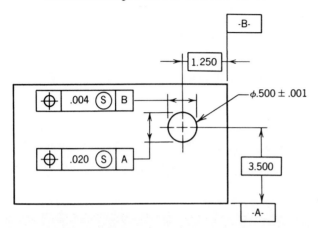

Figure 3.98   Example of a bidirectional position tolerance.

In Figure 3.98 the tolerance zone for each basic dimension is a different width. This forms the same rectangular zone as the old system.

Figure 3.99   Tolerance zone.

Now, the feature is controlled in both directions in the same way as the coordinate system. The old system has been translated to ANSI tolerancing. The only exception is that we could use MMC now to get bonus tolerance, and the datums are clearly specified (Figure 3.99).

One must remember that in the same way the old coordinate system produced a rectangular zone, the new system also created this type of zone. The old system (Figure 3.100) is ±0.002, and the new system (Figure 3.101) is a 0.004 total wide zone.

Figure 3.100   Tolerance zone.   **Figure 3.101**   Tolerance zone.

THREADS (NOMENCLATURE)

Screw threads are a very important part of industry. In the old days, threads were not standardized. One manufacturer would make the bolt and another manufacturer would make the nut and they might not fit together. Today, threads are standardized and interchangeable with each other no matter who makes them.

Threads have names for each geometrical shape and dimension that make them up. The *unified thread form* is one that is discussed here. There are many other forms of threads such as square, acme, buttress, whitworth, and knuckle, which are not covered in this book.

Unified thread form (external)

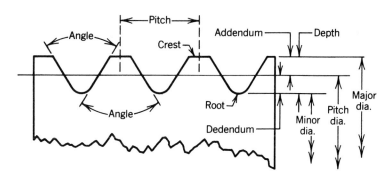

Figure 3.102 Unified external thread. (L.S. Starrett Co. Reprinted by permission.)

The drawing in Figure 3.102 shows a sectional view of a thread and names the various parts. The most important parts of the thread that should be known are shown here and are listed as follows.

Pitch. The distance from a point on one thread to the same point on the next thread.

Crest. The top of the thread.

Root. The bottom of the thread.

Angle. The included angle between the two flanks of the thread.

Minor diameter. The distance from root to root.

Major diameter. The distance from crest to crest.

Depth. The distance from crest to root.

Addendum. The distance from the pitch diameter to the crest on an external thread.

Dedendum. The distance from the pitch diameter to the root on an external thread.

Pitch diameter. The imaginary cylinder throughout the thread.

Screw threads are described on drawings as follows: thread nominal size; how many threads per inch; the type of thread; the class of thread, A external, or B internal; and the direction of the thread (left-hand is always stated, and right-hand is understood).

Examples of Thread Callouts

See Figures 3.103 and 3.104.

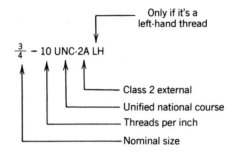

Only if it's a
left-hand thread

$\frac{3}{4}$ – 10 UNC-2A LH

Class 2 external
Unified national course
Threads per inch
Nominal size

Figure 3.103   English thread callouts.

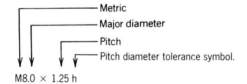

Metric
Major diameter
Pitch
Pitch diameter tolerance symbol.

M8.0 × 1.25 h

Figure 3.104   Metric thread callouts.

Class Numbers.   There are basically class 1, 2, and 3 threads.

- Class 1 is the loosest fit for quick assembly.
- Class 2 is tighter toleranced for torque applications.
- Class 3 is very tightly toleranced for close-fitting requirements.

Pitch (or Lead).   On a single thread, the pitch and lead are the same. The formula for the pitch or lead of a single thread is

$$P = \frac{1}{\text{number of threads per inch}}$$

CHAMFERS

Chamfers are commonly referred to as the "unimportant dimension." However, they can become important for the follow reasons: (1) They can break burrs and sharp edges; (2) they aid in easy assembly of component parts; and (3) in some cases, an oversize chamfer, or one with the wrong angle, will cause problems in fit or function of the part. The examples in Figure 3.105 show what kind of callouts may be used for a chamfer.

Example 1

On an outer edge, or outside diameter.

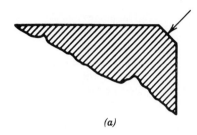

(a)

Example 2

Inside a diameter.

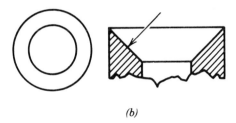

(b)

Example 3

A countersink.

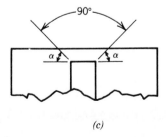

(c)

Figure 3.105 Various chamfers: *(a)* on an outer edge or outside diameter; *(b)* inside a diameter; *(c)* a countersink.

Example 4

(Figure 3.106)

As Shown on Blueprint	This Means	To Measure It

(a) | The angle should be 45 degrees and the depth should be 0.050 within certain tolerance limits. | A protractor of some sort.

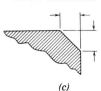

(b) | 45 degrees can be measured in either direction.

(c) |

Figure 3.106   A 45-degree chamfer.

Example 5

(Figure 3.107)

As Shown on Blueprint	This Means	To Measure It
	The angle of cut should be 45 degrees and the depth must be 0.100 within certain tolerances.	The part is still measured with some type of protractor, but now you must measure it in the specified direction. This is true when it is called out as shown, or with any angle other than 45 degrees.

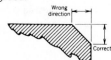 |

Figure 3.107

Care should always be taken to ensure that chamfers are **measured as they are shown on the blueprint.**

POSITION: FIXED AND FLOATING FASTENER CASE

When using position tolerances on mating parts it is necessary to understand fixed and floating fastener cases.

Fixed fastener case. This relationship between mating parts is simply defined by saying that the fastener (or bolt) that connects these parts is restrained in one of them or both. An example of this is when one of the parts has a tapped hole. This tapped hole would restrict the bolt from moving around. This restriction would mean that the holes must be lined up more accurately to fit.

Floating fastener case. This relationship between mating parts is much easier to work with since the fastener (or bolt) is not restricted in either part. Since it is not restricted, the clearance can be put to use in tolerance for the location of each hole position.

Floating Fastener Case

The formula for calculation of the position tolerance using the floating fastener case is

$$T = H - F$$

T = position tolerance diameter

H = MMC of the holes

F = MMC of the fasteners (bolts)

When you have an established position tolerance, and you must calculate the other sizes, the following formulas can be used:

$$H = T + F \qquad F = H - T$$

Example

$$\text{MMC (hole)} = 0.380$$

$$\text{MMC (fastener)} = 0.375$$

What should the positional tolerance be for both mating parts?

Solution $T = H - F$

$$= 0.380 - 0.375$$

$$= 0.005 \quad \textit{Answer}$$

Example: Floating Fastener Case

$$\text{Position tolerance} = 0.010$$

$$\text{MMC (hole)} = 0.510$$

What is the MMC of the fasteners you should use?

Solution $F = H - T$

$$= 0.510 - 0.010$$

$$= 0.500 \quad \textit{Answer}$$

Fixed Fastener Case

The formula for the fixed fastener case to calculate the position tolerance is

$$T = \frac{H - F}{2} \quad \text{(the tolerance for both parts)}$$

This formula can be changed to also calculate the hole size or fastener size with an existing position tolerance as follows:

$$H = F + 2T \qquad F = H - 2T$$

Example What is the required position tolerance for each mating part if the hole MMC is 0.380 and the fastener MMC is 0.370?

Solution $T = \dfrac{H - F}{2}$

$$= \frac{0.380 - 0.370}{2}$$

$$= \frac{0.010}{2}$$

$$= 0.005 \text{ position tolerance} \quad Answer$$

Understanding the fixed and floating fastener cases makes position tolerance calculation much easier (using the MMC concept per ANSI Y14.5M–1982). With the old square tolerance zone for hole positions, this was not possible.

THE DIMENSION ORIGIN SYMBOL

There are certain instances on drawings where there is a need to specify the reference surface of a linear dimension. The dimension origin symbol (Figure 3.108) adequately defines the origin (or reference surface) of a dimension using a circle around its extension line.

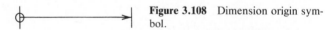

Figure 3.108   Dimension origin symbol.

The example in Figure 3.109 shows one application for the dimension origin symbol.

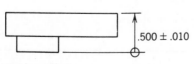

Figure 3.109   Application for the dimension origin symbol.

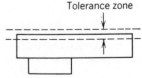

Figure 3.110

As shown in Figure 3.110 the origin of the 0.500 dimension lies at the shorter surface of the part. Therefore, measurement of the 0.500 dimension must be made using the shorter surface as a reference and the tolerance of ±0.010 surrounds the longer surface of the part. This is true because in the final assembly the shorter surface is the interface (or the surface that makes contact with the mating part).

To Measure the Part on the Surface Plate

To measure this part you must locate the short surface on the surface plate and indicate on the long surface as shown in Figure 3.111.

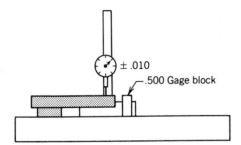

± .010

.500 Gage block

Figure 3.111 Transfer measurement.

The dimension origin symbol clearly specifies a required "datum" reference for a simple linear dimension without having to use datum symbols or a geometric tolerance.

THE CHAIN LINE

The chain line was adopted in the 1982 revision of ANSI Y14.5M specification. It is similar to a centerline except that it is much thicker (see Figure 3.112).

Figure 3.112 Chain line.

The chain line has several uses, most of which are to limit the geometric control of a feature to only certain areas (depending on design requirements).

Chain Line Applications

Chain lines have the following uses:

1. *To define a limited length of control* (Figure 3.113). The chain line shown in Figure 3.113 means that the position tolerance of 0.004 is limited to control only 1 in. of the length of the controlled diameter. This is true because the rest of the diameter is nonfunctional and need not be controlled.

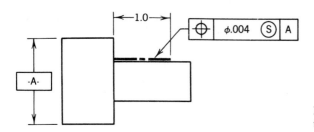

Figure 3.113   Limited length of control.

2. *To define a limited area of control* (Figure 3.114). The area shown in Figure 3.114 is the only area on the part that must be flat within 0.002. The area is bounded by the chain line and filled in with section lines for clarity.

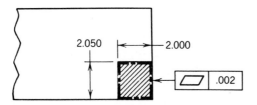

Figure 3.114   Limited area of control.

3. *To define partial datums* (Figure 3.115). The drawing in Figure 3.115 means that datum A is not the entire bottom surface. Only 2 in. inward from the end as shown should be contacted to measure the perpendicularity of 0.003. (This is so because only 2 in. of the bottom surface is contacted by the mating part at assembly).

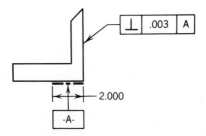

Figure 3.115   Partial datum.

4. *To show the direction of a projected tolerance zone* (Figures 3.116 and 3.117). (Refer to the Appendix for further information on projected tolerance zones.)

Using the chain line a designer can limit the control on a part to exactly where it is needed. As a result, limited control products are more economical to produce without any loss in function. There are many cases in industry where limited control can and should be applied.

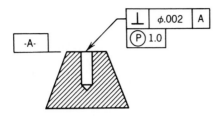

Figure 3.116 Projected zone direction—blind hole. There is no need for the chain line; the direction is obvious.

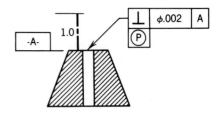

Figure 3.117 Projected zone—through hole. Use the chain line to show the direction and height of the projected zone.

COMPOSITE POSITION TOLERANCES

Composite position tolerances are often used to describe two different controls on a single feature or pattern of features using one geometric symbol for position. The example in Figure 3.118 is a very basic example of composite position tolerancing. There are three datums (C, D, and E). Datum C is the primary datum for the hole. Specifying datum C normally would control the perpendicularity of the hole where datums D and E control its location.

In this case the hole can be located within a 0.010 diametral zone for its location, but must be perpendicular to datum C within the 0.005 diametral zone. To specify this clearly on the drawing, composite position tolerance is used.

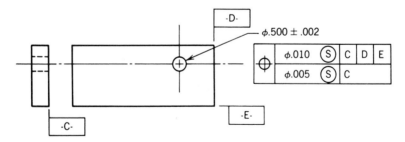

Figure 3.118 Composite position tolerance.

There are two different measurements required here. One must first measure the location of the hole regardless of feature size to datums C, D, and E. Then one must measure the perpendicularity of the hole regardless of feature size to datum C alone. If the controls were at MMC, this would call for two different functional gages (or one gage with two different virtual size pins).

Hole patterns. When composite position tolerances are used on a pattern of holes. The upper portion of the feature control frame locates the pattern to the outside edges of the part, and the lower portion locates the holes to each other.

Holes in line. When composite position tolerances are applied to hole drilled in line. The upper portion of the feature control frame locates each hole to the outside edges of the part, and the lower portion controls the coaxiality of the holes to each other.

HOW TO CONTACT FEATURE DATUMS

There are two basic categories of datums: *feature datums* and *datum targets*. Datum targets are covered in another portion of this book (see the Appendix for datum targets). Feature datums are surfaces, holes, slots, and the like. A feature datum is defined as an actual surface on the part. A size feature datum is defined as an actual size feature (holes, slots, etc.) on the part. A centerline cannot be called a datum.

The ANSI standard states that feature datums (surfaces or holes) must be fully contacted prior to the measurement of a geometric tolerance wherever datums apply (the only exception is targets). These feature datums are contacted regardless of their condition (rough, out of round, etc.). This is why a designer must apply a form tolerance on a datum, such as flatness, when it is necessary to make sure that the datum surface is flat enough to reference the measurement.

Simulated Datum Planes and Features

Simulated datum planes are any surface that is accurately flat within gage tolerances. Examples of simulated datum planes are *surface plates, toolmakers flats,* and *fixtures*. Simulated datum features are any size feature that is to size, round, and/or cylindrical within gage tolerances. Examples of simulated datum features are *gage pins, plug gages,* and *ring gages*. The ANSI standard states that datums on the part may be contacted by simulated datum planes to establish the datum plane for the measurement.

Feature Datums (Example: A Surface on the Part)

If a surface or group of surfaces have been identified as a datum on the part they must each be fully contacted against a simulated datum plane prior to any measurement. If the datum is *qualified* by a form tolerance (such as flatness tolerance), the flatness must be measured or verified before the surface is contacted for any other measurement. When the actual part surface is brought in contact with the simulated datum plane, the datum plane has been established and the measurement can begin.

Size Feature Datums (Example: A Hole or Pin)

Size feature datums apply to those measurements that must be related to a size feature on the part. Size feature datums are fully contacted by the most accurate equipment that will do the job.

Example: Datum Is a Hole

> When a hole has been identified as a datum, it must be contacted with the largest true cylinder that will fit the hole (in an RFS situation). This cylinder is often an approved simulated datum feature of size, such as a gage pin or an expanding mandrel.

Example: Datum Is a Pin

> When a pin is a datum it must be contacted with the smallest true cylinder that will fit it. This true cylinder is often a ring gage or a collet.

Inspectors and Datums

It should be clearly understood that whoever does the inspection of the finished product *must* at all times locate the specified datums. The inspection of the product cannot be from any other feature of size or surface on the part even if it is more convenient. If inspection is performed from datums as shown on the drawing, that inspection will be a correct functional inspection of the part.

Datums That Are Technically Not Acceptable Unless Targeted

There are some tools that are often used to locate datums that are technically not acceptable unless datum targets are used. The most widely used simulated datum is the V-block. A V-block only makes contact on two "lines" of the surface at one time. This violates the rule of "full contact unless targeted." Another example is the jaws of a rotary table. These jaws only contact at three areas of the datum diameter. The last example (not the least) is the coordinate measuring machine, which only makes contact on points of the datum surface. This can be corrected, however, by bringing the part in contact with a simulated plane, then entering the simulated plane information into the machine.

POSITION TOLERANCE STATED AT LMC

There are some applications where position tolerances are stated at LMC. LMC is that condition of a feature of size where all of the material that is allowed to be removed (per the size tolerance) has been removed. The LMC is simply remembered as the largest allowable hole, or the smallest allowable shaft.

The 1982 version of the geometric tolerancing specification ANSI Y14.5M allows position tolerances to be stated at LMC condition. Whenever the position tolerance is stated at LMC in the feature control frame, it means that the tolerance shown only applies when the feature is produced at LMC size. As the feature departs from LMC size (toward MMC) the stated tolerance gets larger by the same amount of that departure.

One application for position at LMC could be a boss that takes a drilled hole. If the boss and the hole are both located by the same basic dimensions, and both are in position at LMC, the edge distance (or break out distance) between them can be guaranteed at a certain minimum. This is true because the tolerance applies when the hole is its largest, and the boss is its smallest size. Another example application follows.

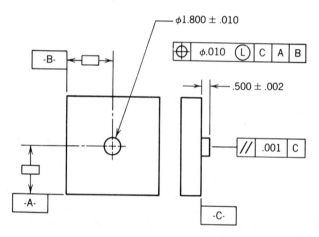

Figure 3.119 Position at LMC.

As shown in Figure 3.119, there is a 1.800 diameter pad on this part that is used as a locator for the part. It must be in position such that it will make contact with a fixed locator on the machine. Position at LMC is applicable here because as the diameter of the pad increases it can afford to be farther out of position as it will still make contact with the fixed locator on the machine (Figure 3.120).

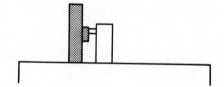

Figure 3.120 LMC application.

The LMC of the pad is 1.790 and at this size, and the position tolerance is 0.010. If the pad is larger than LMC size, you have bonus tolerance equal to that amount. The table in Figure 3.121 shows the position tolerance allowed because of the actual size of a produced pad.

Actual Feature Size	Position Tolerance Allowed
1.790 Ⓛ	.010
1.791	.011
1.792	.012
1.793	.013
↓	↓
1.810 Ⓜ	.030

Figure 3.121 Table of tolerances.

ZERO POSITION TOLERANCE AT MMC

Position tolerancing has been around for a long time, but in many cases, the best method of positioning features (zero tolerance at MMC) is not used. One of the main reasons it is not used is that it is misunderstood. The thought of putting zero (or no tolerance) on a blueprint strikes fear in the minds of many. Zero tolerance at MMC is not taking tolerance away from the part. In fact, it does not change the tolerance value that is traditionally allowed for a part. The part in Figure 3.122 is an example of zero tolerance at MMC.

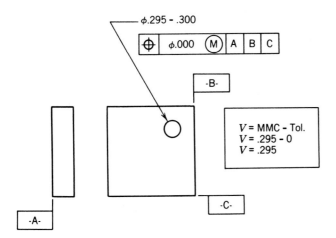

Figure 3.122 Zero tolerance at MMC.

In Figure 3.122, the position tolerance allowed for the 0.300–0.295 hole is solely dependent on the actual size of the produced hole. If it is produced at MMC size (0.295) it must be in perfect position. That position is described by the basic dimensions (not shown) that locate the hole. The only position tolerance allowed is the amount that is the difference between the actual size and the MMC

Actual Hole Size	Position Tolerance Allowed
.295 (M)	0
.296	.001
.297	.002
.298	.003
.299	.004
.300 (L)	.005

Figure 3.123   Table of tolerances.

size. For example, if the hole is produced to 0.298 diameter, a 0.003 diametral tolerance zone is allowed for the position of the hole (see Figure 3.123).

The part could be traditionally toleranced as:

Hole size range: 0.300–0.298

Position tolerance = 0.003 at MMC

Notice that the size range of the hole had to be tightened. The maximum hole size has not changed, and the virtual size of the hole has not changed (it is 0.295 virtual size either way). Let's also say that a *functional gage* has been made to check the hole position. The functional gage would have a 0.295 nominal pin located precisely at the basic dimensions of the hole.

We can also say that on the traditional part some parts were made with undersized holes (0.297) but these holes were in perfect position. The functional gage would accept every one of these parts (because they are good parts), but since the hole is undersize, they must be rejected. This is why zero tolerance at MMC is so effective. The functional gage also *doubles as a Go size gage*. No functional parts can be rejected simply because of hole size, as with traditional position tolerancing.

CHAIN, BASELINE, AND DIRECT DIMENSIONING

There are times when a row (or chain) of features must be dimensioned to control their location. An example of this is a set of holes in a straight line as shown in Figure 3.124. Careful selection of the type of dimensioning to use on these holes will assist you in getting the maximum tolerances allowed with a specific build up of tolerance over the entire chain of holes. It must be remembered that with any type of dimensioning there is no *tolerance buildup* when you use *basic dimensions*. The examples to follow all use linear plus and minus dimensions, and will have tolerance buildup (or accumulation).

There are three methods in which a "chain" of size features can be dimensioned: *chain dimensioning, baseline dimensioning,* and *direct dimensioning.*

Chain Dimensioning

Figure 3.124 shows an example of chain dimensioning with plus and minus tolerances. When chain dimensions are used, the maximum tolerance accumulation overall is equal to the sum of the intermediate tolerances. Since all three tolerances over distance A are ±0.005, the maximum tolerance for distance A is ±0.015 (the sum of the three intermediate tolerances).

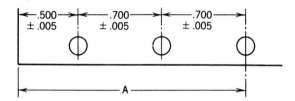

Figure 3.124 Chain dimensioning.

Baseline Dimensioning

Using baseline dimensioning (preferred in most cases), the maximum buildup tolerance for A dimension is equal to the sum of the two dimensions' tolerances that make up A dimension (they are the 0.500 ± 0.005 and the 1.300 ± 0.010, and add up to ±0.015) (Figure 3.125). This is the greatest tolerance accumulation.

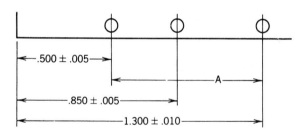

Figure 3.125 Baseline dimensioning.

Direct Dimensioning

Direct dimensioning allows you to limit the tolerance buildup in a dimension to the stated tolerance on that dimension. Figure 3.126 shows an example of direct dimensioning where the tolerance on A dimension is as stated.

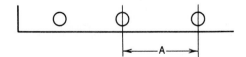

Figure 3.126 Direct dimensioning.

REVIEW QUESTIONS

1. Describe two uses for a phantom line on a blueprint.
2. What is the most important orthographic view on a blueprint?
3. What type of tolerance is 0.500–0.505?
4. What is the name of the fit where two mating parts can easily be assembled?
5. What does FIR mean?
6. What is the MMC size of a hole that has a size tolerance of 0.500–0.510?
7. What is the LMC size of a shaft that has a size tolerance of 0.250–0.255?
8. What is the virtual condition of a 0.500 diameter shaft that is bent on axis by 0.004 in.?
9. Draw the symbol for functional datum A.
10. Is there allowable bonus position tolerance when the tolerance is stated at RFS?
11. When evaluating flatness, is a datum surface used?
12. Which of the general rules of ANSI Y14.5 states that size controls form?
13. A basic dimension has no tolerance. It locates a geometric tolerance

    _____.

14. What is the minimum points necessary to contact a primary datum?
15. What is the function of a secondary datum?
16. A zero position tolerance at MMC means that the tolerance for position is equal to the

    amount of _____ tolerance allowed.
17. When pattern location and hole-to-hole location tolerances differ, the applicable posi-

    tion tolerance for this application is _____ position tolerance.
18. An example of the floating fastener case for position tolerance means that both mating
    parts have clearance holes. (True or False)

4

Common Measuring
Tools and Measurements

MEASURING TOOLS

Some very basic measuring tools are covered in this chapter. Although it does not discuss all of the basic measuring tools, it does lead you through the ones that are commonly used. Helpful hints are given on the many causes of errors in measurement and certain ways to avoid such errors. It is worthwhile to study the possible errors related to each tool and keep them in mind when using the tools.

References and Measured Surfaces

One of the first things to consider with these measuring tools is which surface of the tool is the reference surface, and which surface is the measured surface. The reference surface is fixed (not movable), and the measured surface is movable. Proper measurement with these tools requires that you fix the reference surface of the tool on the part and take your measurement at the movable (measured) surface.

Reading the Tools

It is also necessary to become proficient at reading the various tools, for example, *micrometers, indicators,* and *verniers.*

There are different micrometer models (such as those that discriminate to

0.001 in. and those that discriminate to 0.0001 in.). There are also different *vernier scales* (such as each line division being spaced at 0.025 in. or 0.050 in. apart.

Selecting the Proper Tool for the Measurement

The rule of thumb is to select a measuring tool that is ten times more accurate than the total tolerance to be measured (or can discriminate to one-tenth of the total tolerance). For example, if you are measuring a part with a 0.010 in. total tolerance, use a tool with 0.001 in. discrimination. If the part is 0.001 in. total tolerance, use a tool with 0.0001 in. discrimination capability.

Unnecessary Accuracy

It is important to select the proper accuracy of the tool for the measurement. An example of unnecessary accuracy is measuring a casting with a tolerance of ±0.030 in. with a dial snap gage graduated in 0.0001 in. Unnecessary accuracy can take considerable time in inspection of parts that is just not necessary or economical. A more appropriate accuracy of the tool would be graduated in 0.001 in.

Transfer Tools

It is also necessary to understand *transfer tools*. These are tools that cannot be read directly. They make contact with the part (on the dimension being measured); they are then locked in the measured position; and they are measured with another tool (such as a micrometer). Therefore, the dimension has been "transferred" from the part, to the transfer tool, to the scale on the tool used to make the measurement.

Attribute Gages

Attribute gages are simply gages that are designed to either *Go* or *Not Go* in the dimension of a part. Attribute gages tell you if a part is "good" or "bad" but do not provide a reading to tell you how good or how bad. There are many attribute gages. Two of the most widely used ones are *plug gages* and *ring gages*. The plug gage is designed with two plugs (one at either end). One is the Go member and the other is the NoGo member. If the Go member goes into the hole, and the NoGo member does not, then the hole is within its limits, and acceptable. The ring gage does about the same thing only it gages an outside diameter on a Go–NoGo basis. Attribute gages are used for quick verification of component parts, but limit you severely in your capabilities of measurement. They will not tell you how good or how bad a part is. They also cannot be adjusted in any way. They can only be reworked into another size gage.

Not being able to tell how good or how bad the part is can become troublesome during production or inspection. The operators cannot tell when their machines are going out of tolerance; it is not possible to maintain control charts for averages and range. Inspection must use another method of measurement when rejecting a part to get its actual size (actual size is the measured size with a variable tool).

Variable Gages

These gages provide actual size information and this can be very helpful in spotting the need for adjustments. Variable gages allow inspectors to measure and record actual sizes of parts. Variable gages, for the most part, use indicators and travel mechanisms to show actual movement so a dimension can be compared to the master from which the gage was set.

Accuracy and Precision

Accuracy means an unbiased true value, and *precision* means getting consistent results, repeatedly. Accuracy and precision are two terms that must be understood in all measurement. Accuracy is something that can be drastically reduced by measurement errors. Precision can be reduced by mechanical means. The difference between accuracy and precision can be explained by an example.

A dimension is to be measured with a gage block stack. Upon stacking gage blocks that are of different thicknesses, the stack will be exact to whatever block you put into it. If a person chooses the wrong blocks to make his stack, it will be inaccurate, although measurements taken from the stack will still be precise (every measurement will repeat the same "wrong" results). On the other hand, if the height of the stack is the unbiased true value, the measurements will be both accurate and precise.

Feel

Measuring tools require the person using them to obtain a *feel* for the proper measuring pressure. You cannot use a micrometer as a C-clamp, nor can you have it too loose on the part. Measuring pressure should be light enough to avoid squeezing the part or damaging the tool, yet heavy enough to make good contact with the part surfaces. At no time should the pressure be heavy enough to secure parts in the tool. Most often, the feel for hand tools is improved with experience. For good practice with feel, use a pressure micrometer. This micrometer has an indicator for proper measuring pressure and it can be used to increase your ability to feel.

Care of Tools

Hand tools are all susceptible to damage, rust, wear, miscalibration, and other things that can cause them to fail in their function and reduce the accuracy of the measurement. Precision tools are generally sturdy but they do require careful handling and proper care.

A light film of oil will protect tools that are susceptible to rusting. Avoid rough handling and dropping tools. If a tool is dropped, check it for damage and calibration. Examine them frequently for wear on the measuring surfaces and try to keep tools calibrated. Keep tools protected in a toolbox and avoid storing them on top of each other. Know the particular precautions necessary to keep your tools in good condition. Well-cared-for tools and effective inspectors are an unbeatable combination.

MEASUREMENTS AND ACCURACY

When making measurements, there should be one goal in mind—to be as accurate as possible. Due to many variables, no measurement in the shop is ever perfectly accurate; we can only come very close. The accuracy of any measurement made in the shop is largely dependent on a few select variables. These variables are *calibration, manipulation, reading, workpiece geometry, measuring pressure,* and *dirt* and *burrs*.

Calibration

If a measuring tool is not properly calibrated the accuracy of the measurement will be poor. There are many conditions that can cause a tool to be out of calibration, including wear, dropping the tool, improper standard used, and improper pressure (too much or not enough). Measuring tools should be checked periodically for calibration to make sure that this variable is not a problem.

Manipulation

The person using the tool must know how to manipulate (use) it to perform the measurement. He must be aware of those measurements that require centering, or avoid cocking (or tilting) the tool the wrong way. An example of manipulation error is when a micrometer is used to measure a diameter. The micrometer faces (anvil and spindle) must be 180 degrees apart so that the diameter, and not a chord, is measured. Manipulation errors are easily made unless the inspector understands that it is essential to locate the tool properly on the part.

Reading

Regardless of the other variables, if a person is not well versed in how to read the scale of the tool, accuracy cannot be achieved. Reading errors that are made (on any scale) are of two basic types—*lack of knowledge* and *parallax*. A person must know how to read the scale of the tool, and even then, must avoid *parallax error*. Parallax error occurs when a person takes a scale reading from an angle instead of looking directly at the scale. Looking at any scale from an angle will cause the person to make an error because the person cannot see the reading exactly.

Workpiece Geometry

Many measurements are not accurate because the workpiece has excessive geometry errors. The piece may be tapered or out-of-round, may have surface waviness, or be barrel-shaped, hour-glass-shaped, and so forth. If the person making the measurement does not understand these conditions and does not search for them, accuracy cannot be achieved.

Measuring Pressure

Measuring pressure is a large factor when considering accuracy. If the pressure is too heavy or too light, the measurement will not be accurate. A key to measuring pressure is to attempt to use a *constant* pressure, which is preferably the same pressure that was used to calibrate the tool. This is where "ratchet stops" and friction thimbles come in handy. Strive to develop the capability to achieve constant and proper pressure in your measurements.

Dirt and Burrs

Dirty measuring tools or parts, or parts with machining burrs are always a problem. Tools should always be kept clean, and parts should be cleaned before any attempt is made to measure them. Dirt and burrs will cause errors in measurements as follows. On an external feature (such as a shaft), dirt and burrs will cause your measurement to be larger than the actual size of the shaft. On an internal feature (such as a hole), dirt and burrs will cause your measurement to be smaller than the actual size of the hole.

There are many other variables that cause inaccuracy, for example, heat (which will cause a part or a tool to expand in size). Understanding these possible errors and how to avoid them will give you a much better chance of obtaining an accurate measurement. Remember, accuracy means getting an unbiased true value. This definition, when simplified, will say: "Accuracy means the ability to measure the true size of an object as closely as you possibly can."

THE STEEL RULE

A common measuring tool that is used in the shop for a variety of measurements is the steel rule. The standard rule is graduated on both sides in inches and fractional parts of inches. These graduations are shown in Figure 4.1.

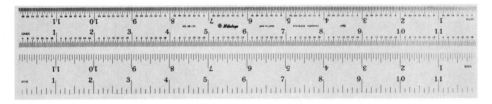

Figure 4.1   Steel rules. (M.T.I. Corp. Reprinted by permission.)

The *finest* divisions that are on a steel rule (sixty-fourths on the one at the top of Figure 4.1) establish its *discrimination*. This is the greatest accuracy that can be achieved on this particular steel rule. Some steel rules discriminate to $\frac{1}{10}$ in. and $\frac{1}{100}$ in.

Use of the Steel Rule

To avoid errors, the steel rule must be used properly. The first thing to consider is where a steel rule can be used and where it cannot. Obviously, it cannot be used on any dimension that is tighter in tolerance than its discrimination. For example, you would not use the steel rule to measure a dimension that has a tolerance of 0.010 total, because the best discrimination of this tool is $\frac{1}{64}$ (or almost 0.016 in.). However, if you are measuring a casting or forging with a tolerance of, for example, 0.060 total, the steel rule will do very well.

Proper reference is important

Steel rules should be referenced in one of two ways. Either the edge of the rule should be firmly fixed, or your measurement should begin with the 1-in. line. These two conditions are shown in Figure 4.2.

Hook Rules.   Some steel rules have a hook built in the reference end that is designed to provide an accurate reference on the scale when applied to a part (Figure 4.3). The hook must be watched for looseness or wear, to avoid errors. The hook is designed to line up with the end of the steel rule (or 0 in.). Applications for the hook rule are shown in Figures 4.4 and 4.5.

Steel rules are a necessary measuring tool but must be used for what they were intended for—measurements within the limits of their discrimination.

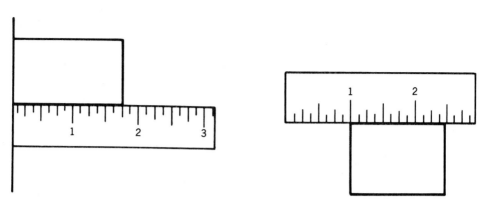

Figure 4.2 Referencing the steel rule.

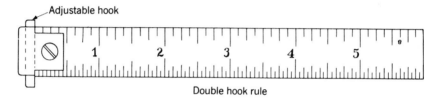

Double hook rule

Figure 4.3 Hook rule. (L.S. Starrett Co. Reprinted by permission.)

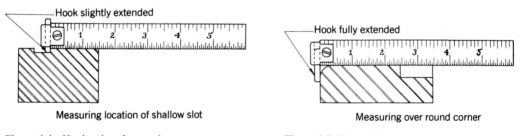

Measuring location of shallow slot

Measuring over round corner

Figure 4.4 Hook rule referenced. (L.S. Starrett Co. Reprinted by permission.)

Figure 4.5 Measuring over a round corner. (L.S. Starrett Co. Reprinted by permission.)

SPRING CALIPERS

Spring calipers are adjustable calipers that are used to *transfer* a dimension to a line-graduated instrument (such as the steel rule). There are inside and outside spring calipers as shown in Figure 4.6. Dividers are also shown.

Inside calipers are used for measuring inside dimensions and outside calipers are used on outside dimensions. Spring calipers are a transfer tool. First, you

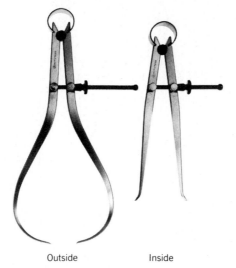

Outside   Inside

Figure 4.6 Spring calipers. (L.S. Starrett Co. Reprinted by permission.)

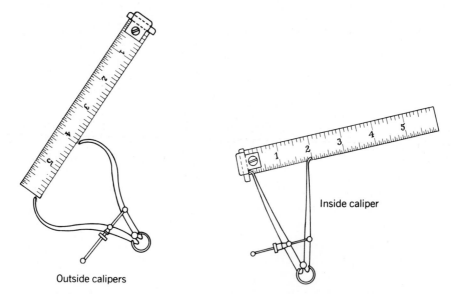

Inside caliper

Outside calipers

Figure 4.7 (*a*) Setting an outside caliper. (*b*) Setting an inside caliper. (L.S. Starrett Co. Reprinted by permission.)

locate them on the diameter (or length) to be measured and adjust them by feel as the contacts rub against the surfaces of the dimension. Then you transfer the measurement to the line graduated tool to find the actual size of the part. Spring calipers are generally used on wide tolerances, such as castings, or forgings, and generally involve the use of a steel rule for the measurement (see Figure 4.7).

GAGE BLOCKS

Gage blocks are used in almost every shop that manufactures a product that requires mechanical inspection. They have many purposes. Some of these are: to set up a length dimension for a transfer measurement; to set (or calibrate) fixed gages, such as snap gages; and to be used with special attachments for various applications. Gage blocks come in three standard shapes: *round, square,* and *rectangular* (Figure 4.8).

Gage blocks are made of hardened tool steel and their surfaces are square, parallel, and flat to very accurate tolerances. There are three grades of gage blocks, with approximate accuracies as shown.

1. Master blocks, 0.000002 in.
2. Inspection blocks, 0.000005 in.
3. Working blocks, 0.000010 in.

There are many methods to clean gage blocks before use; one is with kerosene and a chamois. It is best to filter the kerosene through the chamois. Note,

Figure 4.8   Gage blocks. (Federal Products Corp. Reprinted by permission.)

however, that perfectly clean blocks will not *wring* together. It is necessary for a light oil film to be on them.

Gage blocks are used by wringing them together to obtain the length you need for the measurement. Wringing or rubbing them together squeezes the air out from between them, and since their surfaces are accurately flat, a "cementing" action occurs. Therefore, the blocks will stick together. The correct way to wring blocks together is shown in Figure 4.9. Position the blocks as shown at (*a*), and with a clockwise, twisting action, slide them together.

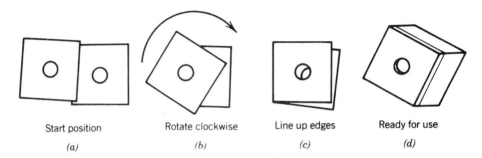

Start position	Rotate clockwise	Line up edges	Ready for use
(a)	(b)	(c)	(d)

Figure 4.9   Wringing gage blocks.

Wear Blocks

Wear blocks are frequently used on both ends of a gage block stack. They come in two sizes, 0.050 in. and 0.100 in. thick. Their purpose is to protect the rest of the blocks from wear. Using wear blocks is simple. You always put them on the ends of a stack, and always the same side out. Remember to include their length in the stack size you need.

To maintain the accuracy of gage blocks, they must be treated as carefully as a gage; therefore, delicate handling is important at all times.

General guidelines

1. Avoid using the gage blocks if they are not required for the accuracy of the measurement.
2. If there are no wear blocks in the set then pick two small blocks and use them as wear blocks (or order wear blocks).
3. Always use the least amount of blocks in a stack as possible.

Gage Block Sets

A typical gage block set (Figure 4.10) contains various increments of block sizes. Listed below is an 81-piece set.

Ten-Thousandths Blocks		**Fifty-Thousandths Blocks**	
0.1001		0.050	0.500
0.1002		0.100	0.550
0.1003		0.150	0.600
0.1004		0.200	0.650
0.1005		0.250	0.700
0.1006		0.300	0.750
0.1007		0.350	0.800
0.1008		0.400	0.850
0.1009		0.450	0.900
			0.950

One-Inch Blocks	**One-Thousandths Blocks**				
1.000	0.101	0.111	0.121	0.131	0.141
2.000	0.102	0.112	0.122	0.132	0.142
3.000	0.103	0.113	0.123	0.133	0.143
4.000	0.104	0.114	0.124	0.134	0.144
	0.105	0.115	0.125	0.135	0.145
	0.106	0.116	0.126	0.136	0.146
	0.107	0.117	0.127	0.137	0.147
	0.108	0.118	0.128	0.138	0.148
	0.109	0.119	0.129	0.139	0.149
	0.110	0.120	0.130	0.140	

Figure 4.10 Square gage block set. (M.T.I. Corp. Reprinted by permission.)

Note. Usually included in the set are two wear blocks. These blocks are either 0.050 in. or 0.100 in.

For practice in the selection of blocks, try this example.

Example

Stack up 3.6824 in.

Solution   3.6824
          − 0.1004   (use this block)
          ‾‾‾‾‾‾‾
           3.582
          − 0.132   (use this block)
          ‾‾‾‾‾‾‾
           3.450   (use 0.450 and 3.000 blocks)

SURFACE PLATES

Many reference surfaces on component parts or assemblies are established by *planes*. The ideal plane for dimensional measurement should be perfectly flat, but since nothing is perfect, we must settle for the next best thing. Surface plates provide a true, flat reference surface for dimensional measurement (see Figure 4.11). They are a simulated datum plane per ANSI Y14.5M−1982.

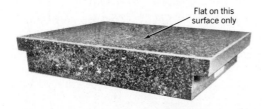

Flat on this
surface only

Figure 4.11 Granite surface plate. (L.S. Starrett Co. Reprinted by permission.)

There are different materials from which surface plates are made. Two of the main materials are cast iron and granite (stone). Cast-iron surface plates have their good and bad points. Two good points are that they are magnetic and usually have the configuration that allows parts to be clamped down during the measurement. A bad point is that if they are damaged, it produces raised material on the surface that could cause a considerable amount of error in measurements. Another bad point is that since they are iron, they will easily rust.

Granite surface plates are far superior to those of cast iron because:

1. Damaged areas produce *no* raised material.
2. They are lower in price.
3. They retain their flatness longer than cast iron.
4. They will not rust.

Whichever type of plate is used, it is important to be sure that the plate is level when in use. Dimensions measured on surface plates are taken *from the plate up* since the plate is the reference surface (see Figure 4.12).

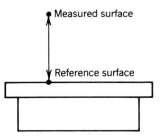

Figure 4.12 Reference and measured surfaces.

One widely used accessory to the surface plate is a gage block stack (see Figure 4.13). Gage blocks reflect the actual distance from the plate up, so that comparisons can be made to the part being measured with a high degree of accuracy.

As shown in Figure 4.14, a dimension that must be located by three perpendicular planes is not a problem because of another accessory called an *angle plate* (or knee). This is also an approved simulated datum plane.

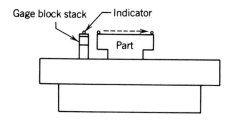

Figure 4.13 Transfer measurement on a surface plate using gage blocks.

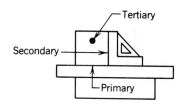

Figure 4.14 Locating primary, secondary, and tertiary datums on a surface plate.

Some surface plate accessories are shown in Figure 4.15. There are many more accessories that can be obtained to aid in setting up a wide variety of measurements.

Angle plates

(a)   **Figure 4.15a** Angle plate (knee).

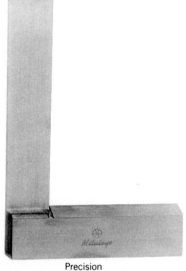

Precision
squares
(b)

Figure 4.15*b* Precision square.
(M.T.I. Corp. Reprinted by permission.)

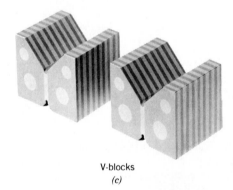

V-blocks
(c)

Figure 4.15*c* Matched V-blocks.
(M.T.I. Corp. Reprinted by permission.)

Gage
blocks
(d)

Figure 4.15*d* Square gage blocks.
(M.T.I. Corp. Reprinted by permission.)

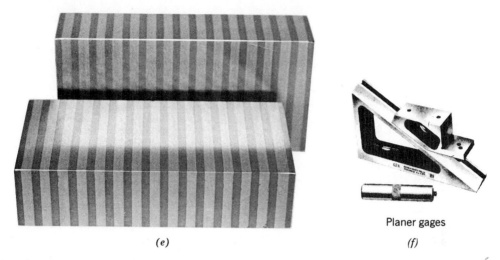

Planer gages

(e)

(f)

Figure 4.15e Fixed parallels. (M.T.I.
Corp. Reprinted by permission.)

Figure 4.15f Planer gage. (Brown &
Sharp. Reprinted by permission.)

Care of Surface Plates

Listed below are some hints that are helpful in taking good care of a surface plate.

1. Do not abuse surface plates; they are a precision "tool."
2. Do not allow unnecessary objects to be placed on the surface plate.
3. Do not store tools on the surface plate. Keep on the plate only what you are using at any given time.
4. Clean the surface plate before and after use.
5. Keep the surface plate covered when it is not in use.

The proper method of putting items on, or removing items from, a surface plate is shown in Figure 4.16.

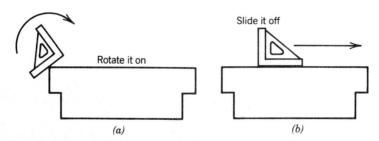

Rotate it on

Slide it off

(a)

(b)

Figure 4.16 *(a)* Proper method of putting parts on a surface plate. *(b)* Proper method of removing parts from a surface plate.

To avoid damage to the surface plate, the methods described above should be used. By putting items on the plate in this manner, you avoid the possibility of damage to the plate because you ensure that the surface of the item and the surface of the plate meet evenly. You will also avoid damage when removing items by always sliding the item from the plate.

MECHANICAL INDICATORS

Mechanical indicators are a widely used *variable* tool in manufacturing and inspection areas. They are called a variable tool because of their ability to detect the actual variation (or difference) between a dimension and a reference standard. Indicators are used in many applications such as making surface plate measurements and checking parts on the machine during production. Figure 4.17 shows an example of a mechanical indicator.

Amplification

Indicators are a tool that amplifies the actual movement of the tip to the dial face so that this movement can be measured (see Figure 4.18). The discrimination of the common indicators used today is 0.00005 in., 0.0001 in., 0.0005 in., and 0.001 in. You can determine an indicator's discrimination by looking at the dial face. For example, if the indicator is 0.001 in. discrimination, the dial face would show (have lines at) 0.001 in. On this indicator, the distance between each line on the dial face represents 0.001 in. of movement (or travel) at the tip. Because of their internal mechanisms, an advantage of using dial indicators is *constant measuring pressure*. This adds to the accuracy of any measurement.

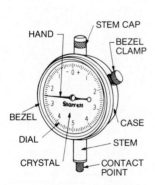

Figure 4.17 Parts of a dial indicator. (L.S. Starrett Co. Reprinted by permission.)

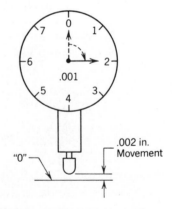

Figure 4.18 Dial indicators amplify distance.

Types of Dials

There are two main types of dials used on indicators: *balanced* and *continuous* dials (see Figure 4.19).

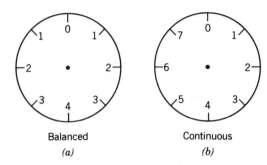

Balanced
(a)

Continuous
(b)

Figure 4.19 Types of dial faces: (*a*) balanced; (*b*) continuous.

A balanced dial simply means that the indicator can be read directly in either direction from 0. Balanced dials are used mostly to measure dimensions with bilateral tolerances. The continuous dials starts at 0 and continues in number sequence all the way around until you reach zero again. This type of dial is used mostly to measure dimensions with unilateral tolerances.

Revolution Counters

Some indicators are equipped with another small built-in indicator called a *revolution counter* (see Figure 4.20). It tells how many full revolution the large indicator needle has made. In some applications, use of the revolution counter is very helpful in keeping track of the total travel during a measurement.

Choosing Which Indicator to Use

There are two important items to consider when choosing what indicator to use for a given measurement—*range* and *discrimination*. The range of an indicator is simply the total travel (or linear movement of the tip or plunger) of which the indicator is capable. An example is a 0.0001-in. indicator with a 0.004-in. range. In this example, 0.0001 in. is the discrimination and 0.004 in. is the range. This indicator could be used on a part tolerance of ±0.001 or so, but you could not use it on a tolerance of ±0.005 (it is outside the 0.004-in. range). Be sure you choose an indicator that has the range that will envelop the entire tolerance of the dimension you are measuring, and that will be accurate enough (have adequate discrimination) to do the job.

Revolution
counter

Figure 4.20   Dial indicator with a
revolution counter. (M.T.I. Corp.
Reprinted by permission.)

The Rule of Discrimination

It is bad practice to use an indicator on a tolerance where you would have to read
between the lines on the dial face. An example might be where the tolerance on
the dimension is ±0.0002 in. and an indicator with only 0.001-in. discrimination is
erroneously chosen. Since this indicator can only measure to 0.001 accuracy, you
cannot possibly measure a tolerance of 0.0002 in. without trying to estimate by
reading between the lines. Do not estimate—use an indicator that will do the job.

Indicator Travel

When using any indicator, always make sure that you have enough plus and minus
travel to measure the entire tolerance you are trying to measure (Figure 4.21).
Many errors occur because the indicator was set up to have travel in only *one*
direction (not both) from the nominal size of the part (0 on the indicator dial).

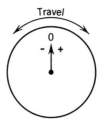

Figure 4.21   Travel on an indicator.

Contact Tips

Dial indicators can be used with a variety of contact tips for many different types of measurement (see Figure 4.22). You should use the proper type tip for the feature you are measuring. Shown in Figure 4.23 are examples of correct and incorrect choices of contact tips.

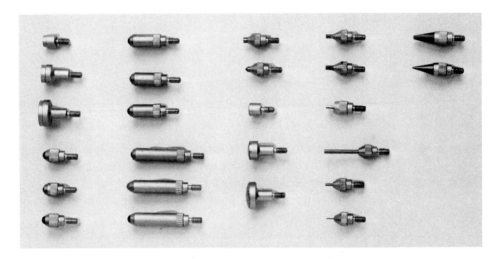

Figure 4.22   Various dial indicator tips. (M.T.I. Corp. Reprinted by permission.)

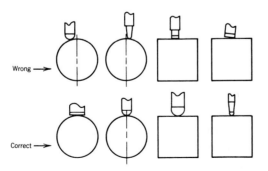

Figure 4.23   Proper selection of indicator tips.

Checking for Repeatability and Accuracy

Indicators should be checked frequently for repeatability and accuracy. Repeatability can be checked by measuring the same gage block over and over. Accuracy within the indicator's range can be checked by using a series of gage blocks within the range of the indicator; the discrimination (tolerance) of the blocks should match the discrimination of the indicator. You should also be very cautious not to use an indicator that is erratic (jerks or is sluggish) in its movement.

Care of Indicators

Indicators should be treated as a delicate precision instrument.

1. Never bump or bang them around.
2. Avoid dropping them.
3. Never blow them with air pressure.
4. Store them in a safe, dry place.

Errors in Using Indicators

Below are listed just a few of the many possible errors in using mechanical indicators.

1. *Cosine error* (discussed later). Error caused when an indicator tip is at an angle to the surface being measured.
2. *Mistaken travel.* A person using an indicator that makes more than one revolution must be aware of how many revolutions the pointer has made when making the measurement.
3. *Parallax.* Caused by not looking straight (90 degrees) at the face of the indicator when reading it. The indicator pointer can look like it is right on a line from the side view when it really is not.
4. *Travel.* Indicator pointer should be moved at least one-fourth revolution from *dead* position when used. Dead position is the position of the pointer when the indicator is not being used.

COSINE ERROR IN INDICATORS

When using indicators in making linear measurements, it is necessary to understand and avoid cosine error. Cosine error is simply defined as that linear measurement error that occurs when an indicator tip is at an angle to the surface of the workpiece being measured. Note, at this point, that the position of the body of the indicator makes no difference. It is the indicator tip itself, when at an angle, that causes cosine error. As shown in Figure 4.24, indicator tips should be parallel to the workpiece, because the proper movement of the tip is when it moves directly up and down (or in a plane perpendicular to its axis).

Movement in any other direction has the same effect as using a tip that is too short.

Example of Cosine Error

In Figure 4.25 you can see that the indicator tip is at an angle to the surface being measured and therefore causes cosine error. Figure 4.26 shows the correct methods (remember, only the tip must be parallel).

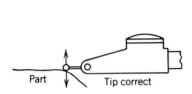

Figure 4.24   Indicator tip movement.

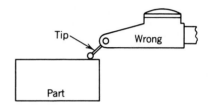

Figure 4.25   Angle between the indicator tip and the workpiece—cosine error.

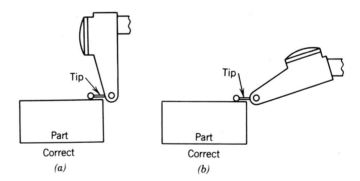

Figure 4.26   No cosine errors here.

Correcting for Cosine Error

There are times when, because of the shape of a part, you must use the indicator tip at an angle to perform the measurement. At these times, there is a formula for correcting for the cosine error if it is necessary to do so.

   Cosine error can be calculated. The answer is expressed as a % (percentage). This is a percentage of the reading seen when making the measurement.

With cosine error, the reading will be too high, and you compensate by decreasing the reading by whatever percentage you get by using the formula.

First, know the meaning of the following symbols:

- ψ means exact angle of the tip.
- Delta (Δ) means the change (in percent) that the reading must be decreased.

The formula is

$$\Delta = \left(\frac{1}{\text{cosine } \psi} - 1 \right) \times 100$$

This reads: *Delta is equal to 1 divided by the cosine of the angle, minus 1.* That answer is multiplied by 100.

Example

Your measurement is made with the indicator tip at an angle of 15°, and the reading (variation) on the indicator shows 0.010. The error is

$$\Delta \, (\%) = \left(\frac{1}{\text{cosine } 15°} - 1 \right) \times 100$$

$$= \left(\frac{1}{0.9659} - 1 \right) \times 100$$

$$= (1.035 - 1) \times 100$$

$$= 0.035 \times 100$$

$$= 3.5\% \text{ error}$$

Now you must multiply the reading (variation 0.010) times 0.035 (or 3.5 percent written as a decimal) to get the actual error. Then subtract the actual error from the original 0.010. So

$$0.010 \times 0.035 = 0.00035 \quad \text{(the cosine error)}$$

and

$$0.010 - 0.00035 = 0.0097 \quad \text{(the true reading or variation)}$$

This tells you that instead of 0.010 variation as shown on the indicator dial, there was only 0.0097 variation. The 0.00035 error occurred because of the indicator tip being at an angle. The cosine error here is therefore 0.00035.

Note. In most measurements cosine error is relatively negligible, but it is still a measurement problem of which we should be aware.

Remember. **The larger the tip angle, the larger the error becomes.**

DIAL INDICATING GAGES

Dial indicating gages are often referred to as *variable gages*. This is so because they are gages that use a dial indicator to gage a part on a variable basis (variable meaning that the gage will detect the actual size of the part, not just whether or not it is in, or out of tolerance).

Dial indicating gages work on a *comparison basis*. The gage is usually preset to indicate zero at the nominal size of the part with a set master. Then it is applied to the part, and the difference between the master size and the actual part size is shown on the indicator. For example, you have a dial indicating snap gage that is preset on zero with a 0.5000-in. cylindrical master. You use a cylindrical master because you are going to measure a shaft diameter with the gage. The dial snap gage is then applied to the shaft you are measuring, and the indicator reads a plus 0.003 in. Knowing that zero set on the gage is the same as 0.5000 in., the actual size of the part is 0.003 in. larger, or 0.503 in. If the dial read a minus 0.003 in. (for example), the actual size of the part would be 0.5000 in. − 0.003 in. = 0.497 in.

As you can see, dial gages are simply hand-held mechanical comparators. Unlike the micrometer (where you can read the actual size directly), the dial gage only shows the "difference" between the master size and the actual size. You must use simple mathematics to add or subtract the reading from the set master size.

Cautions When Using Dial Gages

1. *Allow for indicator travel.* There must always be enough plus and minus travel from zero set to allow for the tolerance limits you are measuring. It would be poor practice to measure a dimension with a tolerance of 0.005 in. total with a gage that has a total travel of 0.002 in. There must be enough travel in the gage to indicate the actual size of the part, no matter what that size is within the tolerance band.

2. *Use the proper discrimination.* The dial gage used should have an indicator that has the necessary discrimination to measure the part. For example, a shaft with a tolerance of 0.5000 in. ± 0.0004 in. *should not* be measured with a gage that discriminates to 0.001. It should be measured with a gage that discriminates within 0.0001 in. Therefore, the gage must have a ten-thousandths indicator on it and at least enough travel to measure the part within the tolerance band.

3. *Know the set master size.* Most set masters have the size marked on them and, ideally, they are the nominal size of the dimension being checked. For example, if the nominal size of the dimension is 0.500 in., the set master should be 0.5000 in. This avoids confusion when using the gage. Always be sure of the set master size before setting the gage.

4. *Watch indicator travel when the gage is applied to the part.* Sometimes, when a part is out of specification, the indicator can make a complete revolution quicker than the eye can see. In this situation, the part can seem to be within tolerance when, in fact, it is considerably undersize or oversize. Care should be taken to ensure that the same amount of pointer travel from null position, when the gage is set, is duplicated when the gage is applied to the part.

5. *Reference the gage.* Whatever the design of the gage, there is always a reference surface. This reference surface should be kept clean and free of damage. When the gage is being used, care should be taken to make sure the reference surface is securely located and in the proper place. The reference surface of any gage is one that is not movable, whether it be a bar of some kind, an anvil, or pins. The only movable part of a dial indicating gage is the indicator rod itself.

6. *Beware of indicator problems.* When using a dial indicating gage, regardless of how often it is calibrated, the user should always be on the lookout for indicator problems. Indicator pointers that move in a sluggish manner, jerk from one point to another, or stick, are a sign of indicator malfunction, and the gage should be checked out prior to use. Always make certain that the indicator will repeat.

7. *Avoid parallax error.* When using a dial indicating gage, or any other instrument that has a moving pointer and line divisions, make sure you are looking directly (90°) at the dial during the measurement. Avoid looking at the dial from an angle at all times. Look at the dial directly, so that you can accurately line up the pointer to the scale line.

8. *Use revolution counters on long-range dial gages.* If you use a dial gage that has the ability (within the indicator) to measure long ranges (such as 2 in., etc.), it is better to use an indicator with a revolution counter. The revolution counter involves a small pointer on the indicator face that tells you how many revolutions the large pointer has made during a measurement. Since the large pointer moves very quickly during the measurement, carefully count the revolutions to avoid making a serious error in the measurement.

9. *Always be aware of what is minus and what is plus.* All indicating gages have a *null position*. This is the "free" position of the pointer when the indicator is not being used. Generally, this position is at nine o'clock on the dial face. Depending on the type of gage and the dimension being measured, a plus or minus value can be on either side of the zero setting. While setting the gage with the set master, you must compare the measuring jaws of the gage to the direction of the indicator movement. An example is a dial snap gage. As the jaws open (or the gap gets larger), the indicator moves clockwise from null position. In this case, a reading clockwise from zero is a plus reading and must be added to the set master size to obtain the actual size of the part. In some dial indicating gages, it is exactly the opposite. Just re-

member to compare the motion of the measuring jaws to the motion of the indicator pointer, and this error can be avoided.

10. *Make sure that all clamps are tight.* All indicating gages have some sort of clamping device to hold the indicator to the gage. Make sure that this clamp is not loose, or your measurement could be in error. Hand-tight will usually suffice.

Examples of Dial Indicating Gages

There are a wide variety of dial indicating gages on the market today and this book does not intend to cover them all. There are, however, a few types that are seen more often on the shop floor. Some of these are as follows.

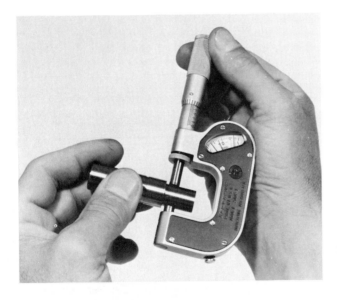

Figure 4.27 Dial indicating micrometer. (Federal Products Corp. Reprinted by permission.)

Dial indicating micrometers

The dial indicating micrometer has a constant pressure anvil that transmits movement to the dial face which is built into the frame (refer to Figure 4.27). The dial indicating micrometer is the same as any other micrometer with the exception of this feature. When using the dial indicating micrometer, you read the micrometer on the thimble and sleeve for 0.001-in. increments, then read the 0.0001 in. directly from the dial face. This micrometer is very useful in making accurate measurements because of:

1. Constant measuring pressure (in the anvil)
2. Direct reading in the ten-thousandths range

The dial indicating micrometer can also double as a snap gage for outside measurements. You simply preset the desired size, lock it in, and use the dial for comparisons.

Dial calipers

The dial caliper is very useful for quick and accurate measurements (within the capability of the tool). The dial is there to represent the vernier scale, so that the user does not have to find the coincidence line (see the discussion of vernier equipment on p. 182). The dial caliper is a direct reading tool. You simply read the main scale and add what you see on the dial indicator.

Care should be taken to use the dial caliper slowly so that the teeth on the track are not damaged or jumped. Figure 4.28 shows a typical dial caliper. Figures 4.29 and 4.30 are further examples of dial calipers.

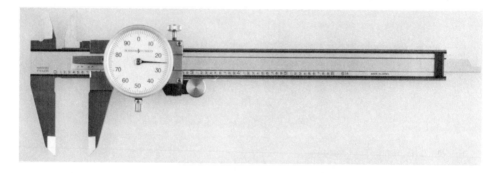

Figure 4.28   Dial caliper. (Scherr Tumico. Reprinted by permission.)

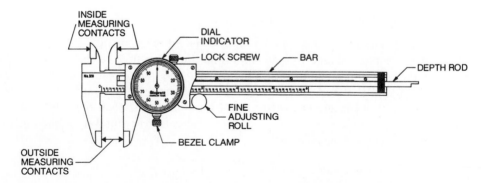

Figure 4.29   Parts of a dial caliper (L.S. Starrett Co. Reprinted by permission.)

OUTSIDE
MEASUREMENT

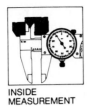
INSIDE
MEASUREMENT

DEPTH
MEASUREMENT

Figure 4.30 Dial caliper in use. (L.S. Starrett Co. Reprinted by permission.)

Dial depth gage

There are two basic types of dial depth gages. The first type is shown in Figure 4.31. This dial depth gage employs a base and a drop indicator with a revolution counter. This gage does not use a set master, but does still require calibration. This is a *direct-reading dial depth gage*. It can be calibrated with various stacks

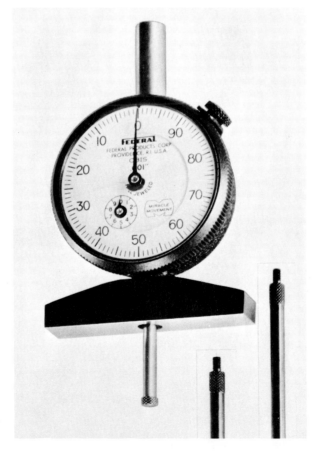

Figure 4.31 Dial depth gage. (Federal Products Corp. Reprinted by permission.)

of gage blocks throughout its measuring range. Once calibrated, you can apply this gage to the part and use the indicator itself to establish the depth of the part.

The second type of dial depth gage is one that "compares" the depth of a part to the depth of a set master. The gage and master are shown in Figure 4.32. This gage generally employs an indicator with a *balances dial* face (one that reads directly in the plus or minus direction), and the user must compare the plus or minus reading to the set master depth.

Figure 4.32 Dial depth gage. (Federal Products Corp. Reprinted by permission.)

Dial bore gages

All dial bore gages employ some kind of mechanism that translates movement from the contact points to the indicator. There are a variety of different contact points used in dial bore gages. These contact points may expand because of a tapered shaft inside, or may utilize an internal lever mechanism. The mechanisms that translate movement in a dial bore gage are not covered in this book.

There are two basic methods in which dial bore gages make contact: (1) two-point contact, and (2) three-point contact. Whenever possible (because of the size of the bore) three-point contact is preferred. This is so because three-point contact centralizes the gage in the bore automatically.

Centralizing the gage avoids *chordal* error. This error is often found when using two points of contact to measure any inside diameter. Two- and three-point contact are shown in Figure 4.33.

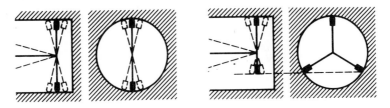

Figure 4.33   Positioning a dial bore gage. (M.T.I. Corp. Reprinted by permission.)

When using a dial bore gage, care should be taken to insert the gage into the bore so that the bore is not damaged. Rock the gage to find the true diameter. Some dial bore gages have a built-in *trigger* to retract the contacts for easy entry into the bore. Once inside the bore and rocked into position, the difference between bore size and the size of the set master can be seen on the dial face. The set master for a dial bore gage is usually a ring gage. Care must also be taken to centralize the gage in the set master. Dial bore gages can be very accurate, depending on the indicator used, the set master, and the manipulation of the gage inside the bore (see Figure 4.34).

Dial snap gages

Dial snap gages come in a variety of different shapes and sizes, but all share the same purpose—to measure an external feature on a variable basis. Most dial snap gages are used to measure the following:

1. Outside diameters.
2. External pitch diameters of threads.

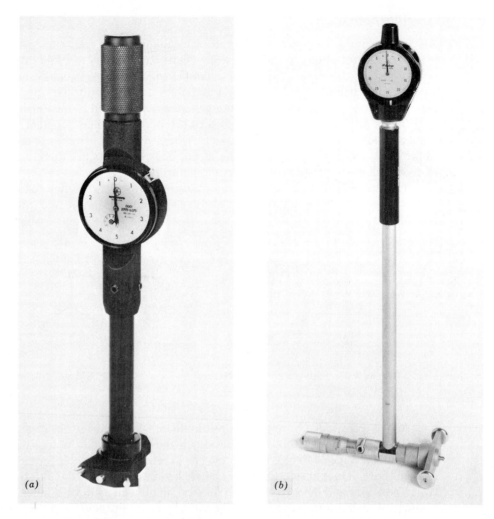

(a)   *(b)*

Figure 4.34   Two models of dial bore gage. (M.T.I. Corp. Reprinted by permission.)

3. External groove diameters.
4. Other external features.

Some examples of dial snap gages are shown in Figure 4.35.

Dial snap gages employ a reference anvil and a measuring face. The reference anvil is fixed in the frame and should always be fixed (not moved) during the measurement. The measuring face is equipped with spring loaded tension and transmits the movement to the dial indicator. There is one more feature of a dial

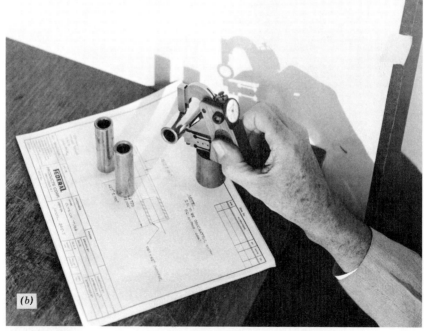

Figure 4.35 (a) and (b) Dial snap gage in use. (Federal Products Corp. Reprinted by permission.)

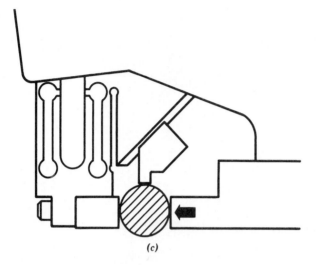

(c)

Figure 4.35c Positioning the part in the dial snap gage. (Federal Products Corp. Reprinted by permission.)

(d)

Figure 4.35d Dial snap gage. (Federal Products Corp. Reprinted by permission.)

snap gage that should be discussed. This is the centralizing anvil. The centralizing anvil is usually movable to allow you to adjust it to any size part that you are checking with the snap gage. Its purpose is to be a positive stop for the part when it is placed in the gage. This anvil should be set so that the part being measured rests on it in such a way that the actual diameter of the part is centralized between the measuring faces of the gage.

Using the dial snap gage is simply comparing the size of the set master to the actual part and reading the difference on the dial face (plus or minus).

In-Process Dial Gages

There are various kinds of dial gages that are used directly on the machine during production (see Figure 4.36). These gages are similar to other dial gages with the exception of mounting brackets and special indicators. The special indicators are simply indicators that have *dampers* built into them so that they can indicate a piece that is being machined and will not have erratic pointer movements. During machining there is considerable vibration, and the damper on the indicator is necessary to enable the operator to read the indicator. The mounting brackets are there so that the gage can be mounted to the machine in such a way that its measuring faces can rest on the part during machining.

In process gages are very useful because they tell the machine operator when the part is at its desired size. There are some in-process gages that are equipped with electrical connections that will shut the machine off automatically when the part is to size. In-process gages should be checked periodically for calibration, and reset, if necessary, to avoid running defective parts.

Comparing Fixed Gages to Dial Indicating Gages

Almost all fixed and indicating gages are used during production to determine whether parts are to the size specified (within tolerances). There are some major drawbacks and benefits to using either type of gage. Some of these are listed in Table 4.1. Remember that fixed gages are only used for *one* dimension and tolerance.

TABLE 4.1   Drawbacks and Benefits of Fixed and Indicating Gages

Characteristics	Fixed Gage	Indicating Gage
Cost	Less expensive	More expensive
Ease of use	Easier to use	More difficult to use
Maintenance costs	Less cost	More cost
Calibration time	Less time	More time
Information per piece inspected	Less information	More information

Figure 4.36   In-process gage in use. (Federal Products Corp. Reprinted by permission.)

The drawbacks and benefits noted in Table 4.1 should be weighed against your needs in gaging. In my opinion, the indicating gage is the best gage to use (even though it is more costly). There are several reasons why an indicating gage is suggested for use in most applications of gaging in the shop.

1. The indicating gage gives you more information about the part being inspected. It not only tells you whether the part is good or bad, but it tells you *how good* or *how bad* the part is.

2. The indicating gage must be used when you are using preferred statistical quality control techniques. The most widely used control chart is the X-bar and range chart.

3. Indicating gages allow you to inspect a variety of different dimensions and tolerances. To plot an X-bar and range chart, you must have variable gages (see Chapter 10 on statistical process control).

4. The information given by an indicating gage can help the operator tell when the machine is nearing the high or low limit of the tolerance.

5. The indicating gage enables the operator to make more precise setups than when using the attribute (or fixed) gage. When the setup is precise, there is greater chance of producing good-quality parts in the long production run.

6. The indicating gage is useful when parts must be rejected. Most material review boards require a rejection that clearly states how bad the part is, so that a decision can be made on what to do with the part. On rejections, you cannot simply say that the part is bad; you must say how bad.

Fixed gages do find considerable use in the shop and at the assembly line, but it must be remembered that they only tell you that the part is in tolerance or out, and nothing more. If your only need is to tell whether or not the part is in tolerance, fixed gages should be used because they are less expensive and easier to use and maintain.

Variable gages (such as dial indicating gages) are recommended for use because of the information they can supply about the part. Even though a variable gage is a little more expensive to purchase and maintain, they can often save you considerable costs in preventing defective products from being made.

SURFACE GAGES

Surface gages (Figure 4.37) are used in many measurement applications, but since they do not have a direct-reading scale, they are used in transfer (or comparison) measurement. In Figure 4.38, you see the surface gage transferring a dimension established by a stack of gage blocks to the part in order to measure it. This is simply a comparison of a known standard to the part.

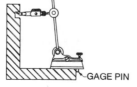
GAGE PIN

Figure 4.37 Surface gage and indicator in use. (L.S. Starrett Co. Reprinted by permission.)

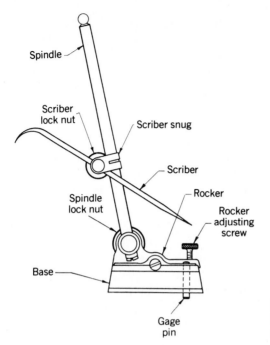

Spindle

Scriber
lock nut

Scriber snug

Scriber

Rocker

Spindle
lock nut

Rocker
adjusting
screw

Base

Gage
pin

Figure 4.38 Parts of a surface gage. (L.S. Starrett Co. Reprinted by permission.)

Remember. All methods of measurements are comparisons to known standards. Gage blocks are a good example of a standard traceable to the National Bureau of Standards.

Uses of the Surface Gage

Surface gages have many uses, including the transfer measurement shown above, scribing parallel lines (using the two pins provided), inspecting runout, concentricity, and perpendicularity. These are just a few uses. Many more are learned by obtaining experience with the tool.

Possible Errors in Using the Surface Gage

Although the surface gage is a very worthwhile tool for use in production and inspection, there are a few points that must be remembered to avoid errors. The causes of errors include (1) loose clamps, (2) a too long indicator support rod, (3) an upright post at an angle to the base, and (4) completely loose fine adjustment bolts (Figure 4.39).

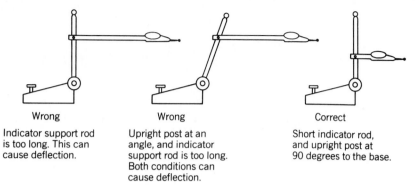

Wrong	Wrong	Correct
Indicator support rod is too long. This can cause deflection.	Upright post at an angle, and indicator support rod is too long. Both conditions can cause deflection.	Short indicator rod, and upright post at 90 degrees to the base.

Figure 4.39   Possible errors using a surface gage. (L.S. Starrett Co. Reprinted by permission.)

The Cantilever Effect

An example of a cantilever is a beam that is only supported at one end. Because of this single support, and length, it is very unstable and can bend with weight on the other end. How much it bends depends on how much weight is on it and how far the weight is from the supported end. The farther the weight is away from the supported end, the more the beam will bend (see Figure 4.40). This deflection, resulting from long support rods, can cause measurement errors.

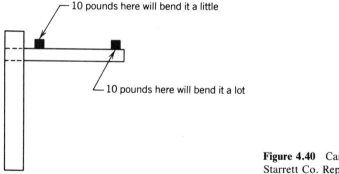

Figure 4.40   Cantilever effect. (L.S. Starrett Co. Reprinted by permission.)

MICROMETERS

One of the most widely used basic tools in the shop is the micrometer (or ''mic'' as it is sometimes called). There are many kinds of micrometers, but one thing that is common to all of them is the way they are read. Before going further, study Figures 4.41 and 4.42.

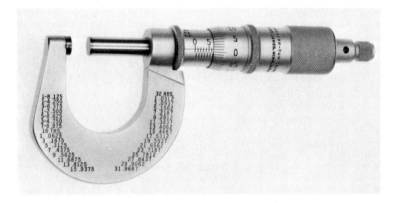

Figure 4.41   Micrometer. (Scherr Tumico. Reprinted by permission.)

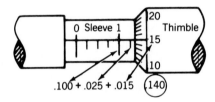

Figure 4.42   Micrometer reading.

Reading the Micrometer

The scale of a micrometer ranges from 0 to 1 in. and is usually graduated in "thousandths" of an inch (think of 1 in. as $\frac{1000}{1000}$). The sleeve of the mike shows the numbers 0 1 2 3 4 5 6 7 8 9 0. Each of these numbers represents 0.100 (or 100 thousandths) in travel. In between each number, there are four spaces each representing 0.025 (or 25 thousandths) of an inch. So, you can see that the micrometer *sleeve* divides 1 in. into 0.025 increments.

Now look at the thimble. Each line on the thimble represents 0.001 and there are numbers every five lines (or 0.005). One complete turn (revolution) of the thimble is equal to 0.025. So each revolution of the thimble (0.025) is equal to one division on the sleeve (0.025). Following the same logic, one revolution of the thimble moves the thimble one line along the sleeve. When the edge of the thimble lines up with one of the divisions on this sleeve and the thimble shows 0 in line with the horizontal line on the sleeve, you simply add what you see on the sleeve.

Example

The edge of the thimble is past the 0.100 mark and in line with the first line after the 0.100 mark. This reading would be 0.100 + 0.025 = 0.125.

The example in Figure 4.42 shows what happens when the 0 on the thimble does not line up with a division on the sleeve. In this example, you can see that the edge of the thimble is past one of the 0.025 marks, but not quite to the next 0.025 mark. Also, the horizontal line on the sleeve is lined up with the 15 mark on the thimble. To read this, you simply add what you see on the sleeve and the thimble, as shown. Keep in mind, that every line on the thimble is equal to 0.001 and the numbers on the thimble are in increments of 0.005. The reading, as shown in Figure 4.42 would be

$$0.100 + 0.025 + 0.015 = .140 \text{ in.}$$

Another way to look at it is in dollars

If 1 in. = $1000, each number on the sleeve (1, 2, 3, etc.) is $100, each small division on the sleeve is $25, and each division on the thimble is $1. Now the previous reading is $140.

Parts of the Micrometer (Figure 4.43)

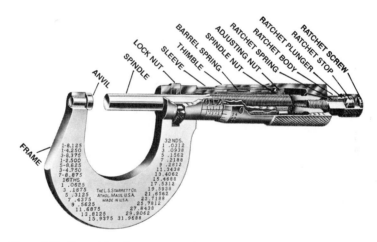

Figure 4.43   Parts of a micrometer. (L.S. Starrett Co. Reprinted by permission.)

Types of Micrometers

Following are listed just a few of many types.

> **Depth micrometer.** The micrometer shown in Figure 4.44 reads the same way as the one in Figure 4.46 but watch it because the thimble hides the scale. You must know the scale well to read the "mic" in Figure 4.45.
>
> **Hub micrometer.** This micrometer is very useful in tight places where the conventional micrometer cannot reach because of its frame (see Figure 4.46).

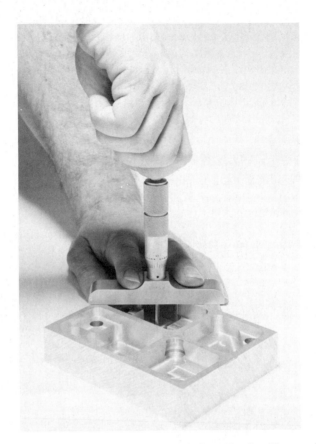

Figure 4.44   Depth micrometer in use. (M.T.I. Corp. Reprinted by permission.)

V-anvil micrometer. This "mic" is very good for measuring diameters or detecting out-of-roundness in a part (Figure 4.47).

Inside micrometer set. Sets of inside micrometers come in handy for measuring a variety of inside diameters (Figure 4.48). They can be used to measure ranges from about 2 in. to 12 in. simply by changing the rods (Figure 4.49).

Unimike. This tool has flat and round changeable anvils. You can remove the anvil to measure the height of a step (Figure 4.50).

There are many more special micrometers that range from 0 to $\frac{1}{2}$ in., 0 to 1 in., up to very large sizes.

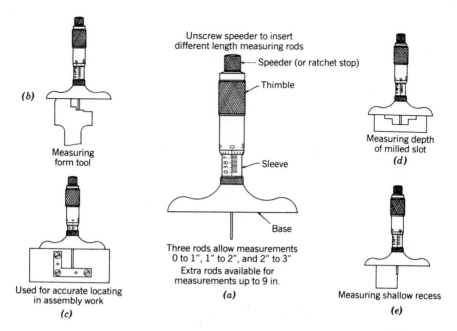

(b)

Measuring
form tool

Used for accurate locating
in assembly work
(c)

Unscrew speeder to insert
different length measuring rods

— Speeder (or ratchet stop)

— Thimble

— Sleeve

— Base

Three rods allow measurements
0 to 1″, 1″ to 2″, and 2″ to 3″
Extra rods available for
measurements up to 9 in.
(a)

Measuring depth
of milled slot
(d)

Measuring shallow recess
(e)

Figure 4.45   Various uses for a depth micrometer. (L.S. Starrett Co. Reprinted
by permission.)

Figure 4.46   Hub micrometer. (M.T.I. Corp. Reprinted by permission.)

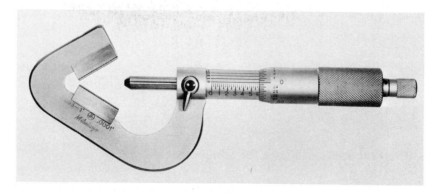

Figure 4.47   V-anvil micrometer. (M.T.I. Corp. Reprinted by permission.)

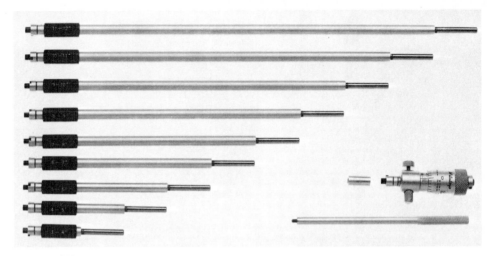

Figure 4.48   Inside micrometer set. (M.T.I. Corp. Reprinted by permission.)

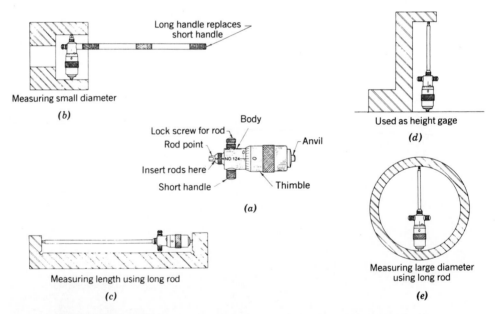

Figure 4.49   Inside micrometer measurements. (L.S. Starrett Co. Reprinted by permission.)

The Proper Way to Care for a Micrometer

The correct way to care for a micrometer is shown in Figures 4.51 to 4.53. Keep it clean, oiled, and safe. Treat it like an expensive tool, because it is.

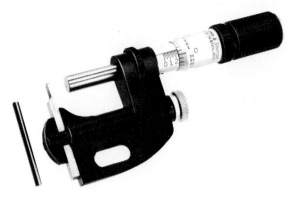

Figure 4.50 Multianvil micrometer. (Brown & Sharp. Reprinted by permission.)

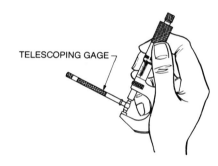

TELESCOPING GAGE

Figure 4.51 Proper way to hold a micrometer. (L.S. Starrett Co. Reprinted by permission.)

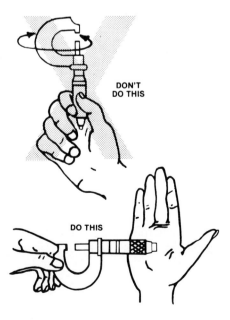

DON'T DO THIS

DO THIS

Figure 4.52 Opening and closing a micrometer. (Scherr Tumico. Reprinted by permission.)

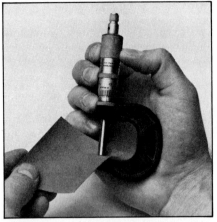

A handy method to clean the contact surfaces is to close the micrometer lightly on a piece of soft paper. It is then withdrawn. The paper will probably leave fuzz and lint on the surfaces. Blow this out with lung power—not the air hose. Never use compressed air to clean any precision instrument. The high velocity forces abrasive particles into the mechanism as well as away from it.

Figure 4.53 Cleaning micrometer contact surfaces. (Scherr Tumico. Reprinted by permission.)

Errors in Using Micrometers

1. Misreading the tool (most common misreading is by 0.025, incorrectly counting sleeve divisions).
2. Wear on the anvil and spindle faces.
3. Miscalibration.
4. Damage.
5. Heat.
6. Pressure too tight.
7. Pressure too loose.

Various Micrometers

Figure 4.54 shows various micrometers that are used in industry.

Calibrating Micrometers

For effective results there are certain techniques and tools to use when calibrating the micrometer. In calibrating the micrometer, you are looking for some specific problems that may cause inaccuracy. These problem areas are

1. Lead error in the threads.
2. Out of parallelism of anvil face to spindle face.

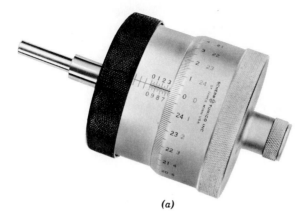

(a)

Figure 4.54a Micrometer head. (Scherr Tumico. Reprinted by permission.)

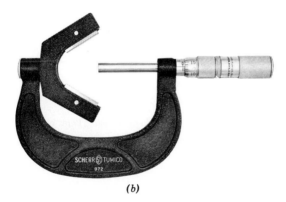

(b)

Figure 4.54b V-anvil micrometer. (Scherr Tumico. Reprinted by permission.)

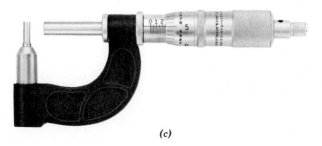

(c)

Figure 4.54c Hole location micrometer. (Scherr Tumico. Reprinted by permission.)

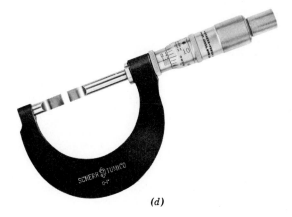

(d)

Figure 4.54*d* Blade micrometer. (Scherr Tumico. Reprinted by permission.)

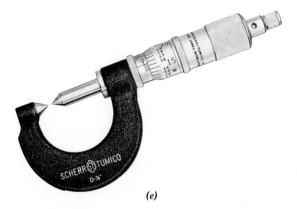

(e)

Figure 4.54*e* Point micrometer. (Scherr Tumico. Reprinted by permission.)

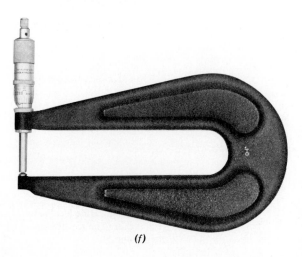

(f)

Figure 4.54*f* Deep throat micrometer. (Scherr Tumico. Reprinted by permission.)

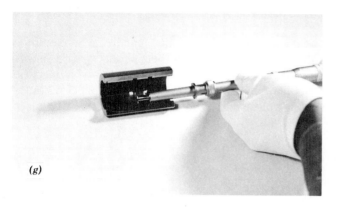

(g)

Figure 4.54g Multimicrometer in use. (M.T.I. Corp. Reprinted by permission.)

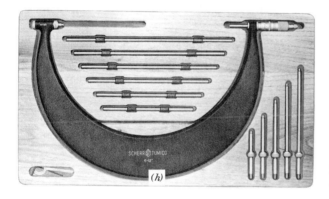

(h)

Figure 4.54h Micrometer set. (L.S. Starrett Co. Reprinted by permission.)

3. Worn measuring faces.

4. Measuring faces not perpendicular to the sleeve and thimble centerlines.

5. Zero setting not set at zero.

The conditions above must be checked and verified to be acceptable before a micrometer is truly said to be in calibration. *Lead error* can cause a micrometer to reflect good calibration in one area of its range, but there is measurement error in another area of its range (or travel). The *parallelism of the anvil face to the spindle face* can cause errors of contact, and the workpiece will not give the same (accurate) reading. *Worn measuring faces* have the same effect as parallelism. *Measuring faces that are not at right angles to the centerline* of the spindle and thimble can cause the micrometer not to be accurate on sides of the measuring faces. An *incorrect zero setting* can cause many problems. When the micrometer measuring faces are clean and closed, the 0 lines on the thimble and the sleeve should coincide.

Cleaning micrometer faces for calibration and use

Before using or calibrating the micrometer, the anvil and spindle faces should be cleaned. A good method for doing this is with a clean sheet of paper. You simply close the micrometer on a clean piece of paper, pull the paper out slowly, open the micrometer, and softly blow on the faces to make sure there are no paper particles left on them. This should be done before calibration and, often, before using it in the shop.

Standards for calibrating the micrometer

There are two basic standards used in calibrating the micrometer—*gage blocks* and *micrometer end standards*. Gage blocks are used more often than end standards because they have the range capability to check for lead error, and they are flat so that complete contact is made with the anvil and spindle faces. End standards are used too, but they are rod-shaped with spherical ends that make only single point contact with the measuring faces. End standards that come with micrometers are generally available only in 1 in. increments. With either standard, the micrometer should read the correct size when the measuring faces are closed and the ratchet stop (if there is one) is used. The ratchet stop, or the friction thimble, has one purpose. This is to maintain constant pressure that is adequate for measurement.

A good practice is to calibrate the micrometer with the ratchet stop and use it with the same movement of the ratchet stop. In other words, if the ratchet stop clicks twice in calibration, you should click it twice in use.

Calibration technique

The basic steps in calibrating the micrometer are outlined below:

Step 1. Visually inspect the micrometer for damaged parts.
Step 2. Clamp the micrometer to a micrometer stand during calibration. It is not recommended to hold it in the hand.
Step 3. Select the standard (gage blocks or end standards).
Step 4. Clean the micrometer faces (anvil and spindle) with the method prescribed earlier.
Step 5. Use a variety of gage blocks to check for lead error. A standard sequence to follow in size order is

0.105 in.	0.210 in.	0.315 in.	0.420 in.	0.500 in.
0.605 in.	0.710 in.	0.815 in.	0.920 in.	1.000 in.

The micrometer should be accurate throughout this range if there is no lead error.

Step 6. Check the zero setting. When the anvil and spindle faces are closed, the zero setting should be correct. If not, the sleeve can be rotated with a key wrench (usually supplied with the micrometer when purchased) (see Figure 4.55).

Step 7. Check the parallelism of the micrometer's measuring faces with an optical parallel. Note, however, that the measuring faces could be parallel to each other, but could be out of square with the centerline (of the spindle or anvil).

Step 8. Check the flatness of the measuring faces with an optical flat. A perfectly flat measuring surface, when observed through an optical flat, will show a fringe pattern of straight lines that are equally spaced. Other patterns, such as circular patterns, may be seen. These indicate convex or concave conditions.

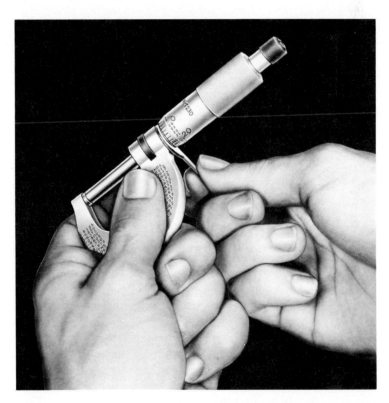

Lining up zeros on thimble
and sleeve

Figure 4.55 Setting the micrometer. (L.S. Starrett Co. Reprinted by permission.)

The calibration techniques stated above should be performed with standards, usually maintained in the metrology lab, and are traceable to national standards.

Along with the calibration system, which will be discussed later, the user of the micrometer should perform checks on calibration during use. The user can spot check the micrometer by using shop standards or gage blocks. Simply apply the micrometer to the standard (after cleaning the surfaces) and keep a close check on the zero setting during use.

THE VERNIER SCALE

Many measurements today are made with tools that require the ability to read the vernier scale. Examples of these tools are *height gages; vernier calipers;* and *protractors*. In Figure 4.56, notice the parts of the height gage (particularly the *bar* and the *plate*). This is where the vernier scale is "read."

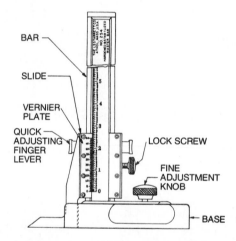

Figure 4.56 Parts of a vernier height gage. (L.S. Starrett Co. Reprinted by permission.)

Each large number on the bar is 1 in. The small numbers in between the large numbers represent 0.100 in. each. The 0 line on the *plate* is where you begin to read the vernier.

Also on the plate, you will see numbers from 0 to 50 in increments of 5. The space from 0 to 50 on the plate is equal to one small line on the bar, which is 0.050 in. Without the 0 to 50 vernier scale on the plate, the closest you could measure something is only 0.050 in. accuracy. The starting point, again, is the line 0 on the plate. This line is the first thing you look at when reading the vernier.

Figure 4.57 is an example of one vernier reading. Following are steps in how to obtain this reading.

Step 1. Look at the zero line on the plate to see where you start. In this case, it is not quite at 1 in., and the zero line is at 0.950 in. (plus something).

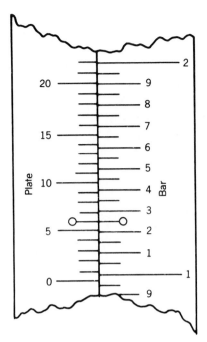

Figure 4.57 Vernier reading.

Step 2. Remember (or write down) 0.950.

Step 3. Now you must go to the vernier scale on the plate to see how much to add to 0.950 to obtain the correct reading.

The vernier scale has 50 lines on it. However, only one of those 50 lines will line up perfectly with a line on the bar. *This is the line you are looking for.*

Step 4. When you find the one line that lines up with any line on the bar, you simply add the amount shown on the plate (at that line) to the 0.950 you already have. In this case, the line is 0.006 as shown by the small circles drawn on the scale.

Once again, the zero line showed that your reading starts at 0.950 plus something. The vernier scale shows 0.006 line. So, the answer is 0.956.

Another tool that uses the vernier scale is the vernier caliper. This is a hand-held tool where feel is as important as it is with the micrometer. The vernier caliper measures three different characteristics: *inside dimensions, outside dimensions,* and *depth dimensions.*

The Universal Bevel Protractor

One of the best tools available for measuring angles (not angularity) is the universal bevel protractor (Figure 4.58). Other small protractors are only accurate

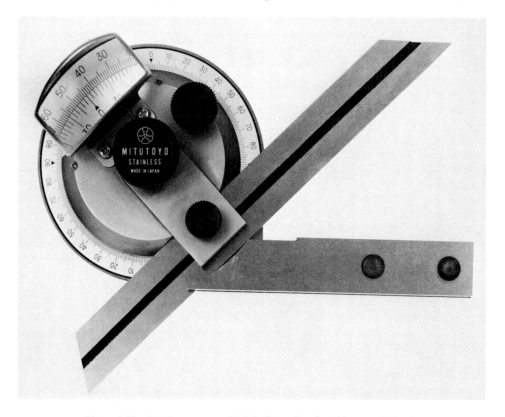

Figure 4.58   Bevel protractor (M.T.I. Corp. Reprinted by permission.)

within 1 degree, but the universal bevel protractor has a vernier scale that gives you accuracy within 5 minutes of a degree. For example, an angle of 20 degrees ± 30 minutes cannot be measured with a simple protractor. It must be measured with the universal bevel protractor because of the vernier capability in minutes.

Remember. There are 360 degrees in a circle and there are 60 minutes in each 1 degree.

The top scale of the universal bevel protractor shows you the measurement in degrees, but the vernier scale obtains the accuracy in minutes.

The picture in Figure 4.58 shows a universal bevel protractor, which by the use of an attached magnifying glass, allows the vernier scale to be read easily.

Examples of the various uses of the universal bevel protractor (see Figure 4.59)

Reading the vernier scale of the universal bevel protractor

Note, at this point, that the basic principle of reading the protractor vernier scale is the same as reading the vernier caliper.

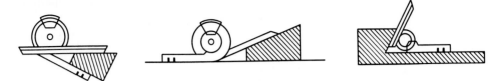

Figure 4.59 Bevel protractor applications.

Figure 4.60 shows one example of how to take a vernier reading.

Step 1. Use the 0 line on the main scale and obtain the reading in degrees. This is 12 degrees, (almost 13 degrees) here.

Step 2. Look to the right of the 0 line and see what line (between 0 and 60) lines up precisely with any line on the main scale. In this case the 50-minute line is the one.

Note. Read the vernier scale in the same direction from zero as the main scale reading.

Step 3. Therefore, the reading you get is 12 degrees and 50 minutes (12° 50″). Remember, the 50-minute line is the only one of all the graduations that line up.

Remember. The selection of what vernier scale to use is simple. If the numbers on the main scale are getting larger to the right, use the right vernier scale.

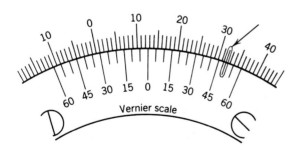

Figure 4.60 Bevel protractor reading.

TELESCOPING GAGES

There are many tools used to measure inside diameters. One of these tools is the telescoping gage. A telescoping gage is a tool that *transfers* an inside diameter to an outside diameter so that it can be measured with a tool such as an outside micrometer. Care must be taken when using the telescoping gage, because (if it is not used correctly) it can be very inaccurate. These gages generally come in full sets ranging in size, so that you may choose the correct size gage for the particular hole size you are measuring. The usual set ranges from $\frac{5}{16}$ in. to 6 in. capability (see Figure 4.61).

The telescoping gage, quite simply explained, is a tool with one spring-loaded cylinder and one fixed cylinder. As shown in Figure 4.62, you choose the gage you need, depress the spring-loaded cylinder (so that it will go inside the diameter), and then position it correctly in the diameter and lock it in with the lock provided. Then measure over the ends of the two cylinders with a micrometer.

After locking the gage in (Figure 4.63) to ensure accuracy (due to positioning the gage properly), you next have to "mike" the gage to obtain the actual size of the diameter being measured.

To mic the telescoping gage requires certain steps to ensure a close degree of accuracy (see Figure 4.64).

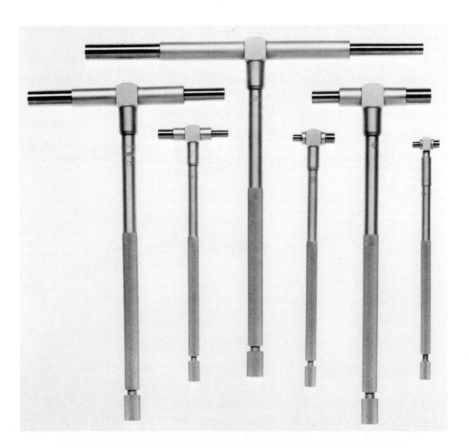

Figure 4.61   Telescoping gages. (M.T.I. Corp. Reprinted by permission.)

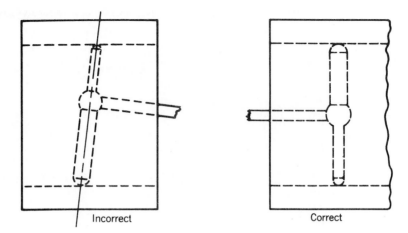

Figure 4.62 Positioning telescoping gages.

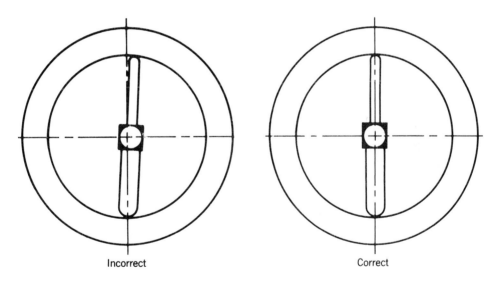

Figure 4.63 Positioning telescoping gages.

Step 1. Fix the fixed cylinder end at the anvil of the micrometer.

Step 2. Slowly bring the micrometer spindle down on the spring-loaded cylinder end until you feel a slight rub between the telescoping gage and the micrometer spindle.

Step 3. Read the micrometer. The feel of this rubbing when you obtain the actual size comes with experience.

Figure 4.64 Telescoping gage in use. (L.S. Starrett Co. Reprinted by permission.)

Step 4. Finally, as shown in Figure 4.65, the need to hold the micrometer properly with one hand arises, because the other hand is occupied in holding the telescoping gage. Never allow the micrometer to be run down far enough to actually hold the gage on its own. Always go by "feel."

Caution.   It is important to remember that transfer tools, such as telescoping gages and small hole gages, reduce the reliability of a measurement because of the buildup of possible errors when transferring measurements from tool to tool.

SMALL HOLE GAGES

There are many ways to measure small holes (holes that are 0.400 in. diameter or less). One way is with *small hole gages*. A small hole gage (Figure 4.66) is simply a tool with two rounded contact stems, a "spreader," and an adjustment knob. They usually come in sets of four gages. Each gage has a range of approximately 0.100 in. Small hole gage sets measure hole sizes from about 0.125 diameter to 0.400 diameter. These gages are similar to telescoping gages in that they are adjustable and are transfer-type gages. A micrometer (or other tool) must be used

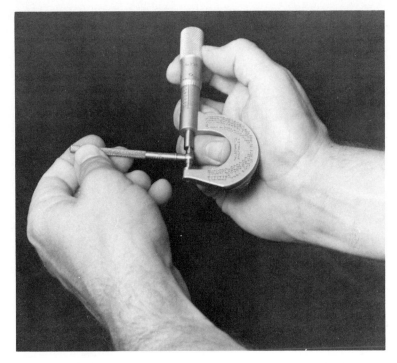

Figure 4.65 Transfer measurements with a telescoping gage and micrometer. (L.S. Starrett Co. Reprinted by permission.)

in conjunction with this tool to read the measurement. Also, as with telescoping gages, a certain feel is involved for accuracy.

Using Small Hole Gages

To use small hole gages, follow the steps below.

Step 1. Select the gage that has the range for the hole you want to measure.

Step 2. Put the gage in the hole with the adjustment knob loose.

Step 3. Moving the gage up and down for a short distance, tighten the adjustment knob until you "feel" the contact points "rub" in the hole (Figure 4.67).

When measuring the small hole gage with a micrometer, make sure that the rounded contact points of the gage are on the same line as the centerline of the anvil and spindle (Figure 4.68). Again, you must have a feel for knowing when to stop tightening the micrometer.

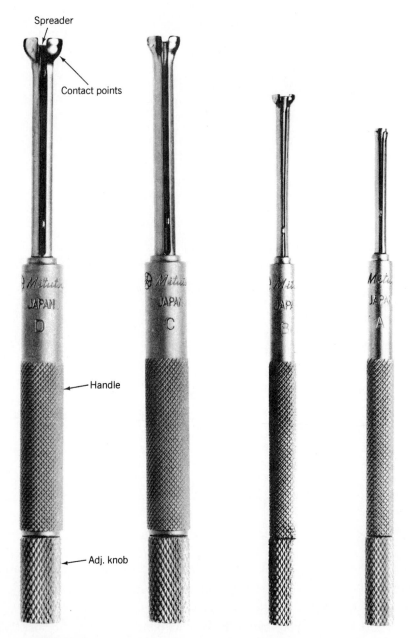

Figure 4.66   Small hole gages. (M.T.I. Corp. Reprinted by permission.)

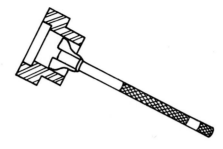

Figure 4.67 Small hole gage application. (L.S. Starrett Co. Reprinted by permission.)

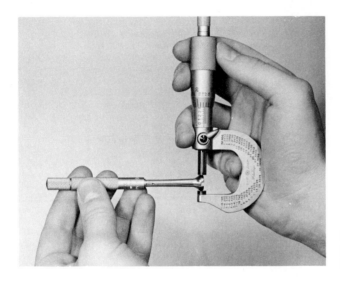

Figure 4.68 Measuring a small hole gage. (M.T.I. Corp. Reprinted by permission.)

There are other ways to measure small holes such as Go–NoGo plug gages (attribute gages). However, these gages only tell if the hole is in print size or not.

RADIUS GAGES

Radius gages (Figure 4.69) are used frequently during manufacturing processes to measure *radii* when the radii are toleranced so that fractional gages are applicable. Radius gage sets come in sizes from $\frac{1}{64}$ to $\frac{1}{2}$ in. in increments of 64ths: for example, $\frac{1}{64}$, $\frac{1}{32}$, $\frac{3}{64}$, and so forth. Each radius gage has five different measuring surfaces to measure various radii as shown in Figure 4.70. Three of the surfaces are for measuring outside radii, and two are for inside radii, as Figure 4.70 shows.

Note. The smallest gage in a set is usually $\frac{1}{64}$ in. (or 0.016 in. in decimals). Therefore, since the radius is 0.016, this gage set cannot be used to measure a radius such as 0.010 or 0.008. Nor can any radius larger than $\frac{1}{2}$ in. (or 0.500) be measured, since the largest gage in the set is $\frac{1}{2}$ in.

Radius gages are visual gages. After selecting which one to use, you apply the gage to the part. If the actual radius (on the part) is smaller than the gage, you will see the conditions shown in Figure 4.71a and b. If the actual radius is larger than the gage, you will see the conditions shown in Figure 4.71c and d.

Radius gages are also a handy tool for Go–NoGo use. You simply choose two gages that are the high and low limit of a radius tolerance and apply them. But if you need to find the actual radius of a part and do not have a radius gage that fits perfectly, you must use other methods.

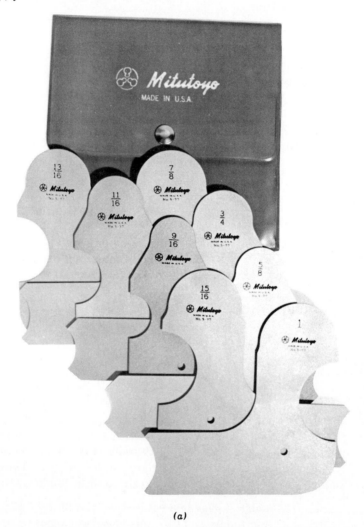

(a)

Figure 4.69a Radius gage set. (M.T.I. Corp. Reprinted by permission.)

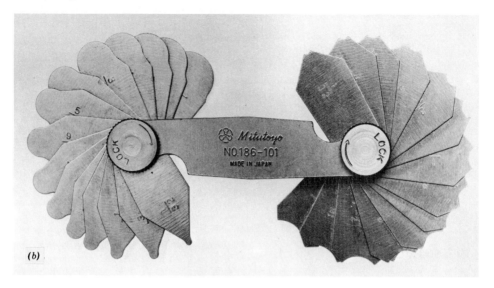

Figure 4.69*b* Radius gage set. (M.T.I. Corp. Reprinted by permission.)

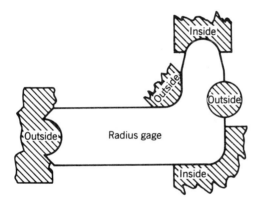

Figure 4.70 Radius gage applications.

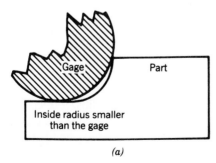

(a)

Figure 4.71*a* Inside radius is smaller than the gage.

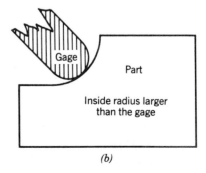

(b)

Figure 4.71*b* Inside radius is larger than the gage.

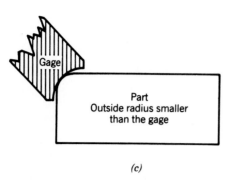

(c)

(d)

Figure 4.71c Outside radius is smaller than the gage.

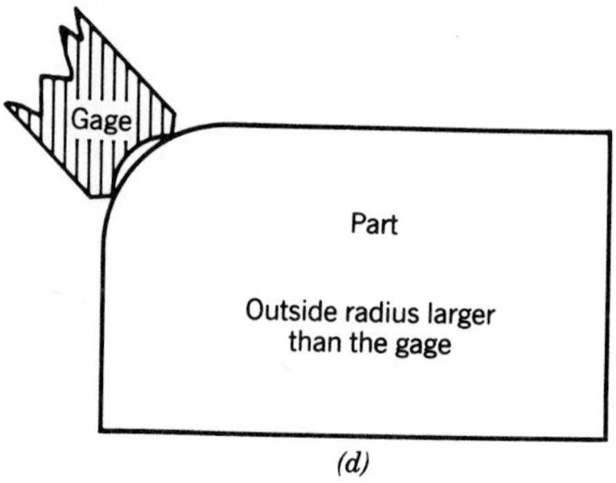

Figure 4.71d Outside radius is larger than the gage.

ATTRIBUTE GAGES

There are several different kinds of *attribute* (or Go–NoGo) gages. These gages are designed to tell you only that a dimension is or is not within blueprint limits. Remember that attribute means that a condition does or does not exist. Attribute gages do not tell you the *actual* size of a dimension, just whether it is good or bad.

Gages discussed here are the ones most widely used in manufacturing and inspecting components. They include *plug gages, ring gages, flush pin gages,* and *snap gages.*

Plug Gages

These are simply hardened steel "pins," that are accurate in size, with a handle. There are usually two *members,* the Go and the NoGo (see Figure 4.72). They are used to check inside diameters such as drilled holes on an accept, or reject basis. There are three basic kinds of plug gages:

> **Single purpose.** A Go or NoGo member only (not both).
>
> **Double end.** Those who have both a Go and NoGo.
>
> **Progressive.** Those where the Go and NoGo are on the same side (made from the same piece of steel). The Go is the first part used.

Application of the plug gage is very simple. Start the member at a slight angle into the hole, rotate it to a 90-degree position, and allow it to go into the hole (if it will) under its own weight. Do not try to force it in. If the Go goes in and the NoGo does not, the hole is to print. If the Go will not go in, the hole is undersize. If the NoGo will go in, the hole is oversize. How much undersized or oversized a hole is must be determined with another tool (such as a small hole gage and micrometer).

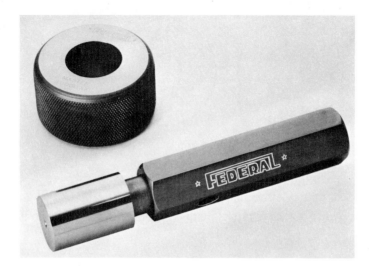

Figure 4.72 Ring gage (upper). Plug gage (lower). (Federal Products Corp. Reprinted by permission.)

Plug gage limitations

There are two main limitations of plug gages.

1. They only tell you that the hole is bad or good, not *how bad* or *how good*.
2. They will not detect out-of-roundness, taper, or barrel-shaped or bell-mouthed holes (which can be a problem).

Thread Plug Gages

There are also thread plug gages (Figure 4.73) that will measure (on a collective basis) internal threads such as tapped holes. They work on the same principle; the Go must go and the NoGo must not. Note, however, that the gage is designed so that the NoGo will at least start (one full turn, or so) but should not go farther.

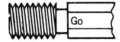

Figure 4.73 Thread plug gage.

Ring Gages

Like plug gages, ring gages (Figure 4.74) are hardened steel with very accurate inside diameters. They are also *attribute* gages. They come in two basic ways.

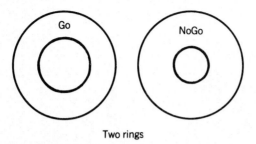

Two rings

Figure 4.74 Two ring gages (Go–NoGo).

1. Two different rings (one a Go and the other a NoGo). (*Note:* NoGo has a slot outside the ring.)
2. One progressive ring (Go and NoGo) with diameters in the same ring (Figure 4.75).

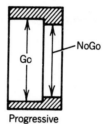

Progressive

Figure 4.75 Progressive ring gage.

Ring gages are used to measure outside diameters on a Go–NoGo basis. Their uses and limitations are about the same as a plug gage, except that they are used on outside diameters instead of inside diameters. When they are a set of two rings, the NoGo ring will have a groove on the outside. This groove is simply there to readily identify it as the NoGo ring.

There are also thread ring gages for Go and NoGo applications on external threads.

Flush Pin Gages

Flush pin gages (sometimes called fingernail gages) are attribute gages that provide additional information similar to that from a variable gage. This is so because the position of the pin (its step) gives you a rough idea of whether the part dimension being checked is near the high limit, low limit, or nominal dimension (Figure 4.76).

Flush pin gages work in two different ways:

1. The tolerance step is on the bar of the tool (Figure 4.76*a*).
2. The tolerance step is on the pin itself (Figure 4.76*b*).

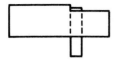

Figure 4.76a Tolerance step is on the bar.

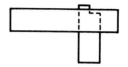

Figure 4.76b Tolerance step is on the pin.

In these figures, the step (on the bar, or on the pin) is the exact height of the total tolerance of the dimension being checked.

Figure 4.77 shows an example of 0.010 total tolerance. If the part is not within the tolerance limits, the pin will be below the lowest step, or above the highest step. You must look closely at the gage and pin to determine if the part is oversize or undersize.

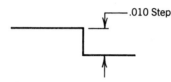

Figure 4.77 Tolerance step—flush pin gage.

Figure 4.78 shows examples of determination of *oversize* or *undersize* conditions.

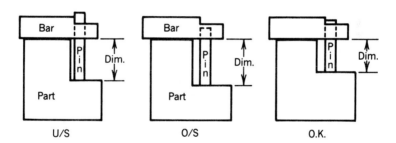

Figure 4.78 Determination of oversize/undersize.

Snap Gages

Snap gages are quite often used in Go–NoGo applications. They are usually shaped like a "C" as far as the frame goes, with adjustable measuring *jaws* (Figure 4.79). They generally have one solid jaw on one side, with adjustable jaws on the other. The snap gage works on the basis that the first jaw is adjusted to be the Go, while the second jaw is progressively adjusted to be the NoGo. Snap gages are applied to the dimension carefully (not quickly) in a rotating motion. The Go should go on the diameter and the NoGo should not.

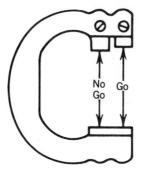

Figure 4.79   Snap gage (adjustable).

Whenever using any of the above attribute gages, you must remember that they will only tell you "good or bad" and nothing else.

Specific Precautions When Using Attribute Gages (Ring Gages, Plug Gages, and Snap Gages)

Ring gages

Ring gages will not detect geometric problems within the feature you are measuring. Examples of these geometric problems are shown in Figure 4.80.

(a)   *(b)*

Figure 4.80   Geometric problems with fixed gages: *(a)* shaft is out-of-round; *(b)* shaft is tapered.

Plug gages

Plug gages will not detect geometric problems within the feature (hole) you are measuring. Examples of these geometric problems in holes are shown in Figure 4.81.

Snap gages

Snap gages must be applied using the reference surface. Make sure that the piece being measured is straight (not cocked) in the gage (see Figure 4.82).

At all times when using ring, plug, snap, or any other type of gaging equipment, make sure that you have the correct gage for the dimension, that the part is clean, that the gage is clean, and that you know the limitations of the gage.

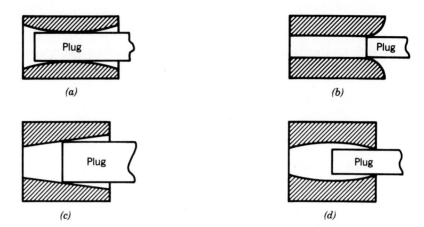

Figure 4.81 Geometric problems with fixed gages: *(a)* hole is hourglass-shaped; *(b)* hole is bell-mouthed; *(c)* hole is tapered; *(d)* hole is barrel-shaped.

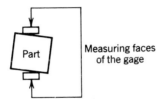

Figure 4.82 Manipulative error.

Application

All gages have a particular way of application to the part, or application of the part to them. The rule of thumb is very simple.

1. If the gage can be held in one hand and the part in the other, there is not a problem.
2. If the gage is heavier than the part, apply the part to the gage.
3. If the gage is lighter than the part, apply the gage to the part.
4. In all cases, use good common sense.
5. Avoid any application that may cause damage to the part or the gage.

FIXED PARALLELS

Fixed parallels are precision tools that are often used in production and inspection applications. These applications vary from machine setups to surface plate set-ups.

Fixed parallels are rectangular in shape and come in various sizes (see Figure 4.83). They are usually purchased in matched pairs. These parallels are made to be square and parallel on all sides to within tooling tolerances.

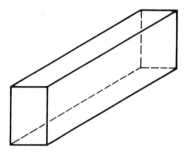

Figure 4.83   Fixed parallel extends the surface plate.

Their main purpose is simple to *extend* a reference surface where it is necessary.

This reference surface can either be on a machine that is producing the part or on the surface plate during the inspection of the part.

Fixed parallels can also be used in some sorting tasks. For example, some washers must be sorted to their *virtual thickness*. Here, matched parallels could be used. They are separated by feeler stock (or gage blocks if applicable) to the maximum virtual thickness allowed on the washers. Next, they are clamped securely, and the feeler stock is removed. Now, the washers can be rolled through them. Any washer that is beyond the virtual size allowed will not go through.

In Figure 4.84*a*, we see another example of the uses of fixed parallels. This part must be located on datum C to inspect it for parallelism, but datum C cannot be directly located on the surface plate because there is a diameter in the way.

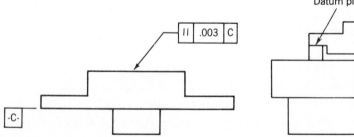
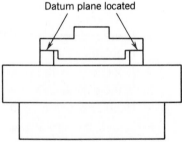

Figure 4.84*a*   Parallelism—interrupted datum.   **Figure 4.84*b***   Part located using fixed parallels.

As shown in Figure 4.84*b*, with parallels, locating datum C for inspection is no problem. The parallels extend the surface plate so that the diameter that was once in the way, is now clear from the plate.

In production and inspection fixed parallels have a wide variety of uses, most of which come with experience.

TAPERED (ADJUSTABLE) PARALLELS

Tapered parallels are a very simple tool in design, yet they are widely used in the production and inspection of a product. They are simply two pieces that are inclined and joined together by a dovetail slide and a lock screw (see Figure 4.85).

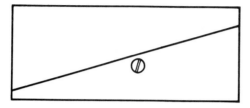

Figure 4.85 Adjustable parallel.

They are joined together in such a way that as the two pieces are moved in opposite directions, the sides spread wider apart but still remain parallel to each other (as shown in Figure 4.86).

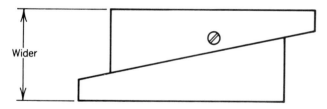

Figure 4.86 Parallel expands due to the angle.

Tapered parallels come in various sizes each with its own specific range. They can be set to a specific width with a micrometer (see Figure 4.87).

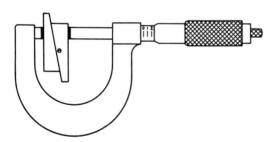

Figure 4.87 Measuring an adjustable parallel.

After being set, they can be used for machine setups, inspection setups, virtual size gages, Go–NoGo gages, and even measurement of certain size diameters (if the edges had radii). As shown in Figure 4.88, they can be used to measure

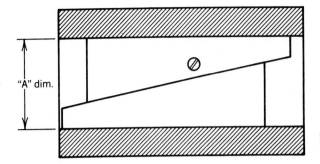

Figure 4.88 Adjustable parallel used to transfer a measurement.

the width of a slot. Simply loosen the lock screw, insert the parallel into the slot, spread it to meet the surfaces of the slot (with correct measuring pressure), lock it in, and remove and measure it with a micrometer (or other tool, depending on the tolerance you are working with).

Tapered parallels can also be set and locked in as a Go–NoGo gage for sorting tasks. Another use for them is the same as for fixed parallels: to extend a reference surface (such as the surface plate for inspection). Where tolerances permit, they can be used in conjunction with a micrometer to replace the need for gage blocks. Tapered parallels are used for more applications than fixed parallels simply because they are adjustable and can work through a larger dimensional range than fixed parallels.

CENTERLINE ATTACHMENTS

There are various methods of measuring the *centerline to centerline distance* between holes. One of the quickest ways to do this (and still retain accuracy) is to use centerline measuring attachments. There are several kinds of centerline attachments made today, but the two types covered in this book are the most popular. These two types are the *centerline gage,* shown in Figure 4.89, and the *center master,* shown in Figure 4.90.

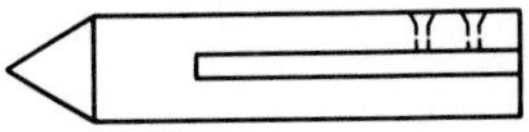

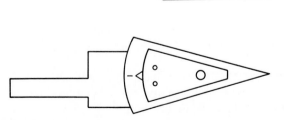

Figure 4.89 Centerline attachment for a vernier caliper.

Figure 4.90 Center master (used on a height gage).

The Centerline Gage

Centerline gages are very simple in their construction. They are made such that when you clamp them onto a vernier caliper, the center of the cone is in line with the measuring face of the caliper (see Figure 4.91).

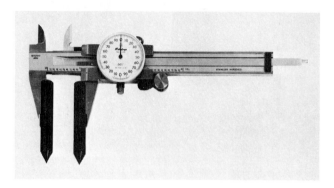

Figure 4.91 Centerline attachments on a caliper. (M.T.I. Corp. Reprinted by permission.)

The cone-shaped tip of the gage will automatically find the centerline of any hole that is not larger than the gage itself. Once the gages are properly clamped to the caliper, you can measure the distance between two holes and read the caliper directly.

Figure 4.92 shows the centerline gages in use. Care must be taken when attaching these gages to the caliper. Both must be attached at the same depth unless you desire to measure the distance between two holes that are offset in depth.

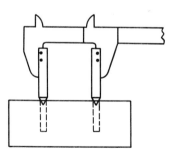

Figure 4.92 Application for centerline attachments.

The next type of centerline attachment, called the *center master,* is shown in Figure 4.93. It is used with a height gage and surface plate and attaches directly to the height gage. Using the center master is very easy. The center master has an indicating arrow and a single line on it that tell you when the blade in is line with the centerline of the hole.

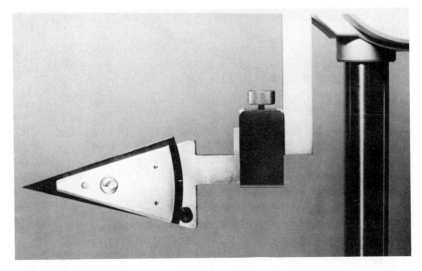

Figure 4.93   Center master on a height gage. (M.T.I. Corp. Reprinted by permission.)

Using a *vernier height gage,* the steps to finding the distance between centerlines of two holes are quite simple (see Figure 4.94).

Step 1. Insert the center master into the bottom hole and line it up. Read the vernier scale of the height gage.

Figure 4.94   Center master helps measure hole to hole. (M.T.I. Corp. Reprinted by permission.)

Step 2. Insert the center master into the top hole and line it up. Read the vernier scale again.

Step 3. Subtract the reading at the bottom hole from the reading at the top hole. This is the centerline-to-centerline distance.

ROUNDNESS: DIAMETRIC MEASUREMENT

Roundness is another form tolerance of great importance. The ANSI symbol for roundness is a circle. As mating parts go together, the roundness of each part at the features where they connect is controlled directly by the tolerance of that feature. For example, if a diameter is toleranced 0.502–0.498, this part can be out-of-round up to 0.004 (provided that it is not over 0.502 or under 0.498). Therefore, it has an *implied* roundness tolerance of 0.004 on the diameter.

Note that when measuring any diameter, you should always look for out-of-roundness, because that condition may cause the diameter to be defective. This is true especially with an out-of-round and undersized (ID) or an out-of-round and oversized (OD).

Example 1

The structure in Figure 4.95 means that this diameter must be round within two ten-thousandths of an inch. To measure this diameter, use a tool with an accuracy capable of measuring in ten-thousandths (or tenths as it is sometimes called). This tool could be either a tenths micrometer or a dial snap gage (Figure 4.96). With a dial snap gage, you simply rotate the part and watch the *total indicator reading* (TIR).

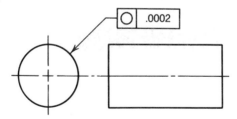

Figure 4.95 Shaft diameter.

Example 2

The information shown in Figure 4.97 means that there is no roundness callout on the part but that there is an implied roundness tolerance. The total tolerance of the part is 7.004 − 7.002 = 0.002; therefore, the diameter must not exceed these limits and must be round within 0.002 on the diameter.

There are a variety of tools that can be used to measure this. One is a *dial indicating gage* with *master*.

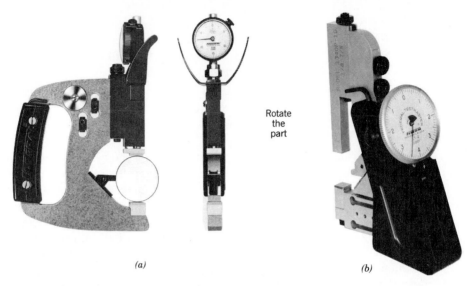

(a)   *(b)*

Rotate
the
part

Figure 4.96   Dial snap gage. (Federal Products Corp. Reprinted by permission.)

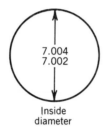

7.004
7.002

Inside
diameter   **Figure 4.97**   Inside diameter.

The dial indicating gage and master, shown in Figure 4.98, is simply a bar with an indicator and a fixed end located, plus a movable end locator. The movable end functions based on its spring pantograph.

The important thing is to remember that a dial indicating gage simply compares the accurate length of the master with an actual length of a part, and the indicator shows the difference. Therefore, to measure this part for the 7.004 – 7.002 dimension, we use the dial indicating gage (with the master being 7.003), and if the part is within plus or minus 0.001, it is to print.

However, there is one more thing that should be noted about a dial indicating gage. You should *always* locate the fixed end of the gage in the diameter and *move only* the movable end (or the end just below the indicator). If the measurement is taken any other way, it is likely to be very inaccurate (see Figure 4.99).

One last thing to remember when measuring *any* diameter with any variable tool is that it is good practice to measure that diameter in more than one direction. This will tell you if the diameter is out-of-round. Out-of-roundness can

Figure 4.98 Dial indicating gage. (Federal Products Corp. Reprinted by permission.)

Figure 4.99 Dial indicating gage and master ring. (Federal Products Corp. Reprinted by permission.)

cause problems in the function or assembly of the unit. Therefore, I personally recommend that for the best results, all diameters be measured in four different directions, as shown in Figure 4.100.

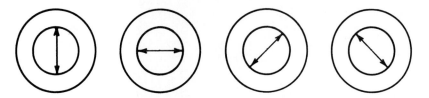

Figure 4.100   Exploring for roundness error.

TRUE CIRCULARITY (ROUNDNESS) MEASUREMENT

Roundness is a characteristic that is not easily measured. Features assume a variety of out-of-roundness errors while the part is being made, depending on the type of process and several other variables. These different "shapes" are caused by lobes or high areas around the circumference of the part. The lobes can be symmetrical (180 degrees apart) or nonsymmetrical (equally spaced, but not opposite each other). There are three types of lobing of out-of-round parts: *odd, even,* and *random* lobes, as shown in Figure 4.101.

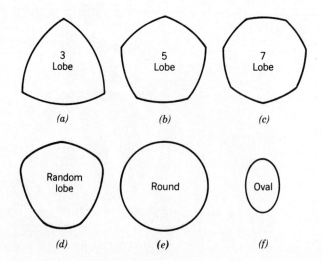

Figure 4.101   *(a)* Three-lobe; *(b)* 5-lobe; *(c)* 7-lobe; *(d)* random lobe; *(e)* round; *(f)* oval.

Diametrical Measurement of Roundness

A part with *nonsymmetrical* (odd) lobes can be very deceiving when measured with a tool such as a micrometer, vernier caliper, snap gage, or V-block. The reason is that no matter in which direction you measure the part, it can look perfectly round to a diameter measuring tool, when it is actually not. This is shown in Figure 4.102 with 3-, 5-, and 7-lobe examples. Note that in these odd-number lobes, the diameter measurement is the same in all directions.

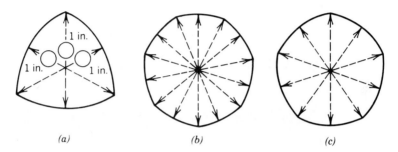

(a)   *(b)*   *(c)*

Figure 4.102 Measurements are the same in all directions with odd lobes.

However, parts with *symmetrical* (or even-numbered) lobes can be measured with diametrical tools because they are not deceiving to the tool. An example of symmetrical lobes is the oval part shown in Figure 4.103*a*.

Note. Vee blocks are effective in measuring roundness if they are the correct angle:

$$\text{V angle} = 180° - \left[\frac{360°}{n}\right]$$

where *n* is the number of equally spaced lobes on the part (see Figure 4.103*b*).

The Virtual Size

Another interesting fact is that you cannot measure odd-lobed parts for size with a diameter measuring tool (other than a ring gage). Because of the spacing and height of the lobes, a part can be virtually larger (if it is a shaft), or smaller (if it is a hole). About the worst virtual size to be found is the 3-lobe part. It can be virtually much larger than the tool measures it to be.

One method for measuring the virtual size of an out-of-round shaft is a *ring gage* (use a plug gage for a hole). When matching a shaft to a ring gage, the virtual size can be detected (see Figure 4.104).

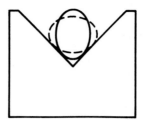

Figure 4.103a   Oval part in V-block.

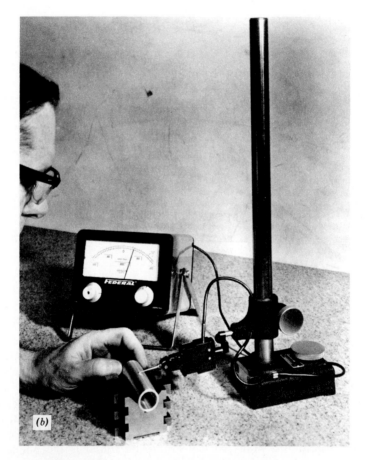

Figure 4.103b   V-block used for roundness measurement.

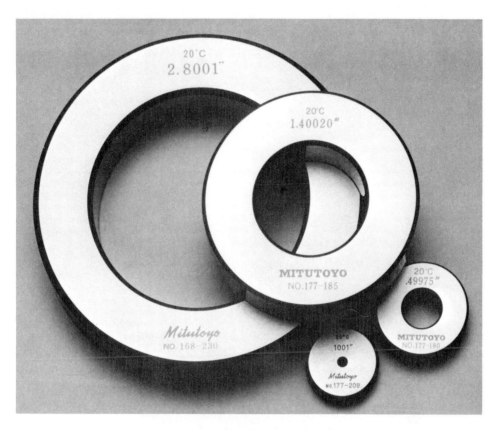

Figure 4.104 Ring gages measure the effective size. (M.T.I. Corp. Reprinted by permission.)

Radial Measurement of Roundness: The Preferred Method

A good method for measurement of out-of-roundness, regardless of the number of lobes, is radial measurement. An example of radial measurement is the precision spindle (see Figure 4.105). This device either has a rotating base with a fixed indicator or a fixed base with a rotating indicator. Either way, the device rotates "true" and allows you to compare the actual part to two concentric circles that are established by the true rotation of the spindle. Or, simply said, the spindle rotates in such a way that if you put a truly round part in it, you will get no movement on the indicator. The indicator (or probe) will detect the amount, height, and spacing of the lobes, giving you a true picture of the roundness of the part.

The interpretation of roundness measurement according to ANSI Y14.5, "Geometrical and Positional Tolerancing," is radial, with the tolerance zone being two concentric circles. This requires radial measurement of odd-lobed parts

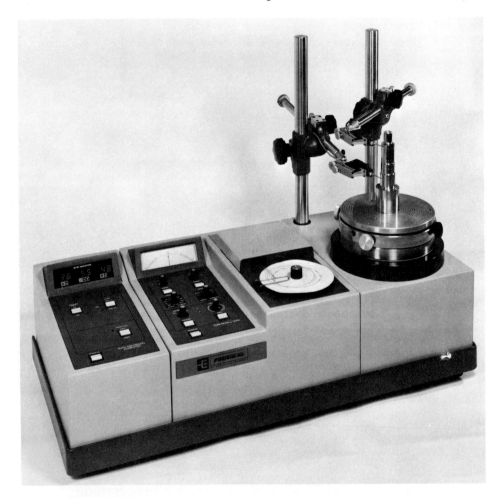

Figure 4.105   Formscan 3000 uses a precision air bearing spindle. (Federal Products Corp. Reprinted by permission.)

and accepts diametrical measurement of even-lobed parts. However, for all purposes where roundness is very important to the function or fit of component parts, radial measurement is preferred.

CYLINDRICITY

Cylindricity is a *form* tolerance that may be considered as an extension of circularity (roundness). The ANSI symbol for cylindricity is a circle with tangent lines.

The only difference between *cylindricity* and *circularity* (roundness) is that cylindricity controls the roundness of a feature throughout its entire length. The tolerance zone for cylindricity is two concentric imaginary cylinders, as shown in Figure 4.106.

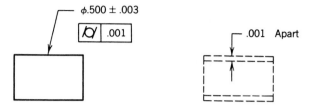 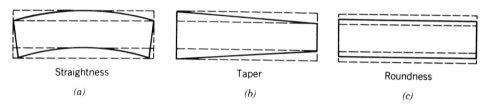

Figure 4.106 Cylindricity callout and tolerance zone.

The surfaces of the shaft at all elements must lie between the two concentric cylinders 0.001 in. apart. Cylindricity controls the feature throughout its entire length. When you apply cylindricity, it controls other characteristics at the same time, as shown in Figure 4.107.

Figure 4.107 Cylindricity controls taper, roundness, and straightness: *(a)* straightness; *(b)* taper; *(c)* roundness.

Measurement of cylindricity is best accomplished using radial techniques such as those described in the earlier section on true circularity (roundness) measurement. The precision spindle (refer to Figure 4.105) may be used in the same way, except that a precision vertical post must be used so that a vertical sweep may be made on the part. This is depicted in Figure 4.108.

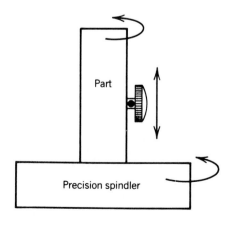

Figure 4.108 A precision spindle with a precision vertical slide will measure cylindricity.

Cylindricity is very difficult to measure; it requires sophisticated equipment for accuracy. In cylindricity, you are making measurements that are affected by the collective geometry of the part. Roundness of circular elements, straightness, taper, surface flaws, and surface texture are all combined to add to the difficulty of the measurement.

FREE STATE VARIATION PER ANSI Y14.5M–1982

There are certain components in industry where free state variation is applicable. *Free state variation* simply refers to that variation seen in flexible parts when they are released by the holding device of the machine. For example, when a thin wall aluminum casting is located in a chuck on the lathe, and the casting is located in the jaws of the chuck, the diameter that is turned will be round; however, when the casting is released from the pressure of the holding device (the chuck jaws) it returns to its original form, causing the diameter that was machined to be out-of-round. Due to this variation, some geometric tolerances applied to flexible parts can exceed the size tolerance of the feature being machined. Note that General Rule 1 of ANSI Y14.5 does not apply to free state parts.

Example

A diameter has a size tolerance of 1.500 − 1.495 and a roundness tolerance of 0.015 free state.

When tolerances are shown on a drawing in the free state, the term "free state" will be shown under the feature control frame, and often the term "avg. dia." will be shown next to the size tolerance, as shown in Figure 4.109.

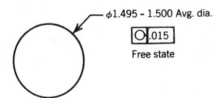

Figure 4.109   Free state callout.

To measure the diameter in Figure 4.109, an inspection is required to measure the diameter in four different directions, to find the average of the four measurements, and to check that the average diameter is within the size tolerance, as shown in Figure 4.110.

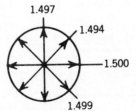

Figure 4.110   At least four measurements should be taken.

Example

Four measurements are taken in four different directions. The average of the four measurements is

$$
\begin{array}{ll}
1.497 & \underline{1.4975} = \text{average diameter} \\
1.494 & 4)\overline{5.990} \\
1.500 & \\
+\ 1.499 & \\
\hline
5.990 &
\end{array}
$$

Note. The average diameter (1.4975) is within the size limits.

Example

Four measurements are taken on another part (Figure 4.111). The average of these four measurements is

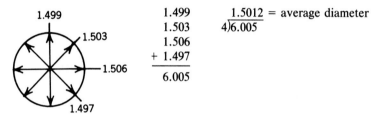

$$
\begin{array}{ll}
1.499 & \underline{1.5012} = \text{average diameter} \\
1.503 & 4)\overline{6.005} \\
1.506 & \\
+\ 1.497 & \\
\hline
6.005 &
\end{array}
$$

Figure 4.111   Measurements on another part.

Note. On this part the average diameter is not within size limits and should be rejected.

It must be noted that the average diameter rule cannot be applied to any part unless "avg. dia." and "free state" are shown on the drawing. If these notations are not shown, size tolerances must be met. Also note that General Rule 1 (ANSI Y14.5–1982) does not apply to parts subject to free state variation in the unrestrained condition.

THE SINE BAR

On some component parts, the term *angularity* is used, relating to the accuracy of an angle cut. The sections on the bevel protractor discussed how angles are measured in degrees when the tolerances are angular. But in angularity, the tolerances are in inches.

Example

Angularity to be within 0.010 in. This type of tolerancing on an angle rules out the use of a bevel protractor, because it can only measure in angular dimensions, such as degrees and minutes. Therefore, to measure angularity, you need a sine bar (or sine plate). The sine bar, or plate (Figure 4.112), is a simple tool based on right triangle

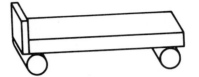

Figure 4.112a   Sine bar.

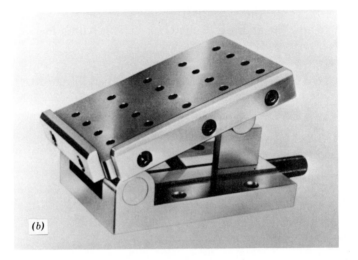

Figure 4.112b   Sine plate. (Grinding Technology, Inc. Reprinted by permission.)

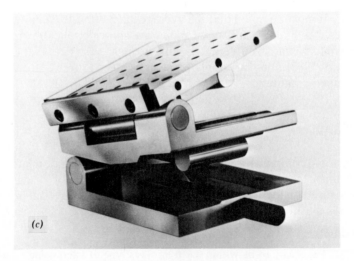

Figure 4.112c   Compound sine plate. (Grinding Technology, Inc. Reprinted by permission.)

trigonometry (see Chapter 2). The sine bar sets up the "perfect" angle you want, so that an indicator can be used to measure the variation in inches.

The example in Figure 4.113 shows a part that has a 30-degree "basic" angle from surface A and must be within an angularity tolerance of 0.015, as shown. To inspect this part, you could use a sine bar set up at 30 degrees. For example purposes only, the length of the sine bar here is 10 in.

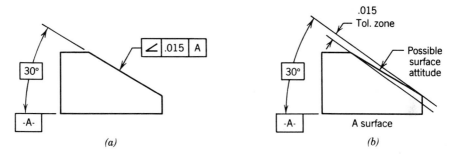

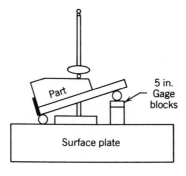

Figure 4.113 (a) Angularity-surface to surface callout. (b) Angularity tolerance zone.

Remember. The sine bar forms the hypotenuse of the angle you are working with (see Chapter 2).
So, the angle is 30 degrees. The hypotenuse is 10 in. Now solve for the opposite side (sine 30° = 0.5).

Formula: Sin 30° × hyp. = opp.; or

$$0.5 \times 10 = 5.0 \text{ in.}$$

This tells you that you must put 5 in. of gage blocks under the sine bar to set up a 30-degree angle.

The picture in Figure 4.114 shows how the part is inspected. As you can see, the angled surface of the part becomes parallel to the surface plate, so you can simply run an indicator across it to obtain the *total indicator reading*. If it is less than .015 in., the part is to print.

Figure 4.114 Sine bar measures angularity.

Remember: The length of the sine bar used must be known when you are using the formula. If the bar were only a 5-in. length, you would have to put only 2.5 in. of gage blocks under it to attain a 30-degree angle.

Understanding the right triangle concepts and knowing that the sine bar forms the hypotenuse of the triangle will allow you to use the sine bar effectively to measure angles toleranced by angularity.

Note. A further study of measuring angles will describe alignment datums that may be necessary. If the part is rotated on the sine bar, you can get any reading you want. Secondary alignment datums are not mentioned in any specification but should be considered.

TORQUE MEASUREMENT

Torque measurement is a vital part of today's industrial measurement, especially when the product is held together by nuts and bolts (fasteners). The wrong torque in an assembly can result in the assembly failing due to a number of problems. Two of these problems are: (1) parts are not assembled securely enough for the unit to function properly; and (2) threads are stripped because torque is too high, causing the unit to fail.

Torque is simply described as a force producing rotation about an axis. This force could be in pounds, grams, ounces, or other units. Torque is based on the law of leverage. The formula for torque is

$$\text{torque} = \text{force} \times \text{distance}$$

The distance could be feet, inches, meters, or other units. As shown in Figure 4.115, a distance D, and a force of F, produces torque T.

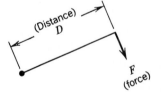

Figure 4.115 Description of torque.

Example 1

A force of 1 lb applied at a distance of 1 ft. So.

$$\begin{aligned}
\text{torque} &= \text{force} \times \text{distance} \\
&= 1 \text{ lb} \times 1 \text{ ft} \\
&= (1 \times 1) \text{ lb-ft} \\
&= 1 \text{ lb-ft}
\end{aligned}$$

Note. This is usually referred to as foot-pounds (ft-lb).

Example 2

A force of 12 lb applied at a distance of 10 in. from the center of a bolt will torque the bolt at 120 lb-in. (pound-inches, usually referred to as inch-pounds).

Measuring Torque

Torque on fasteners is measured by a torque wrench. There have been many types of torque wrenches made. The two types most commonly used are the *flexible beam* and the *rigid frame*. The flexible beam type, shown in Figure 4.116, works on the basic principle that a beam of a certain size, geometry, and material will bend a certain distance at an applied force, and return to the original position after the force has been removed. The flexible beam has a pointer that shows the torque applied. This beam can be round or flat.

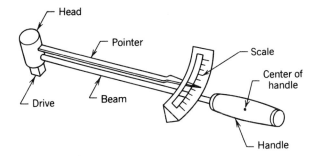

Figure 4.116 Flexible beam torque wrench.

The rigid frame torque wrench works with a torsion bar connected to the frame, as shown in Figure 4.117, and a dial indicator attached so that the torque applied can be read directly. This indicator works using the rack on the lever arm.

Another type of torque wrench (not shown) is the type for which you preset the desired torque, and when you reach the desired torque on the bolt, the wrench

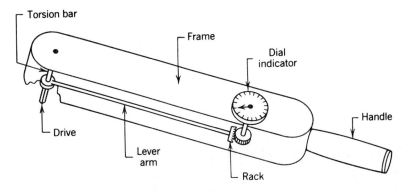

Figure 4.117 Rigid frame torque wrench.

will either make a distinct "clicking" sound or you will simply feel the wrench "give" in the handle.

Cautions in Using Torque Wrenches

The following points should be kept in mind while using torque wrenches.

1. To use the wrench accurately, hold the handle in the center at all times.
2. Apply the force slowly and smoothly; do not jerk the wrench.
3. Hold the wrench steady for a short time after the desired torque has been reached (this is generally considered to be good practice).
4. Handle torque wrenches carefully, especially the pointer.
5. Use torque wrenches for applications only within 80 percent of their range.
6. Beware of **false applications of torque.** *Example:* The bolt is too long, and it bottoms out before the bolt head reaches the surface for tightening.
   OR:
   The threads in the hole are too shallow and the bolt also bottoms out before tightening.
7. Keep torque wrenches calibrated periodically with a torque wrench measuring instrument that is a known standard.
8. If it is necessary to extend a torque wrench, ensure that you compensate for the change in distance. Also ensure that the extension is in line with the centerline of the wrench handle, to avoid *cosine error.*

OPTICAL COMPARATORS

Optical comparators (Figure 4.118) are used throughout the industry for making comparison measurements. The standards used in these measurements vary and may include:

Built-in micrometer heads. To allow you to measure actual movements from a reference point to the measured point.

Graduated screens. To make direct comparisons using the screen itself (such as angular measurements with a screen graduated in degrees and minutes).

Simple overlays. An overlay is made to scale to make comparisons for different shapes.

Pantograph mechanisms. For use in "blind" areas.

The optical comparator projects the profile of a part on the screen after it has been magnified in size.

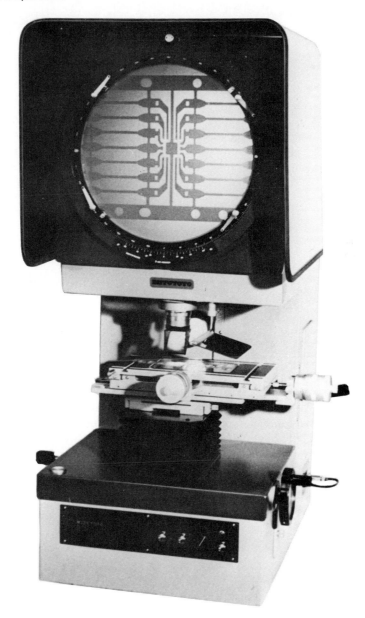

Figure 4.118 Optical comparator. (M.T.I. Corp. Reprinted by permission.)

Example

Using a 10× comparator, if a part is placed in front of the light source, the image on the screen will be 10 times the size of the actual part. If the part has a 0.250-in. diameter, the image on the screen will appear as a 2.50-in. diameter.

The comparator table (or staging area) is capable of movement from left to right, up and down, toward or away from you. The movement toward or away from you allows you to focus the part for a sharp image on the screen.

Crosslines

There are usually crosslines built into the comparator screen for the horizontal and vertical referencing (see Figure 4.119). Direct measurements can be made starting at one of these crosslines and using the micrometer heads to measure actual movement from one surface to another.

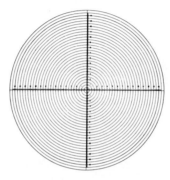

Figure 4.119 Crosslines on the comparator screen. (M.T.I. Corp. Reprinted by permission.)

Overlays

There are various overlays that can be used on a comparator. These overlays are simply a clear material, such as Mylar, that can be clamped onto the comparator screen (usually having their own crosslines for referencing them to the comparator's crosslines).

Most overlays are specially made to compare a particular shape on the product, such as a contour. The overlay will generally have the perfect shape drawn with tolerance lines on each side of it, or sometimes will show just the tolerance lines (see Figure 4.120).

Figure 4.120 Overlay description with tolerance lines.

Angular Measurement

Most comparators have screens that have a built-in scale of 360 degrees with the ability to measure within minutes of a degree using a vernier scale. Crosslines divide the circle into its quadrants (Figure 4.121).

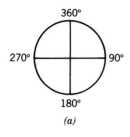

Figure 4.121 Angles and images. (M.T.I. Corp. Reprinted by permission.)

(a)   (b)

With a part on the staging area, the screen can be turned to match the angle of the part, then the angle can easily be determined.

Linear Measurement

Using the crosslines as reference, linear measurements can be made by moving the staging area (using its micrometer heads, or other attachments). By starting at one edge of the part and referencing it on a crossline, then zeroing the micrometer head, the part can be moved to the measured surface, and the micrometer head reading will show the movement.

An advantage of linear measurement with a comparator is that there is no contact made with the part. Therefore, contact pressure and wear on measuring surfaces is not a problem. The comparator is also handy when the reference points are in "space" (not physical points).

Contour Measurement

The comparator cannot see inside a part. If there is an internal contour or shape, there must be some sort of attachment to compare this shape. This attachment is often in the form of a *pantograph* that allows you to physically trace internal shapes. The pantograph system has two parts that move together. The stylus follows the shape of the part being measured, and its counterpart shows this movement on the screen. This is how a pantograph system works. When the stylus moves, the counterpart moves exactly the same way so that you can work easily with internal shapes on a comparator.

Parts of a Comparator (Figure 4.122)

In this book we do not elaborate on the parts of the comparator, but you should have an idea of what makes it work.

   Light source. The part is placed in front of the light source so that its image can be viewed on the screen.

   Lens system. The lens system magnifies and transmits the image to the screen.

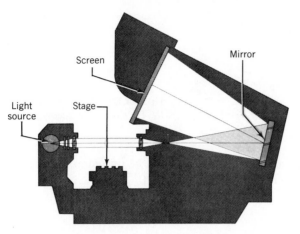

Figure 4.122   Parts of a comparator.
(M.T.I. Corp. Reprinted by permission.)

Screen. The screen must be large enough to show the magnified image of the part.

Holding devices. There must be a staging area for the part to be held in place. The kind of holding device depends on the parts that are to be measured. This can vary from simple fixtures, to centers, or other clamping devices.

Projection of a Surface

Some comparators are equipped with the ability to project a surface (sort of a mirror image) onto the screen (Figure 4.123). This requires a strong light source and the ability of the material to reflect light.

Surface image

Figure 4.123   Projection of a surface.
(M.T.I. Corp. Reprinted by permission.)

Comparators are all a little different. The important thing is to familiarize yourself with how the staging area works, any applicable overlays you must use, the magnification power of the comparator, and how to operate the pantograph (if there is one).

MECHANICAL COMPARATORS

Bench comparators are often used for on-site measurements. They can be carried and used anywhere because they have their own built-in reference surface, indicator stand, and indicator. Some bench comparators use a granite surface plate with a vertical post built into it; others use metal reference surfaces. Some of them have serrated reference surfaces (see Figure 4.124) that will allow dirt or chips to fall between the serrations to avoid errors in the measurements. On all bench comparators there is a method of vertical adjustment and locking it in. Most have a clamp on the vertical post that (when loosened) allows the indicator to be moved up and down to adjust it when necessary. Some have a rack, where turning the knob moves the indicator up or down; then it can be locked into place.

Bench comparators are generally used when quick, but accurate, measurements of height, thickness, and so on, must be made on a variable basis. Examples of this can be seen from time to time on machines for the operator to measure his or her parts during production, or at workbenches when parts must be "screened" or 100 percent inspected.

All bench comparators require the use of some type of standard to establish the size to be compared. More often than not, gage blocks are used as this

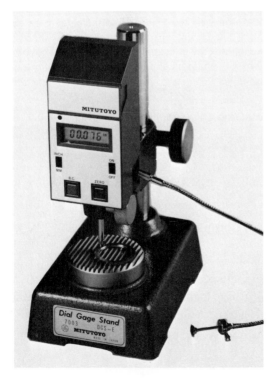

Figure 4.124 Bench comparator with an electronic digital indicator. (M.T.I. Corp. Reprinted by permission.)

standard. For example, the piece (in Figure 4.125) must be measured on a 100 percent basis (each piece must be measured as it is produced). The bench comparator would be set up using a 0.560-in. stack of gage blocks. The indicator would be set at 0 on the stack, and then the part would be passed under the indicator. Any part that was good would not cause a movement on the indicator farther than ± 0.004 in. from 0 set.

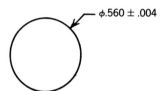

φ.560 ± .004

Figure 4.125   Diameter.

From time to time, the comparator should be checked with the gage block stack for accuracy. Sometimes a representative master may be used, as shown in Figures 4.126 and 4.127.

All the operator has to do is be able to read the device to determine the comparison between the standard and the part.

Example 1

You know that the comparator was set to 0 with a 0.560-in. gage block stack. Therefore, any part that passes under the indicator that is actually 0.560 in. high will cause the indicator to come to 0. If the indicator comes to 0 from a counterclockwise position, plus movements past 0 would mean that the part was higher than 0.560 in.

Example 2

The indicator reads plus 0.003 in. Therefore, the part is 0.560 in. + 0.003 in. = 0.563 in. actual size.

Example 3

The indicator reads minus 0.001 in. Therefore, the part is 0.560 in. − 0.001 in. = 0.559 in. actual size.

Cautions When Using Bench Comparators

1. All clamps should be tight.
2. Make sure that your gage block stack is dimensionally correct.
3. Make sure that the indicator is facing you directly (to avoid parallax error).
4. Make sure that the vertical post is not loose.
5. Make sure that the indicator has the accuracy you require for the measurements you are making.

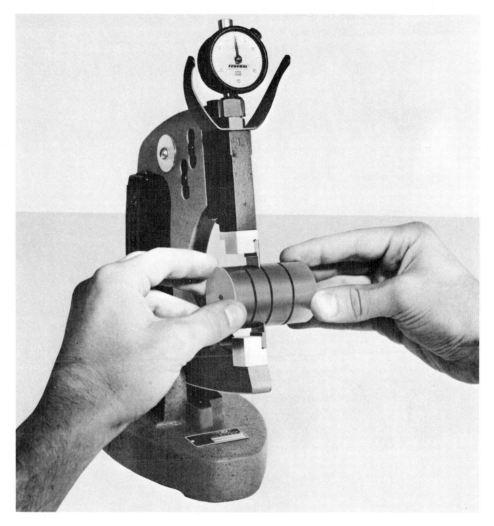

Figure 4.126 Dial snap gage with stand. (Federal Products Corp. Reprinted by permission.)

6. Check the indicator calibration with the gage block stack at regular intervals (especially when being used continuously).

7. Make sure that parts are clean and deburred before passing them under the indicator.

8. To move the bench comparator, use the base; do not lift it with the upright post.

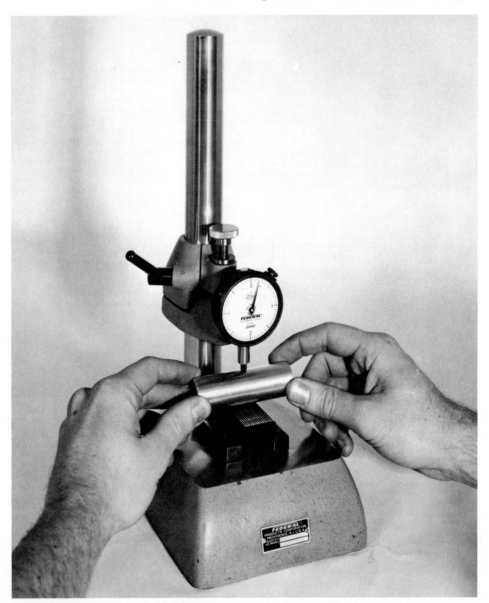

Figure 4.127   Bench comparator in use. (Federal Products Corp. Reprinted by permission.)

9. Avoid bumping the indicator rod from the side. It may be necessary to lift the indicator tip up and then put the part under it.
10. Make sure to use the proper tips on the indicator.
11. Do not use the bench comparator in jerking motions, which could cause miscalibration. Use smooth, even motions to make your measurements.

MEASURING INTERNAL THREADS

Refer to Chapter 3 on internal threads. The nomenclature here changes slightly from that on internal threads in Chapter 3. The *crest* and *root* are looked at in a different manner, as shown in Figure 4.128. The *minor diameter* is actually the original hole (at the crest diameter) that was drilled before tapping. The *major diameter* (after tapping) is now at the root of the thread. (It is at the crest on external threads.) The *pitch diameter*, thread depth, and angle remain the same.

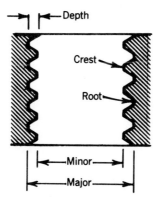

Figure 4.128 Internal threads—nomenclature.

Measurement (in general) includes the pitch diameter, minor diameter, and the depth of the threads in the hole (if required). Sometimes the depth is not a requirement because the hole must be tapped all the way through. The minor diameter can be measured simply with a plug gage. The pitch diameter is usually proved with a *thread plug gage* like the one shown in Figure 4.129. It is a Go–NoGo type of gage.

Figure 4.129 Thread plug gage.

Measuring Depth on Threads

Example

Measurement of the depth of threads can be done in various ways. One of the most accurate ways is the *turn method*. This involves using a thread plug gage's accurate lead to verify the depth of threads (see Figure 4.130).

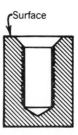

Figure 4.130   Callout for minimum full thread.

Keep in mind that thread plug gages are accurate within gage tolerances and therefore have an accurate lead. The *lead* is the distance that a thread plug gage will travel in *one full turn* (or the distance between two adjoining threads).

To Inspect Depth

If there are 20 threads per inch, each thread is $\frac{1}{20}$ in. or 0.050 in. apart. So 0.050 is the lead. Therefore, if you have an internal thread that is $\frac{1}{4}$ − 20 and is supposed to be 0.600 deep (at the last full thread), and the lead is 0.050 in., then if the threads were actually 0.600 in. deep, the plug would take 12 turns to reach the bottom thread.

$$0.050 \overline{)0.600} \quad \overset{12 \text{ turns}}{}$$

Other Examples

$$\frac{3}{8} \text{ in.} - 16 \times 0.625 \text{ in. deep} = \frac{1}{16} \text{ in.} = 0.0625 \text{ in. lead}$$

and

$$0.625 \text{ deep} \div 0.0625 \text{ lead} = 10 \text{ turns}$$

Let's say that the previous part is measured by the turn method and you get only nine turns. To find out how deep nine turns is, just multiply the nine turns times the lead. Therefore, 9×0.050 in. $= 0.450$ in., and since the minimum depth is 0.600 in. ± 0.010 in., these threads are too shallow.

Remember. When using the turn method, you *must* include the depth of chamfers because the measurement is from surface A to the bottom thread. So the example of 0.450 in. with a 0.030-in. chamfer is actually 0.480 deep.

MEASURING EXTERNAL THREADS

There are various ways of measuring external threads, depending on the application and the desired accuracy. Usually, the most important dimension of threads of any kind is the pitch diameter.

There are various tools that can be used to measure threads. Some are explained in the following examples.

Example 1

One of the quickest ways is the use of a pair of Go–NoGo thread ring gages, one of which is shown in Figure 4.131. These gages are set to the tolerance limits of the part. Their application is simply screwing them onto the part. The Go should go on easily without force over the length of the threads, and the NoGo should start but not go over more than a couple of threads before it drags.

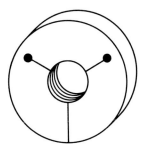

Figure 4.131   Thread ring gage.

Remember. These rings only tell you whether or not the part is in tolerance. They do not tell you what the actual size of the thread is.

Example 2

Another quick method is with the screw thread micrometer shown in Figures 4.132 and 4.133. These "mics" are set up to measure the pitch diameter directly. The only problem with the mics is that they are designed to measure a range of different pitch values; therefore, an incorrect setup could cause error in measurement. *To correct this,* the micrometer should be set with a thread plug gage of known pitch diameter before use.

Example 3: The Three-Wire Method

This method of measuring pitch diameter is relatively quick and more accurate than the others (depending on the accuracy of the wires used). As shown in Figure 4.134, you need three wires of the same specified size, setting up two of them on one side and one on the other. There is a formula that tells you what wires to use and how to arrive at the pitch diameter. However, most wire sets have a direct reading table that will also give you pitch diameter. Make the setup as shown. Measure the dimension over the wires and use the constant number on the table to arrive at the pitch diameter.

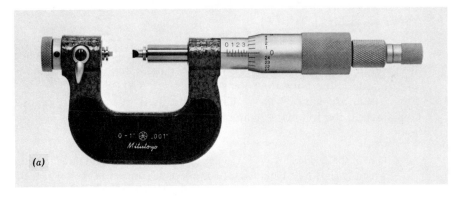

Figure 4.132a Thread micrometer. (M.T.I. Corp. Reprinted by permission.)

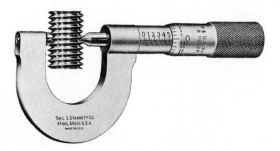

Figure 4.132b Anvils for the thread micrometer. (M.T.I. Corp. Reprinted by permission.)

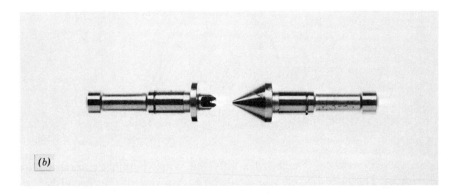

Figure 4.133 Thread micrometer in use. (L.S. Starrett Co. Reprinted by permission.)

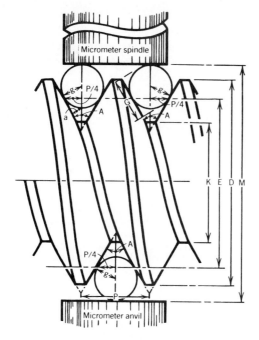

Figure 4.134 Three-wire method. (L.S. Starrett Co. Reprinted by permission.)

Measurement over the wires:

$$M - \text{constant} = \text{pitch diameter}$$

Without a table of constants, you must use a formula that is described later in this chapter.

Caution. Before measuring threads, make sure that they are clean and not damaged in any way that will affect the measurement.

MEASURING THREADS: THE THREE-WIRE METHOD

One of the most important measurements to be made when measuring threads is the *pitch diameter*. The pitch diameter of external threads can be measured in several ways. There are screw thread micrometers, thread comparators, and many other methods, but the most accurate hand-held tools for measuring the pitch diameter of external threads are thread wires, or the three-wire method. In this method, you simply use three wires of the same diameter. Two of these wires are placed on one side of the thread and the other wire on the other side, as shown in Figure 4.134.

Using the three-wire method, any wire size can be used as long as the wire is small enough to make contact with the "flanks" of the threads and large enough to also extend above the crests of the threads.

Best Wire Size

When accuracy is necessary, and to avoid the possibility of error due to the thread angle, you should use a particular *best wire size* for the measurement. The best wire size is calculated with the following formula:

$$\text{best wire size} = 0.57735 \, (p)$$

where 0.57735 is a constant and p is the pitch of the thread you are measuring.

Example

You want to measure a $\frac{1}{4}$ in. $-$ 20 thread.

Solution

Step 1. Find the pitch of the thread.

$$p = \frac{1 \text{ in.}}{\text{number of threads per inch}}$$

$$= \frac{1}{20} = 0.050 \text{ in.}$$

Step 2. Find the best wire size using the pitch above.

$$\text{Best size} = 0.57735(0.050) = 0.0288 \text{ in. (or 0.029-in.-dia. wire)}$$

Remember. Any wire size can be used under the conditions stated above, but try to use the best size, to avoid inaccuracy due to varying thread angles.

Three-Wire Method (Step-by-Step Procedure)

To measure threads using the three-wire method, follow these steps.

Step 1. Calculate the best wire size. (Make sure you use wires that are accurate enough for the threads you are measuring.)

Step 2. Place the wires onto the threads (two on one side and one on the side directly opposite the threads).

Step 3. Using a micrometer (or other instrument), measure the dimension over the wires. (This is called M.) Good contact pressure is required.

Step 4. After finding the measurement M (over the wires), use this formula to calculate the actual pitch diameter:

$$\text{pitch diameter } (E) = M + (0.86603p) - 3W$$

where E is the pitch diameter
m is the measurement over the wires
p is the pitch of the thread
W is the wire size you used
0.86603 is a constant

For example, let's say that M is 0.280 in., p is 0.050, and W is 0.029. So

$$
\begin{aligned}
E &= M + 0.86603p - 3W \\
&= 0.280 \text{ in. } + 0.86603(0.050) \text{ in. } - 3(0.029) \text{ in.} \\
&= 0.280 \text{ in. } + 0.0433 - 0.087 \text{ in.} \\
&= 0.280 \text{ in. } + 0.0433 - 0.087 \text{ in.} \\
&= 0.3233 - 0.087 \\
&= 0.2363 \text{ in.}
\end{aligned}
$$

This is the pitch diameter.

If you find that the pitch diameter is within the tolerance limits on the drawing, the part is acceptable.

Summary

First, find the pitch (p) of the thread.

Next, find the "best wire size."

Next, take the measurement over the wires.

Last, inserting the M, p, and W values in the formula, calculate the actual pitch diameter.

FUNCTIONAL GAGES PER ANSI Y14.5

Functional gages are designed to inspect the collective effects of the size of a feature and the geometric characteristic controlling that feature. Geometric tolerances can be verified with a functional gage only if they utilize the *maximum material condition* (MMC) in the feature control symbol on the drawing. These gages are very useful during production and inspection for quick and accurate feature verification; the gaging is also related to fit and function. It should be noted that functional gages only measure the geometric characteristics, not the actual size of the feature being controlled.

Functional gages are attribute gages, meaning that they will only tell you whether or not the part is acceptable. They will not tell you how good or how bad a part is. Generally, functional gages are used for quick check during production or inspection. When the gage rejects the part, then a "variable" inspection method may be used, if necessary, to find out how bad the part is.

Basics of Functional Gages

Functional gages generally take the shape of the mating part. For example, the feature to be controlled is a shaft (or pin). Here the functional gage would generally take the shape of a ring gage. This is also true for holes. A functional gage to check hole positions would use pins. All functional gages are generally one gage that either goes into the part or does not. They are not two gages (such as Go and NoGo).

Basic Design of Functional Gages

To design functional gages, one must thoroughly understand how to calculate the *virtual size* of any feature (e.g, a hole, slot, shaft, or tab) that is controlled by a geometric tolerance such as angularity, perpendicularity, position, and parallelism.

There are two quick formulas to use in all cases:

1. Virtual size (external feature) = MMC (of the feature)
+ geometric tolerance.
2. Virtual size (internal feature) = MMC (of the feature)
− geometric tolerance.

Example: Virtual Size (External Feature)

A shaft that has a straightness tolerance at MMC of 0.002 and a size tolerance of 0.530 ± 0.003.

$$\text{Virtual size} = \text{MMC} + \text{geometric tolerance}$$
$$= 0.533 \text{ in.} + 0.002 \text{ in.}$$
$$= 0.535 \text{ in.} \quad \textit{Answer}$$

Example: Virtual Size (Internal Feature)

A hole that has a positional tolerance of 0.005 in. diameter at MMC and a size tolerance of 0.310 in. ± 0.003 in.

$$\text{Virtual size} = \text{MMC} - \text{geometric tolerance}$$
$$= 0.307 \text{ in.} - 0.005 \text{ in.}$$
$$= 0.302 \text{ in.} \quad \textit{Answer}$$

Examples of Functional Gages

Example 1: Functional Gage to Measure the Straightness of a Pin at MMC (Figures 4.135 to 4.137)

THIS ON THE DRAWING

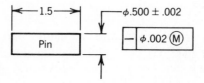

Figure 4.135 Straightness of an axis at MMC callout.

MEANS THIS

The axis of the pin must be straight within a diametral tolerance zone of 0.002 in. diameter only when the pin size is at MMC (which is 0.502 in.). As the pin gets smaller (until you reach LMC or 0.498 in.) the straightness tolerance gets larger automatically by the amount that the pin deviates from MMC. The virtual size of the pin is 0.504 in.

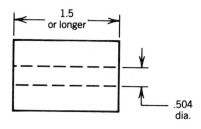

Figure 4.136   Functional gage (a ring).

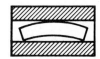

Figure 4.137   Part and gage together.

Example 2: Functional Gage to Measure the Position of Three Holes Equally Spaced on a 6-in. Basic Bolt Circle Diameter at MMC (see Figures 4.138 and 4.139)

   THIS ON THE DRAWING

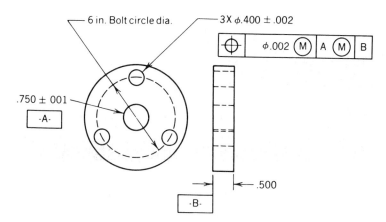

Figure 4.138   Bonus tolerance from pattern and datum.

MEANS THIS

These three holes must lie in true position, with each having a diametral tolerance zone of 0.002 in. (when they are at MMC, which is 0.398 in.). As the holes get larger, so does the tolerance zone. If datum A is larger than MMC size, the pattern of holes is allowed to shift or rotate from its center axis by that much. The 0.002 diameter tolerance zone does not increase in size from the datum bonus tolerance.

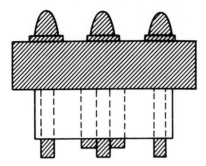

Figure 4.139 Functional gage for pattern.

Example 3: Functional Gage for Perpendicularity of a Hole at MMC (Figure 4.140)

DRAWING

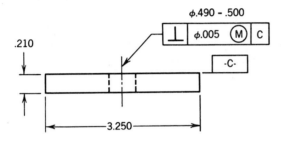

Figure 4.140 Perpendicularity at MMC.

MEANS

The 0.490 − 0.500 hole must be perpendicular to datum plane C within a cylindrical tolerance zone of 0.005 in. diameter when the hole is made at MMC size. The MMC is 0.490.

FUNCTIONAL GAGE

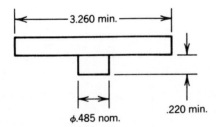

Figure 4.141 Functional gage for perpendicularity at MMC.

MEANS

The functional gage for the part must have a virtual size pin that is 0.485 in. in diameter (which is the MMC of the hole minus the perpendicularity tolerance), and the plate of the gage is large enough to cover datum C completely. The pin must be long enough to go all the way through the hole (see Figure 4.141).

FUNCTIONAL GAGE AND PART

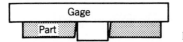

Figure 4.142 Functional gage and part.

MEANS

As shown in Figure 4.142, the functional gage surface must mate with datum C (at its three highest points), and the pin must go through the hole at the same time.

BASIC CALIBRATION TECHNIQUES

Calibration is simply defined as comparing an instrument of known accuracy with another instrument.* For the most part, inspectors, machinists, and operators are concerned with calibration at the company level (which is the company's metrology lab) or at the individual level (which is a person calibrating his or her own tools). At either level, it is important that calibration be performed at certain intervals of time. These calibration intervals depend mainly on

1. How often the tool is used.
2. How often it is found to be out of calibration.

Calibration Intervals

Calibration intervals must be based on both of the above mentioned items, or they will not be effective. The reason for this is that if the calibration intervals are too short, you will be spending too much time calibrating the tool. If the intervals are too far apart, you run the risk of using a tool that is out of calibration. Calibration intervals can (and should) be adjusted based on previous calibration results.

Calibration Systems

In commercial industries, identification systems for calibration vary from company to company. Some companies use the color code system, where tools and instruments are assigned a color (yellow, green, blue, etc.) which identifies when they are due for calibration.

Other companies use calibration stickers on the tool with a variety of information on them. The main pieces of information are

1. When the tool was calibrated.
2. Who calibrated it.
3. When it is next due for calibration.

* Traceable to the National Bureau of Standards.

Companies with large numbers of tools to control use the computer for effectiveness in maintaining calibration intervals. The computer prints out lists of tools that are to be calibrated to meet their particular interval requirements. It can also be programmed to print out those tools that need adjustment at particular intervals.

Some companies have calibration systems conforming to military standards, for example, Mil-Std-45662. If the companies have defense contracts, they may be required to conform to Mil-Std-45662. Some commercial companies elect to conform to military standards because it is an effective system of calibration control.

Calibration Methods

A general rule for calibration is that it should be performed with a standard that is 10 times as accurate as the tool being calibrated. A micrometer, for instance, would be calibrated with a gage block (for length) and checked for wear on the measuring faces of the spindle and anvil with an optical flat. Both the gage block and optical flat are far more accurate than the micrometer.

Calibration Environment

The calibration environment (in most cases) is a clean, temperature-controlled, humidity-controlled room where ''working'' standards are kept for the purpose of the calibration of company tools and instruments. When tools are brought in for calibration, they should be cleaned, if necessary, and set aside until they are the same temperature as the room's temperature (usually 68 degrees Fahrenheit). Then calibration can be performed.

Risks Without a Calibration System

In most mass production companies, there are numerous tools and gages to monitor. Without an effective calibration system of some kind, keeping track of calibration intervals for each tool or gage is next to impossible, let alone retaining the ability to fine-tune each interval to the point where you can be assured that calibrated tools and gages are being used.

If operators and machinists use tools or gages that are *not* calibrated, the risk is *producing parts that are not per specifications*. If an inspector uses a tool that is not calibrated, the risks are *accepting a bad part* and *rejecting a good part*. The largest risk that a company or individual takes without a calibration system is that defective products can be shipped to the customer regardless of any other controls it may use.

Production, inspection, and quality control rely heavily on measurements. Measurements can be subject to a variety of possible errors, such as

Heat   Dirt   Manipulation   Geometry

There are many errors possible in any given measurement. Calibration of the tool should not be allowed to be one of them. Calibration is one thing that can be controlled if a company (or individual) wants to control it.

COORDINATE MEASURING MACHINES

Coordinate measuring machines were originally designed to keep up with the production rate of numerically controlled machines. It can be very time consuming and costly for an inspector to go into a detailed layout of most products that are produced on a numerically controlled machine. These products are usually highly complex, with many features (holes, and the like) that must be inspected for position with reference to datums, planes, or surfaces.

An example of this might be a part with an 18-hole bolt circle that is referenced from three datum planes perpendicular to each other. The numerically controlled machine can produce this part very quickly, but for the inspector to lay out the first piece is another matter. Traditionally, the inspector must establish the three datum planes on the surface plate with right-angle plates and clamps. Each of the 18 holes in the part has two coordinates. The inspector must measure manually 18 coordinates in one direction, then rotate the part 90 degrees and measure 18 more coordinate distances. Performed with a vernier height gage, this job can take a long time to perform, and there are *many* chances for error.

The coordinate measuring machine (CMM) can do this job in a few short minutes, and the production machine can be under way to produce additional parts. This is the difference between coordinate measuring machines and traditional methods. The inspection is more accurate (less chance for errors) and takes a lot less time to perform. When considering how important production time is, the CMM is a tool that pays for itself in a very short period of time and soon starts improving profits because of the time and money saved.

How CMMs Work

The basic principle of a coordinate measuring machine is very simple. The CMM has three basic directions of movement. These are called X, Y, and Z axes. When standing in front of the machine, the X axis is the movement from left to right, the Y axis is the movement toward and away from you, and the Z axis is the movement up and down. Some have a W axis, which is rotation.

CMMs come in various makes and models. They also come with different measuring devices and capabilities. Some are very basic with vernier scales in the X and Y directions only. Some have dials in the X, Y, and Z directions. Some have electronic digital display that can be "zeroed" anywhere, and some have computers attached that can record your measurements and even draw you a picture of the part. All CMMs still work on the same basic principle of movement. An example of this movement is shown in Figure 4.143.

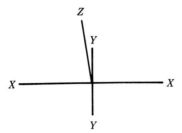

Figure 4.143 X, Y, and Z coordinate directions.

There are various probes used with CMMs, depending on the application. They range from the basic conical style (which will find the center of a hole automatically), to electronic probes that make a "beeping" noise when contact is made with the part. One type of coordinate measuring machine is shown in Figure 4.144.

Functional Measurements with CMMs

One problem with using a coordinate measuring machine is the fact that many users misunderstand its ability. CMMs are fast and accurate, but one cannot forget the basics. There are several examples that could be given with reference to the standard for geometric tolerancing (ANSI Y14.5M–1982). This standard defines particular rules that must be followed if inspections on the product are to be functional.

Locating functional datums

The CMM (with either type of probe) does not have the ability to properly locate a secondary or tertiary datum plane. The functional datums must be contacted at the highest points on the surface. The probes of the CMM can contact points, but not the highest points. The only functional datum that can be contacted by the CMM is the primary datum plane (this is done by the staging area on the machine). The secondary and tertiary datum planes must be contacted by using surface plate accessories such as right-angle plates, parallels, and so forth. One only needs to "beep" the right angle plates (two places for secondary, and one place for tertiary) and then place the part against them. (Refer to pages 126–127 on contacting datums.)

Locating datum target points

Datum target points are easier to locate on a CMM. However, there are still some accessories needed. Datum target points are specified points on a part (see datum targets). The only problem is the primary datum (which must be contacted at three specified points). The staging area of the CMM cannot directly contact these three points. Often, three pins (spherical on the end) are used for the

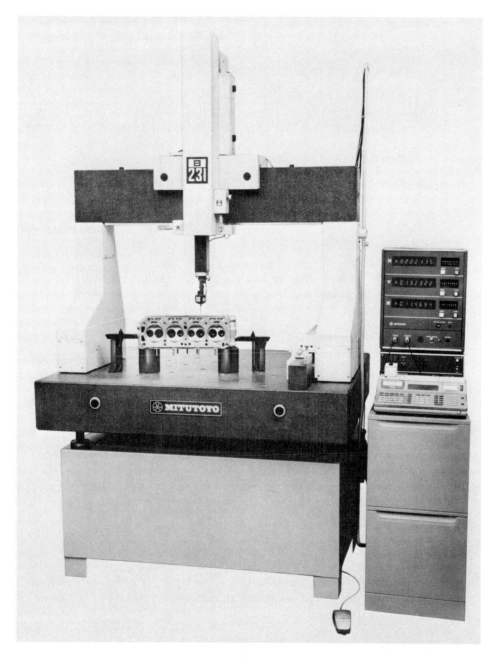

Figure 4.144 Coordinate measuring machine. (M.T.I. Corp. Reprinted by permission.)

primary datum target points. Then the machine's probe can be used to contact the secondary (two specified points) and the tertiary (one specified point).

Locating datum target areas

Datum target areas are specified areas that must be contacted with pins that are flat on the end. The CMM probes cannot directly contact an area, so pins must still be used to assist the CMM in the appropriate contacting of datums.

Measuring hole positions

Hole locations are measured by finding out if the axis of the hole lies within its tolerance zone. To measure the location (position) of a hole, therefore, you must find its axis. The beep probe of the CMM does not find an axis. What the beep probe does find is the position of one point along the axis of the hole. It does not tell you about the actual location of the entire axis. The fixed (hard) conical probe of the CMM only tells you the position of the part of the axis at the end of a hole. Furthermore, it must be remembered that the position tolerance on a hole applies at both ends, so that measurements must be taken twice (the coordinates at one end, then the other).

Coordinate measuring machines are very useful equipment and perform very well. Many labor-hours can be saved without loss of accuracy when they are used. It is very important, however, that they be used properly. Drawing requirements must be functionally verified, or the effort could prove to be a waste of time.

CMM Operations

CMMs are operated in three basic ways: manual, computer assisted, and computer driven. The older manual CMMs, although much better than the alternative of open setup layouts on the surface plate, are not as efficient as computer-assisted or computer-driven CMMs because the operator must read and remember starting points and linear distances, write them down, and then analyze the results. This is a problem specifically in layout and measurement of geometric tolerances (such as position and concentricity). Reporting from the manual CMM must be made on paper after the inspection has been completed, and the proper analysis of position tolerances (see Chapter 6) is difficult. Computer-assisted CMMs are more efficient, in that a computer and a variety of software is added to the CMM for on-line reporting, memorization of zero points, and linear distances. Computer-assisted CMMs are also more efficient in analyzing geometric tolerances, such as position, concentricity, and perpendicularity (to name a few). Software can be added to the computer, which is designed to assist the operator in evaluation of these types of tolerances, and reporting directly from the machine. Computer-driven CMMs are the highest level of efficiency, due to the fact that they can be programmed (one time) to measure a product, and once programmed, can measure subsequent products on its own without help from the operator. The operator can be doing other important things while the measurements are being made automatically.

CMMs Compared to Functional Gages

CMMs are also more versatile than functional gages in measuring products. Functional gages are usually designed to measure one function at a time. The cost of functional gages is limiting, and they must be maintained to avoid wear and damage. CMMs may be advantageous to functional gages because (1) functional gages are designed only to gage the virtual condition (worse case) of products, (2) functional gages cannot be used to gage geometric tolerances specified at LMC or RFS condition (refer to Chapter 3 for LMC/RFS definitions), (3) functional gages are designed with wear allowances and gage tolerances such that they will reject a borderline good product, (4) functional gages usually check only one functional aspect of the product, and (5) functional gages are a Go–NoGo relationship that does not measure the product. By comparison, CMMs can (1) measure all conditions of the product; (2) measure all tolerance modifiers, including LMC and RFS; (3) avoid rejections of borderline good product because they measure actual conditions and are not designed to wear allowances; (4) measure all functional tolerances of the product, not just one; and (5) measure the product and give variable information to the operator for process adjustments and process control purposes.

CMMs Improve Multiple Setups

Another important advantage about CMM use is the avoidance of multiple setups as compared to open surface plate techniques. In many applications the product must be reoriented on various datum structures in order to inspect various functional characteristics. This need for multiple setups adds time and error to the results obtained. CMMs can often measure multiple characteristics in one setup without moving the part.

CMMs and Fixturing

It is important to identify and use the appropriate fixturing (or standard surface plate accessories) as indicated earlier in this chapter. Electronic probes or the fixed probes used on CMMs are limited in the information they provide (especially for establishing functional datum planes). The appropriate fixturing or accessories should be identified, which will allow proper functional setup of the part to be measured. The type of datums on the part should be considered so that the correct datum plane is generated from the contact points. A primary datum target on a surface, for example, is established by contacting a minimum of three specific locations on that surface. Often, the probe on the CMM is able to contact these points, and there are few problems. On the other hand, a primary functional datum plane (for a surface) is established by contacting at least three of the *highest points* on that surface. Fixturing will be needed which will make contact with the highest points so that a plane can be derived from the fixture. Another example of functional inspection on the CMM is the use of gage pins in holes when measuring position, perpendicularity, and other geometric tolerances applied to holes. As with open surface plate setups, there is still no substitute for a snug-fitting gage pin to establish the functional axis of the hole for the purpose of axial measurement. Using the variable and fixed probes of the CMM for axial location can be a problem. The variable probe simulates a dial indicator point of contact. It is often used (when there is no fixturing) in a manner that touches the inside of a hole in a triangle of points (sometimes each point is a little deeper into the hole). This practice is attempting to establish the axis.

Part of the problem here is that holes are out-of-round, tapered, and have other forms of geometric error that can give false signals of the functional orientation and location of an axis. The practical problem, however, is much easier to understand. A probe touching a hole at specific points is not as functional as a full-form cylinder interfacing with the entire hole (when one considers the fact that holes are usually produced in order to put a bolt or a pin through them). The form of a gage pin is representative of the bolt or pin that will be used in assembly. The same thing holds true for the fixed (conical) pins used on CMMs. When used, these pins only find a point of the axis at the end of the hole. They do not identify the effect of the third dimension (depth), nor do they fill the hole as well as a gage pin. The importance of fixturing on CMMs is of growing concern to users. The key to the answer is function. Functional measurement of products is important to take that extra measure and make the CMM results true to the way the product works. A considerable amount of time and money is saved using CMMs compared to open setups. A little more effort should be made to use accessories or fixturing that make the results functional. Refer to examples shown in Chapter 6.

MEASURING SURFACE FINISH

Most component parts today require particular surface textures, depending on how they are to be used. In the old days, this surface roughness was commonly referred to as "coarse" or "fine." Today, the surface texture of a component part is more stringently controlled in many applications and requires more specific classifications than simply coarse or fine.

Before we discuss surface finish measurement, there are some definitions that must be considered. These definitions are listed below.

Roughness. Closely spaced irregularities on the surface.

Waviness. Widely spaced irregularities on the surface.

Lay. The direction of the predominant surface pattern. (Surface finish measurements should be made in a direction that is perpendicular to the lay of the surface pattern.)

Flaws. A defect in the surface that is not part of the surface roughness or lay. This may be a deep tool mark, a void on the surface, or other flaws. The effect of flaws should not be included in roughness average measurements.

Roughness width cutoff. A dimension that distinguishes surface roughness from surface waviness for a particular distance. This cutoff length must be preset prior to using the profilometer. When the cutoff length is not specified on the drawing, you would, in most cases, use a 0.030-in. cutoff length.

Arithmetic average (AA). The average distance between the peaks and valleys of surface roughness.

Cutoff. The electrical response characteristic of the instrument that is selected to limit the spacing of the surface irregularities to be included in the roughness measurement. (0.030 in. if none specified).

Surface Finish and Lay Symbols

There are certain symbols that are used to describe the direction of lay for the surface. These are used in conjunction with the surface finish symbol shown in Figure 4.145.

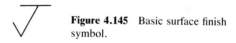

Figure 4.145 Basic surface finish symbol.

The symbols appear under the surface finish symbol (Figure 4.146) and designate the lay of the surface. The symbols are shown below.

//	which means a parallel lay.
⊥	which means a perpendicular lay.
C	which means a circular lay.
R	which means a radial lay.
M	which means a multi-directional lay.
X	which means an angular lay in both directions.

Figure 4.146 Lay symbols.

Examples of each lay (per the symbols in Figure 4.146) are shown in Figure 4.147.

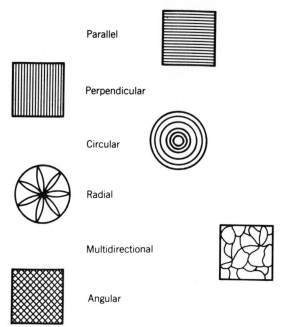

Parallel

Perpendicular

Circular

Radial

Multidirectional

Angular

Figure 4.147 Examples of lay directions.

Figure 4.148 shows other symbols.

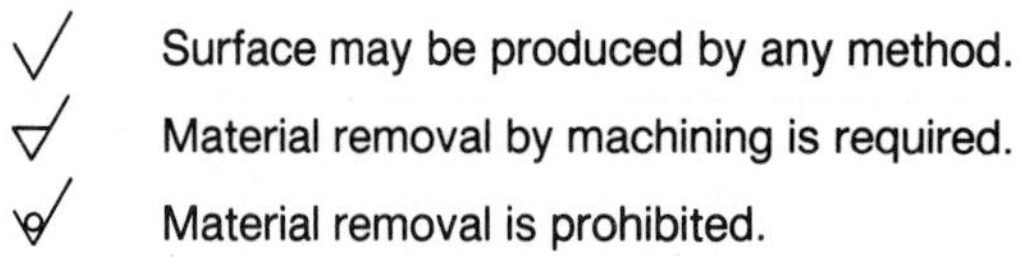

√ Surface may be produced by any method.

√ Material removal by machining is required.

√ Material removal is prohibited.

Figure 1.148 Other surface finish symbols.

It should always be remembered that roughness measurement is usually performed in a direction that is at a 90-degree angle (perpendicular) to the direction of the lay.

The Difference Between Roughness and Waviness

As stated earlier, roughness is considered closely spaced, while waviness represents widely spaced irregularities. Figure 4.149 helps show the difference between roughness and waviness.

Peaks and valleys

There are some other terms that also require definitions. They are not unlike the definitions used when discussing the land around us. In the mountains, a peak is considered to be the highest point and a valley is considered to be the lowest point. It is the same for surface texture.

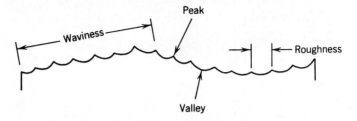

Figure 4.149   Waviness versus roughness.

Surface texture takes the shape of several peaks and valleys. These peaks and valleys, when magnified, look somewhat like threads from a side view (as is shown in Figure 4.149). The measurement of surface roughness is generally the average height of these combined peaks and valleys.

Surface Finish Measurement

In this chapter we do not intend to describe the many methods in which surface finish can be measured. Two methods, however, enjoy wide use in the shop: a *fingernail comparator* and a *profilometer*.

Fingernail comparators (Figure 4.150)

In those situations where surface finish is widely toleranced, fingernail comparators are usually used. These fingernail comparators come in a variety of different styles. One style provides several coins with each coin representing a different surface roughness. Other types are a sheet of metal or plastic that has a variety of roughness samples that you can compare to the part. Each sample is identified with a finish designation, for example, AA32.

They are commonly called fingernail comparators because you run your fingernail across the comparator, then across the surface of the part. Run your fingernail across the comparator (perpendicular to the lay) and you will feel the roughness. Then run your fingernail across the surface of the part (also perpendicular to the lay) and decide which surface is rougher. If the comparator feels rougher than the part, it is likely that the part is acceptable.

Example

The drawing calls for a AA32 (arithmetic average) finish. You scratch the comparator in the area marked AA32 with your fingernail, then scratch the surface of the part. The part surface seems smoother to the fingernail than the comparator was. This means that the part surface is better than an AA32 finish.

Fingernail comparators are widely used (when applicable) because they are simple and inexpensive.

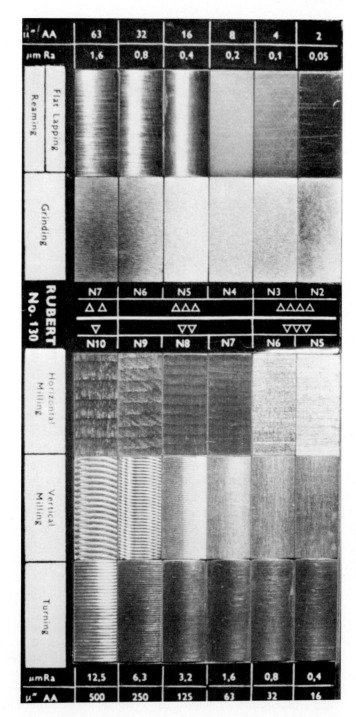

Figure 4.150 Surface roughness comparison standard (fingernail comparator). (Photo by James Pluma.)

Electronic instruments for surface finish measurements

There are several different types of electronic instruments used to measure surface finish. The one that is used primarily is the profilometer (Figure 4.151). This instrument measures surface finish on a *variable* measurement basis; in other words, the actual average surface of a part can be measured. There are several things to consider when using the profilometer.

Figure 4.151 Profilometer in use. (Federal Products Corp. Reprinted by permission.)

1. *Treat the stylus with care.* The stylus of the profilometer in Figure 4.152 is the part that actually follows the surface. It is a diamond-tipped, conical-shaped part that is small enough to get into the peaks and valleys of the surface.

2. *Set the instrument to read your roughness value.* Most profilometers have a switch on the dial to convert the dial to read whatever range of roughness you are measuring, as shown in Figure 4.151. There are generally two scales. The top scale goes from 0 to 30 and the bottom scale goes from 0 to 10. Depending on the setting, the multiplier 30 can mean 300, or 10 can mean 100 finish.

Example

When the selector is set on "10," you can read the bottom scale directly. When the selector is set on 100, you would read the bottom scale as 0, 10, 20, 30, 40, 50, 60, 70, 80, 90 100 (in other words, add a zero to the numbers shown, which are 0, 1, 2, 3, 4, 5, 6, 7, 8, 9, 10).

When the selector is set on 30, read the top scale directly. It is 0, 10, 20, 30. When the selector is set on 300, read the top scale by adding 0 to all of the numbers. So the top scale, 0, 10, 20, 30, now becomes 0, 100, 200, 300.

By selecting the proper position for the application, you can use the same scale for larger readings.

Figure 4.152 Profile tracer stylus. (Federal Products Corp. Reprinted by permission.)

Example

You want to measure a surface finish of AA250. Set the selector at 300 and read the top scale (which is 0, 10, 20, 30). If the top scale reads 20 (and you are set at 300), it means 200.

3. *Make sure that you have set the prescribed cutoff length of the instrument.* Using the cutoff selector (Figure 4.151), you can preset the prescribed cutoff for the measurement. Usually, this cutoff is 0.030 in. but can be other lengths, such as 0.010 in., 0.100 in., and so forth.

4. *Make sure that the stylus is tracing perpendicular to the lay.* The stylus should be tracing in a direction that is perpendicular to the lay of the pattern. In other words, it should cross over the lay (see Figure 4.153).

5. *Make sure that the stylus is set to a stroke that will not cause it to fall off the surface being measured* (Figure 4.154). A poorly set stroke length can damage a stylus by causing it to fall off the surface being measured. There is usually a knob on the instrument that sets the stroke distance the stylus will travel during the measurement. It moves back and forth along this distance. Locate this knob and set the stroke according to your requirements and the part you are measuring. For accuracy, the stroke length must be at least fives times the cutoff length.

6. *Never allow the stylus to drop onto the part; set it on gently.* This is to avoid damaging the stylus.

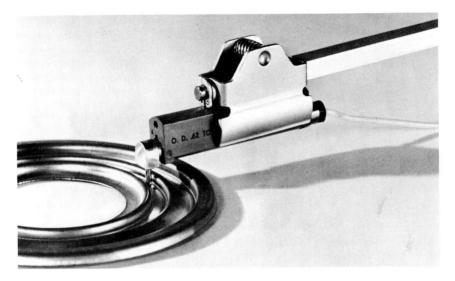

Figure 4.153 Tracing perpendicular to lay. (Federal Products Corp. Reprinted by permission.)

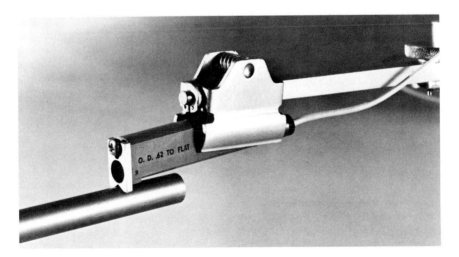

Figure 4.154 Dual skid tracer. (Federal Products Corp. Reprinted by permission.)

7. *The arm holding the stylus should be as parallel as possible to the surface being measured.* This ensures that the stylus is in good contact with the surface.

8. *Locate the workpiece securely* (Figure 4.155). The workpiece should be located so that it will not move around during the measurement. On cylin-

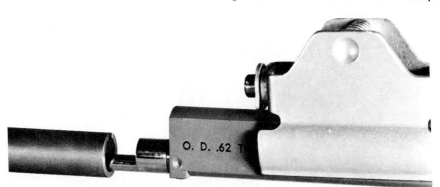

Figure 4.155 Checking surface finish on an inside diameter. (Federal Products Corp. Reprinted by permission.)

drical pieces, it is good practice to use a V-block or, at least, support the part on both sides so that it will not move.

9. *Uses a stylus with the proper type of skids* (Figure 4.156). The *skid* on a profilometer is that part just under the stylus that references (or rides along the surface) so that the stylus can follow the peaks and valleys of the surface. There are basically two types of skids:

   a. *Single skid.* This is basically for flat surfaces, where there is no probable side-to-side movement.

   b. *Double skid.* This is where there are two skids surrounding the stylus. It is used mainly on cylindrical parts to assist the stylus in avoiding side-to-side movement. The dual-skidded tracer is recommended in all cases.

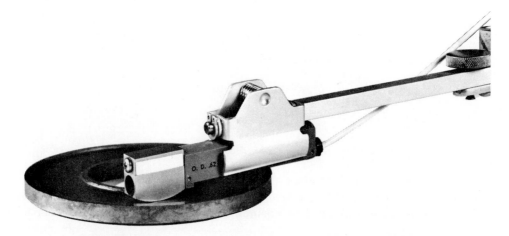

Figure 4.156 Double skid tracer in use. (Federal Products Corp. Reprinted by permission.)

10. *Calibrate the profilometer periodically* (Figure 4.151). There is usually a calibration patch supplied with the profilometer that is used to check calibration of the instrument. This block has a known surface finish, and the profilometer should read very near this finish when the stylus is moved across the patch (within tolerances that are usually stated on the patch). The statistical average of a number of measurements on the calibration patch is recommended.

Basic steps to measuring with a profilometer

Step 1. Turn the profilometer on and let it warm up.

Step 2. Set the prescribed range for the measurement.

Step 3. Set the prescribed cutoff length (if none specified, you use 0.030 in.).

Step 4. Set the stroke length of the stylus according to the part you are measuring. This should be five times the cutoff length.

Step 5. Set the unit of measurement. This is usually AA (which means "arithmetic average").

Step 6. Position (and secure if necessary) the part being measured. Remember, cylindrical parts should not be allowed to roll.

Step 7. Make sure that you have the correct stylus for the finish being measured.

Step 8. Set the arm (holding the stylus) parallel to the surface being measured.

Step 9. Make sure that the stylus is in the *back* position of movement. Holding the stylus up by hand, turn the machine on and stop it when the stylus is back all the way.

Step 10. Gently rest the stylus on the part.

Step 11. Turn the machine on and allow the stylus to trace.

Step 12. Read the dial (at the proper scale) for the surface finish. The average roughness is the mean reading about which the needle tends to dwell, or fluctuate, under small amplitudes.

Surface texture is of prime concern in many industrial applications today. The important thing is to know your particular application and equipment to obtain the desired results.

INSPECTING SIZE TOLERANCES PER ANSI Y14.5 RULE 1

Inspection of size tolerances per ANSI Y14.5 specification (Rule 1) is not a simple task. Traditionally, size tolerances have been inspected using various hand tools such as micrometers, vernier calipers, ring gages, and plug gages. The problem is that each of these hand tools only does half the job.

A size tolerance (per Rule 1) invokes two controls on the feature of size: these are *size* and *form*. Rule 1 states clearly that size features must have *perfect form when produced at MMC size*.

Example

A shaft produced at MMC (its largest allowable size) must be a perfect cylinder.

A hole produced at MMC (its smallest allowable size) must also be a perfect cylinder.

A thickness dimension between two surfaces produced at MMC size must have sides that are perfectly straight and parallel to each other.

None of the conditions above can be verified by using a single hand tool. However, they can be verified using two hand tools.

Example

A plug gage can be used to measure the MMC boundary of perfect form, and a telescoping gage can be used to measure any condition beyond the LMC boundary (Figure 4.157).

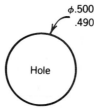

φ.500
.490

Hole

Figure 4.157   Hole diameter.

Example

A ring gage can be used to measure the MMC boundary of perfect form, and a vernier caliper or micrometer can measure any condition beyond the LMC boundary (Figure 4.158).

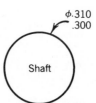

φ.310
.300

Shaft

Figure 4.158   Shaft diameter.

Example

A surface gage and high-limit block stack can set up a boundary of perfect form at MMC where no portion of the surface can exceed. A vernier caliper or micrometer can measure any condition beyond the LMC boundary (Figure 4.159).

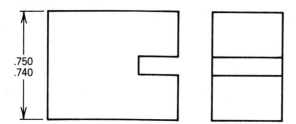

Figure 4.159 Thickness.

HARDNESS TESTING

Industrial materials must often be inspected for hardness, which is simply a material's resistance to penetration. There are many methods of testing material hardness today, but most widely used are the *Brinell and Rockwell hardness tests.* Other tests include Vickers, Knoop, and Scleroscope. In this book we discuss only the Brinell and Rockwell tests. In the following discussion we outline, in step-by-step form, how to perform Brinell and Rockwell hardness tests, and describe some particularly important characteristics, parts, and cautions for each type of test.

Brinell Hardness Testing

Hardness testing is done throughout the industry using several different methods. One of these methods is the Brinell hardness test. Harndess is a vital characteristic of some products because of their function, machinability, strength, or other necessary properties. It can be defined as *the ability of a material to resist penetration.*

The Brinell hardness test is very simple. It is simply a matter of applying a constant load of 500 to 3000 kilograms (kg) (depending on the material) to a 10-millimeter (mm)-diameter steel ball on a flat surface of the part (see Figure 4.160).

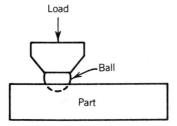

Figure 4.160 Brinell hardness test.

Note. A 500-kg load is usually used for soft materials such as aluminum or copper. A 3000-kg load is usually used for harder materials such as steel or cast iron.

Once the load has been applied for a certain amount of time (15 seconds for hard materials and 30 seconds for soft materials), it is then removed and the diameter of the indentation (dent) is measured in millimeters (mm). This diameter should be the average of two measurements.

Next, the Brinell hardness number must be found. There are tables that can be obtained to convert this number directly, but if not, the following formula can be used:

$$\text{BHN} = \frac{L}{\dfrac{\pi D}{2}\,[D - (D^2 - d^2)^{1/2}]}$$

where L is the load (in kg)

$\quad$ D is the diameter of the steel ball (in mm)

$\quad$ d is the diameter of the "dent" (in mm)

$\quad$ π is 3.1416 (a constant)

Example

The load is 3000 kg, and the diameter of dent is 4.0 mm.

$$\text{BHN} = \frac{3000}{\dfrac{3.1416(10)}{2}\,[10 - (10^2 - 4^2)^{1/2}]}$$

$$= \frac{3000}{15.708(0.835)} = \frac{3000}{13.116} = 229 \text{ BHN} \quad \textit{Answer}$$

When measuring Brinell hardness, it should be noted that the surface preparation is important. The surface should be very smooth so that the diameter of the indentation can be clearly defined and measured. It may be necessary to grind the surface (such as a casting surface) before testing.

In some cases, Brinell testing can be considered a "destructive" test because it leaves an indentation that may be a problem if the surface finish in that area is important. If so, test the part in another area.

Most often, there is a special microscope used to measure the indentation accurately. This microscope has a built-in millimeter scale and a light for easy reading. Care should be taken not to make Brinell indentations too close to the edge of a part, because there is no support for the load at an edge, and the indentation will shift and go out-of-round, giving you a poor reading.

There are many kinds of Brinell hardness testers. Some are manually operated, some automatic, and some are portable. Know your particular tester and how it operates prior to performing any test with it.

Rockwell Hardness Testing

One of the most widely used tests for hardness of materials in industry is the Rockwell hardness test. There are many reasons for its wide use: (1) it is a simple test to perform; (2) hardness is directly read on a dial; (3) it does not require much skill; (4) it leaves a very small indentation on the part (unlike Brinell tests); and (5) it has a very wide range of applications, from soft aluminums to carbides. The Rockwell test also requires that the surface being tested be "prepared" by removing surface roughness or scale. However, some parts are such that they need no preparation.

There are two types of "indenter" used in Rockwell testing. These are the *brale* (which is diamond-cone shaped), and the *ball* indenter (with diameters of $\frac{1}{16}$ in., $\frac{1}{8}$ in., $\frac{1}{4}$ in., and $\frac{1}{2}$ in.) (see Figures 4.161 and 4.162).

Figure 4.161   Brale indenter.   **Figure 4.162**   Ball indenter.

The Rockwell test uses two loads (or forces): a *minor* load to seat the indenter into the material, and then a *major* load. The actual hardness is measured by the additional depth of penetration caused by the major load.

Calibrating the Rockwell tester

This is done with an accurate test block of known hardness. You should take at least five different readings on the test block, and the average of all five readings should match the hardness specified on the test block.

Types of Rockwell tests

There are two types of Rockwell tests: *regular* and *superficial*. Each type of test may require the use of a different tester.

1. *Regular Rockwell test.* Uses a minor load of 10 kg and various major loads of 60 kg, 100 kg, or 150 kg.
2. *Superficial Rockwell test.* Uses a minor load of 3 kg and major loads of 15 kg, 30 kg, or 45 kg.

Precautions prior to testing

1. Check the indenter chosen for damage or wear.
2. The surface to be tested should be flat and smooth.
3. The anvil used should be clean and undamaged.
4. If the part is round, grind a flat area for the test.

5. Do not make the test too near the edge of the part or too near a previous indentation.

Steps in performing a Rockwell test (regular or superficial)

Note. The minor loads are built into the machine.

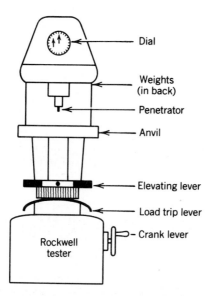

Figure 4.163   Rockwell hardness tester.

Refer to Figure 4.163 when following these steps:

Step 1. Prepare the surface if necessary.

Step 2. Choose the scale to be used (if not already specified) (see Figure 4.164 tables).

Step 3. Choose the major load to be used. This depends on the scale you chose (see Figure 4.164 tables). The load is often shown in the specification. Weights to set the major load are in back of the machine.

Step 4. Choose the penetrator to be used. This is done according to the type of test and the scale used. [For example, regular *C* scale uses the brale indenter (penetrator) (see Figure 4.164 tables)].

Step 5. Choose the proper anvil to use for the part being tested. See Figure 4.165 for various anvils.

Note. The surface being tested must be perpendicular to the centerline of the anvil.

Table of Scales Versus Metals

Scale	Used for:
B	Copper, soft steel, aluminum, malleable iron
C	Steel, hard cast irons, malleable irons, titanium, deep case hardened steels, other metals harder than RB100
A	Carbides, thin steel, shallow case hardened steel
D	Thin steel, medium case hardened steel, malleable iron
E	Cast iron, aluminum, magnesium alloys, bearing materials
F	Annealed copper alloys, thin soft sheet metals
G	Bronze, beryllium, copper, malleable irons
H	Aluminum, zinc, lead
K, L, M, P, R, S, V	Soft bearing materials, other very soft or thin materials

(a)

Normal Rockwell Hardness Scales, Loads, and Indenters

Scale	Indenter	Diameter	Major Load	Dial Color
B	Ball	$\frac{1}{16}$	100 kg	Red
C	Brale	—	150 kg	Black
A	Brale	—	60 kg	Black
D	Brale	—	100 kg	Black
E	Ball	$\frac{1}{8}$	100 kg	Red
F	Ball	$\frac{1}{16}$	60 kg	Red
G	Ball	$\frac{1}{16}$	150 kg	Red
H	Ball	$\frac{1}{8}$	60 kg	Red
K	Ball	$\frac{1}{8}$	150 kg	Red
L	Ball	$\frac{1}{4}$	60 kg	Red
M	Ball	$\frac{1}{4}$	100 kg	Red
P	Ball	$\frac{1}{4}$	150 kg	Red
R	Ball	$\frac{1}{2}$	60 kg	Red
S	Ball	$\frac{1}{2}$	100 kg	Red
V	Ball	$\frac{1}{2}$	150 kg	Red

(b)

Figure 4.164 (a) Table of scales versus metals. (b) Table of Rockwell scales, loads, and indenters. (c) Superficial Rockwell scales, loads, and indenters.

Superficial Rockwell Scales, Loads, and Indenters

Scale	Indenter	Diameter	Major Load
15N	N Brale	—	15 kg
30N	N Brale	—	30 kg
45N	N Brale	—	45 kg
15T	Ball	$\frac{1}{16}$	15 kg
30T	Ball	$\frac{1}{16}$	30 kg
45T	Ball	$\frac{1}{16}$	45 kg
15W	Ball	$\frac{1}{8}$	15 kg
30W	Ball	$\frac{1}{8}$	30 kg
45W	Ball	$\frac{1}{8}$	45 kg
15X	Ball	$\frac{1}{4}$	15 kg
30X	Ball	$\frac{1}{4}$	30 kg
45X	Ball	$\frac{1}{4}$	45 kg
15Y	Ball	$\frac{1}{2}$	15 kg
30Y	Ball	$\frac{1}{2}$	30 kg
45Y	Ball	$\frac{1}{2}$	45 kg

(c)

Figure 4.164   (*continued*)

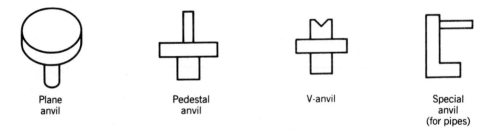

Plane anvil   Pedestal anvil   V-anvil   Special anvil (for pipes)

Figure 4.165   Various anvils.

Step 6. Make sure that the *crank lever* is locked in the "up" position.

Step 7. Carefully install the indenter and anvil you have chosen.

Step 8. Place the part on the anvil and position it for the test. **Avoid bumping the indenter. If you do, it is suggested to inspect the indenter visually under a minimum of 10 × magnification for damage.**

Step 9. Apply the minor load by carefully turning the *elevating lever* clockwise until the part makes contact with the indenter and the small and larger pointers on the dial are pointing straight up.

Step 10. Turn the *knurled adjustment ring* until the "set" arrow (on the dial face) is in line with the large pointer.

Step 11. Apply the major load by pressing down on the *load trip lever* (the crank lever will now move forward).

Step 12. When the *crank lever* has stopped moving forward, carefully pull it back to its locked upright position.

Step 13. Take the reading on the proper scale color while the minor load is still applied.

If *brale* indenter, read the black scale.
If *ball* indenter, read the red scale.

Step 14. Remove the minor load by turning the *elevating lever* counterclockwise to release the part from the indenter.

Step 15. Remove the part **(you must still avoid bumping the indenter).**

A minimum of three readings should be taken on the part, and the average reading for the hardness should be used.

PNEUMATIC COMPARATORS (AIR GAGES)

Pneumatic comparators (commonly called air gages) are often used for tight-tolerance measurements on machined and other precision parts. A pneumatic gage uses an air supply of constant pressure, an indicator, and specific-size nozzles in its operation. There are several kinds of pneumatic gages on the market, and they all have certain common limitations, such as:

1. Gage heads that accompany air gages have limited measuring ranges.
2. The air gage must have a master (such as a master setting ring for measuring diameters).
3. Air gaging requires a specific level of surface finish to be effective. Rough surface finishes (or porous conditions) limit the effectiveness of the gage.
4. The air supply must be clean and dry for the gage to function properly.

Air gaging (Figure 4.166) has several advantages:

1. The training involved in using an air gage is minimal.
2. Variables data can be obtained on several product dimensions quickly and accurately.
3. Air gaging, in the proper environment, can achieve accuracies as small as 0.000010 (ten-millionths) inch.
4. Important part surfaces can be gaged without contacting or damaging the surface, and with little wear on the gage member.

Figure 4.166   Pneumatic comparator. (Courtesy of Federal Products Corporation)

5. The air gage tends to blow loose contaminants from the surface to avoid measurement errors associated with dirt, coolant, and other contaminants.

6. The gage can be rotated (or traversed) when checking dimensions so that geometric error (such as out-of-roundness, taper, and barrel shaped) can be detected.

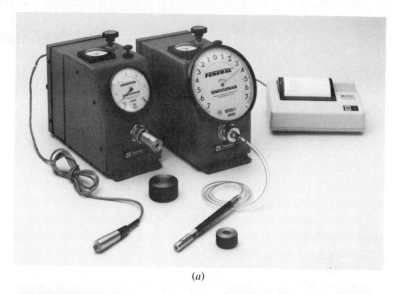

(a)

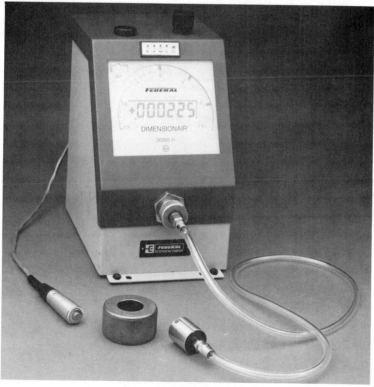

(b)

Figure 4.167 Pneumatic comparator heads. (Courtesy of Federal Products Corporation)

7. Measurement pressure is constant, as opposed to manual gages, which depend on operator skill.

8. One or more air nozzles can be arranged in a variety of locations for effective gaging. An example of this is three air nozzles 120° apart in an air gage for diameters. This configuration is optimal for detecting trilobe diameters. There are air gages today that have been placed into granite surface plates for comparative measurements of flatness.

Air gaging heads can be obtained in many styles (such as those in Figures 4.167a and 4.167b). The common use for air gage heads are plug gaging (for inside diameters), ring gaging, and snap gaging (for outside diameters), and other variations, such as the flatness checking instrument mentioned previously. Several companies have used air gaging heads in multiple-dimension inspection fixtures for fast and accurate inspections. These heads can be designed for direct measurement (no mechanical contact to the part) or indirect measurement (where a mechanical probe or similar device contacts the part).

Each gaging head can have several air jets, depending on the application. Heads with one jet are often used for TIR measurements such as runout, flatness, straightness, and position (or location). They are also used for linear measurements such as length or depth. Heads with two jets are often used for diameters (and geometric errors in diameters such as roundness, taper, barrel-shaped, etc.) and thickness measurements. Based on the limitations and advantages previously stated, pneumatic comparators (air gages) have a wide variety of applications in precision dimensional measurement.

Review Questions

1. (True or False) Precision measurements means getting consistent results repeatedly.
2. What is the primary measurement made on a surface plate?
3. What type of dial indicator has continuous values in a clockwise direction around the dial face?
4. Explain what causes cosine error when using dial indicators.
5. What are the three measurements made by a vernier or dial caliper?
6. What is the amount of common error when using micrometers?
7. Is roundness best measured diametrically or radially?
8. Accurate measurement of angularity is best made using a _____ bar.
9. What is the geometric tolerance that is often measured with optical comparators?
10. Measurements of external thread pitch cylinders are often made using the

    _____ wire method.
11. A gage that measures the collective effects of size and geometric error allowed is a

    _____ gage.

12. What is the typical cutoff value for surface finish measurement?

13. Give one reason for consistent use of wear blocks when using gage blocks.

14. In hardness testing, what does BHN stand for?

15. Accessories to a surface plate that can be used to contact two separate datums in the same plane (multiple datum planes) are _____ _____.

16. The smallest division on a universal bevel protractor is _____ minutes.

5

Surface Plate Inspection Methods

The primary measurement made when using the surface plate is *height* (Figure 5.1). This is true because all measurements made on the surface plate are made from the plate, up. The pages that follow outline specific measurements where a surface plate may be used, including coverage on the surface plate itself. Operators, machinists, and inspectors in the mechanical industry use surface plate techniques during the course of their jobs and should understand these basic techniques. Operators and machinists use surface plate techniques mainly to lay out castings and raw stock, and to measure parts. Inspectors use them for the same purposes, but mostly for making dimensional measurements. In the following pages we discuss the various basic uses of the surface plate and give helpful hints on the many applications of this valuable tool.

POSSIBLE SOURCES OF ERRORS IN MEASUREMENT

Below are some possible sources of errors when using hand tools and the surface plate.

Hand Tools

This list is repeated here for emphasis and because hand tools are used in conjunction with surface plate measurements.

1. *Miscalibration*. Hand tools should be calibrated regularly (according to use).

Figure 5.1 Surface plate equipment (L. S. Starrett Co. Reprinted by permission.)

2. *Dirt*. Parts should be clean before measurement is made.
3. *Wear*. Measuring faces of hand tools should be checked regularly for wear.
4. *Heat*. Hand-held measuring instruments should not be held for an extensive period of time.
5. *Burrs*. Parts should be deburred before measurement.
6. *Training*. The person using the tool must completely understand how to use it (reading, manipulation, etc.).
7. *Manipulation*. Hand-held instruments must be applied properly to the workpiece.
8. *Geometry*. The workpiece may be out-of-round, tapered, or have other geometrical shapes.

Surface Plate Measurements

1. *Gage blocks*. Gage blocks are not properly wrung together, or gage blocks do not include the proper size blocks.
2. *Clamps*. Clamps on height gages, surface gages, and other tools must be tight enough to avoid movement.
3. *Wear*. Surface plates may have excessive wear in high-use areas.

4. *Vibration.* The bench holding the surface plate must be firm and virtually vibration free.

5. *Parallelism.* The workpiece is not parallel, and without "sweep measuring" the workpiece, this would not be discovered.

6. *Dirt.* Surface plates should be kept clean at all times on the working surface.

7. *Cosine error.* Indicator tips should be kept parallel to the workpiece surface being measured.

8. *Support rod length.* Support rods for indicators should not be too long (cantilever effect).

9. *Tip Wear.* The indicator tip is worn flat and does not provide spherical point contact.

10. *Reading.* Errors in reading the vernier scales of height gages occur. (Sometimes they are due to parallax or a person's poor eyesight.)

11. *Parallax.* Reading errors due to parallax can be minimized by holding one's eye perpendicular to the surface being read (for example, the indicator face).

PARALLELISM MEASUREMENT

Parallelism is an important functional tolerance of *orientation* on machined parts. Perfect parallelism would be when two surfaces are exactly the same distance apart throughout their length. But since nothing is perfect, we use tolerances to define the amount of acceptable *nonparallelism* of machined surfaces to each other.

The American National Standards Institute (ANSI) symbol for parallelism is //, signifying two planes the same distance apart (Figure 5.2).

Example 1

THIS ON THE PRINT

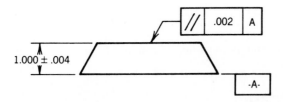

Figure 5.2 Parallelism callout.

MEANS THIS

Datum $\boxed{-A-}$ is the reference plane. If located on datum $\boxed{-A-}$, the other surface must lie within the 0.002 tolerance zone, as shown in Figure 5.3, and still be within the 1.000 ± 0.004 dimension at the same time. Simply locate surface A on a surface plate as in Figure 5.4. Set an indicator at one end of the part on 0, then move the indicator across the surface at random to the other end while watching the total

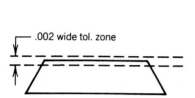

Figure 5.3   Parallelism tolerance zone.

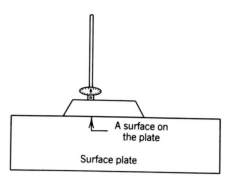

Figure 5.4   Datum is located.

movement of the indicator. This is the total indicator reading (TIR). The TIR shown in Figure 5.5 is 0.003; therefore, the part is *not* to print. It must be 0.002 TIR to be within the specified limits above.

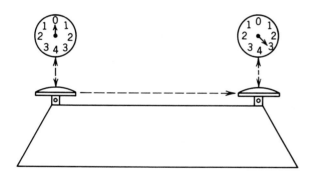

Figure 5.5   Total indicator reading (TIR).

Example 2

THIS ON THE PRINT (FIGURE 5.6)

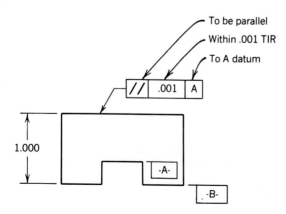

Figure 5.6   Parallelism to a hidden datum.

MEANS THIS

The meaning is the same as in Example 1. The top surface must be parallel to surface A (the datum) within 0.001 TIR. This example, however, is difficult because surface A is not easy to locate.

TO INSPECT THE PART

You must use something to locate surface A, and at the same time, keep surface B from touching the plate. This is often done with a special fixture or a set of parallels; sometimes three pins of equal height will do the job. As shown in Figure 5.7, the ideal tool is a fixture. Note that surface B is clear from the plate.

After this "special setup" to the hidden reference (or datum) surface, the part is inspected in the same way as in Example 1. If the TIR is more than 0.001, the part is not to print.

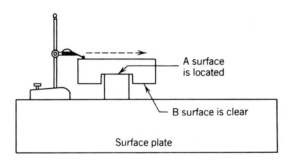

A surface
is located

B surface is clear

Surface plate

Figure 5.7 Fixed parallels locate the hidden datum.

FLATNESS MEASUREMENT

Flatness is another important functional tolerance of *form* on some machined parts. For example, if you had an assembly that required two parts to be bolted or clamped tightly together to form an airtight seal, you would want the two mating surfaces to be flat enough to do so. Therefore, the drawing would indicate a flatness tolerance for the two surfaces. Flatness is also used to "qualify" datum surfaces. Flatness has been described as the geometric characteristic of a reference plane.

The ANSI symbol for flatness is $\diagup\!\!\!\!\diagup$, showing a surface (also, indicating *one* surface) (Figure 5.8).

Notes

1. Flatness only relates to itself; there are no reference surfaces from which to measure it.
2. Flatness measurement is always total indicator reading (TIR) unless otherwise specified, and TIR means the total movement of the indicator.

Example 3

THIS ON THE PRINT

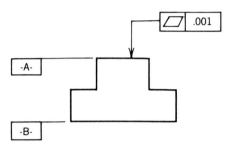

Figure 5.8 Flatness callout.

MEANS THIS

The surface indicated by the arrow must be flat within 0.001 TIR.

Remember. Flatness is to itself. You cannot use surface B for reference. If you did reference surface B, you would be measuring *parallelism*.

TO MEASURE FLATNESS ON THE PART

To measure flatness on a surface plate, set the part on three jack screws (Figure 5.9). (Put surface B on the jack screws.) Then put the indicator surface A and set 0 at three points as shown. This is getting surface A parallel to the surface plate so that the flatness can be measured. After the part is zeroed-in to the surface plate, sweep the indicator randomly over the surface, keeping watch on the TIR.

Jack screws are tools used for setting up the part to be measured. They consist of a base and a threaded locator so that you can set a part on them and adjust the height simply by screwing them up or down.

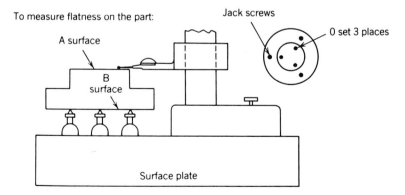

Figure 5.9 Measuring flatness using jack screws to set up the optimum plane.

FLATNESS MEASUREMENT USING OPTICAL FLATS

There are times when flatness must be measured in millionths of an inch. Some examples that require this degree of accuracy are:

1. The measuring faces of gage blocks.
2. The anvil and spindle faces of a micrometer.
3. Other parts with lapped surfaces.

When you are working with surfaces made of material that reflects light and you need flatness measured in millionths, an optical flat may be used. Optical flats (as shown in Figure 5.10) come in various diameters from 1 to 12 in. and various thicknesses from $\frac{1}{4}$ to 1 in. They are generally made from Pyrex glass or fused quartz.

Figure 5.10 Measuring flatness with optical flats. (M.T.I. Corp. Reprinted by permission.)

Optical flats can be purchased in various degrees of accuracy, from a reference grade that is flat about 1 millionth of an inch, to a commercial grade that is flat to about 8 millionths of an inch. You can also get optical parallels which have two measuring faces that must be parallel to each other.

Methods

There are two methods for measuring flatness with optical flats: the *air wedge method* and the *contact method*. In the air wedge method, the flat is held at a very small angle to the surface of the workpiece. In the contact method, the flat is in full contact with the surface being measured.

Visual Measurement

Optical flats measure flatness visually. The observer must be able to interpret interference bands (or fringes) to determine the flatness of a surface. For example, a surface that is perfectly flat will reflect straight, evenly spaced fringes through the optical flat as shown in Figure 5.11.

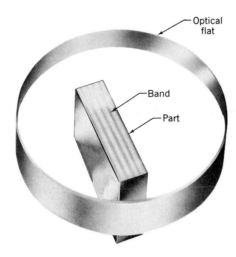

Figure 5.11 Straight and equally spaced bands appear if the surface is flat. (M.T.I. Corp. Reprinted by permission.)

If the bands are curved and not evenly spaced, this indicates that the surface is not flat. Bands should be viewed from a recommended distance of 10 times the diameter of the flat being used and they should be viewed perpendicular to the flat itself. The lighting used should be a monochromatic light (or a light with one wavelength): for example, a light utilizing helium gas. Some optical flats are coated to block out other light sources.

Reading the Bands

When the bands are observed in the recommended manner (perpendicular to the flat), the distance between each band is 11.57 millionths of an inch. In decimal form, this would be 0.00001157 in. The surface in Figure 5.12 represents a surface that is of flat by one band, as shown by the dashed line that just touches the next band. This surface is out-of-flat by 11.57 millionths and the surface is convex.

Another example is if a part were to show a three-band flatness error. The amount of flatness error is then equal to

$$3 \times 0.00001157 \text{ in.} = 0.000035 \text{ in. (or 35 millionths)}$$

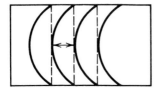

Figure 5.12 A one-band error equals 11.7 millionths if viewed at a 90-degree angle to the flat.

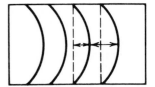

Figure 5.13 A two-band error equals 23.4 millionths.

Figure 5.13 shows a surface that is two bands out of flatness (which means about 23 millionths).

Is the surface concave or convex?

When reading the bands, you must be able to tell if the surface is concave or convex. In the air wedge method, if the bands curve around the point of contact, as shown in Figure 5.14, the surface is convex. If the bands curve around the air wedge itself, as shown in Figure 5.15, the surface is concave.

Basic Steps in Measuring Flatness with an Optical Flat

Step 1. Make sure that the surface of the workpiece has no burrs or nicks.

Step 2. Remove any dust that may be on the workpiece or the flat.

Step 3. Put a clean sheet of paper over the surface to be measured.

Step 4. Place the optical flat on the paper. (This is to protect the flat from possible damage when placing it on the workpiece.)

Step 5. Holding the optical flat in place, carefully slide the paper out from between the flat and the workpiece.

Step 6. Look for interference bands (fringes). If you do not see any, repeat steps 1 through 6.

Step 7. Observe the bands perpendicular to the flat and from a distance of 10 times the diameter of the flat being used.

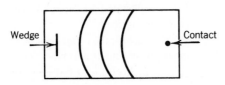

Figure 5.14 Wedge method: surface is convex.

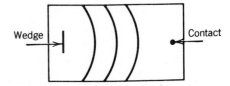

Figure 5.15 Wedge method: surface is concave.

What If Bands Do Not Appear?

If bands do not appear, there may still be dust or nicks on the mating surfaces that are preventing close proximity of the surfaces. Try repeating steps 1 through 6. There may be too much of a wedge (gap) between the workpiece and the flat. Press down on the flat in a uniform manner, and the bands may appear. There may still be moisture or an oil film that is causing the flat to "wring" to the surface. There may also be too great an angle between the workpiece surface and the flat. Applying pressure at different points around the outer edge of the flat may cause bands to appear.

The Distance Between the Bands

Sometimes it is difficult to see the exact distance between the fringes. The thin imaginary line to do this could be simulated by a piece of thread or fine wire attached to the monochromatic light source and lining up the thread or wire with the bands.

Irregular Surfaces: The Contact Method

There are times when surfaces (which are both convex and concave, and have random peaks) are irregular. In such cases the contact method is recommended, and you should not attempt to maintain a wedge. The same steps should be followed as mentioned earlier except that you press down evenly around the flat to allow it to come in full contact with the surface being measured. The flat will make contact at the high points of the surface. These high points will show up as a round band. The flatness error in this method is measured by counting each band (including one side of the high spots) and dividing the total number of bands by 2 (see Figure 5.16).

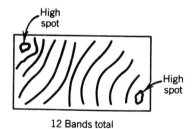

12 Bands total

Figure 5.16 Contact method: 12 bands total.

Cautions When Using Optical Flats

1. The amount of 11.57 millionths (stated earlier) applies only when you view the bands perpendicular to the flat. If you view them at an angle, the amount is different. Refer to measurement handbooks to view bands at an angle.

2. All surfaces of the part and the flat must be clean.

3. Make sure that the flat and the part have reached temperature equilibrium (the same temperature) before making the measurement.

4. Take good care of the optical flat. Avoid damaging it.

5. Use the proper light source (monochromatic light).

6. If bands do not appear, do not wring the optical flat to the surface. It may be that the surface is not flat enough for bands to appear.

When Surfaces Are Both Concave and Convex

When surfaces are both concave and convex, a fringe pattern like the one in Figure 5.17 will appear.

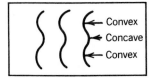

Figure 5.17   Concave–convex.

The fringe pattern in Figure 5.17 is, of course, dependent on where the point of contact and wedge are located.

When they are used properly, optical flats can be an accurate method of measuring flatness. The most frequent source of errors made when using optical flats is related to the technique used and the interpretation of the bands.

Optical Parallels Used in Calibration

Optical parallels come in very handy when calibrating measuring tools such as the micrometer. Errors in measuring with the micrometer can occur if the anvil and spindle faces are worn such that they have not parallel to each other. Optical parallels can measure this directly. One must remember, however, that the anvil and spindle faces could be parallel to each other and still not perpendicular to the axis of measurement.

CONCENTRICITY MEASUREMENT

There are times when cylindrical features must be controlled with respect to their location to each other. The two basic reasons for this are *fit* and *function*. Concentricity is a tolerance of location related to cylindrical features.

The ANSI symbol for concentricity is ◎. This symbol stands for two cylinders that share the same axis (the true definition of concentricity). Since it is impossible for any two cylinders to share exactly the same axis, the word *eccentricity* can be used to describe features that are not concentric to each other.

It must be noted at this point that concentricity is not supposed to be considered on a full indicator movement (FIM) basis. Concentricity is simply the direct linear distance between two different axes. These axes are the datum axis and the axis of the measured feature. Therefore, a concentricity tolerance that is stated on the drawing as FIM is not correct.

The problem in measuring concentricity with one indicator is that you cannot determine the axis of the feature from one side. You must use two indicators and differential measurement to find the *exact* distance between the axes. This is where concentricity and runout differ from each other. Concentricity is an axis-to-axis relationship, and runout is an axis-to-surface relationship. Concentricity is measured with at least two indicators, and differential measurement and runout is measured with one indicator and direct measurement (TIR).

Example of concentricity as shown on the drawing

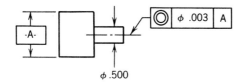

ϕ .500

Figure 5.18 Concentricity callout.

The drawing in Figure 5.18 means that the *axis,* of 0.500 in. diameter, must be concentric to *datum axis* A within a cylindrical zone of 0.003 in.

To Inspect the Part.   First, the datum axis must be established by locating the datum feature and rotating the part. This is normally done with a variety of tools, such as V-blocks, rotary table, precision spindle, centers (if the datums are centers), and others.

For this example we will use a rotary table. Datum diameter A is located in the jaws of the rotary table as shown in Figure 5.19. This will establish the datum axis in rotation. Next, two indicators are placed on the measured diameter (the 0.500-in. diameter) 180 degrees apart. Both indicators are brought to zero on the 0.500-in. diameter. This step is shown in Figure 5.20.

In Figure 5.21 the part is rotated for maximum readings on the indicators. The 0.500-in. diameter axis shows a 0.0015-in. movement in relation to the datum A axis.

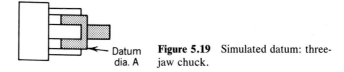

Datum
dia. A

Figure 5.19 Simulated datum: three-jaw chuck.

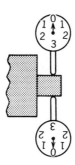

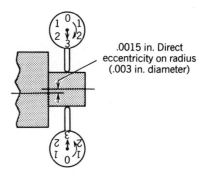

.0015 in. Direct
eccentricity on radius
(.003 in. diameter)

Figure 5.20 Indicators set up for a
differential measurement (should be a
minimum of two indicators).

Figure 5.21 A 0.0015-in. direct eccentricity (or 0.003-in. diametral zone).

Differential Measurement to Establish Actual Eccentricities

Let's say that at one point during the rotation (shown in Figure 5.21), the indicator
on top read +0.003 in. and the opposing indicator on the bottom read −0.003 in.
In this case the actual movement of the axis of the measured feature is 0.003 in.
because both indicators show the same amount of movement. However, in another reading you see that the top indicator shows a +0.004 reading and the
bottom indicator shows a −0.002 reading. The eccentricity is still 0.003 because
the interpolation between the two readings is 0.003.*

The two indicators will find the true axis position related to the datum axis
regardless of problems in the diameter. These problems range from taper to out-
of-roundness and other geometrical conditions that can be misleading to a single
indicator reading.

One should always be sure of what is required in the measurement. Is it
concentricity or runout? They both have a different purpose, symbol, and measurement technique. The odd thing about them is that they are commonly confused with each other.

Note. Keep in mind that the ANSI specification says that the use of concentricity should be carefully considered. In fact, runout can replace concentricity in
several cases.

The inspection of concentricity with two indicators is minimum. More opposing indicators on one part would be better.

---

* +0.004 in. minus −0.002 in. = spread or differential of 0.006 in., and one-half of this total
(0.003 in.) represents axis eccentricity.

RUNOUT TOLERANCES

Two kinds of runout tolerances are used on machined parts. These are shown with their symbols in Figure 5.22.

Circular runout

Total runout

Figure 5.22   Runout symbols.

Runout tolerances apply to features in rotation and are measured with respect to a datum feature. The datum feature is usually located with a tool that will allow it to rotate true (establishing a datum axis of rotation); then the measured feature surface is compared to this axis.

When measuring runout tolerances, an indicator is usually used. This could be a mechanical or electronic indicator, depending on the application. Measuring runout is simply a matter of understanding what the datum is, locating it properly, and using the indicator on the measured surface in the correct manner. An example runout requirement is shown in Figure 5.23.

AS SHOWN ON THE DRAWING

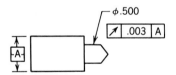

Figure 5.23   Circular runout callout.

THIS MEANS

The 0.500-in. diameter must not runout over 0.003 in. TIR. This 0.003-in. TIR is measured at individual points (if circular runout), or across the entire diameter in one sweep (if total runout).

To Inspect the Part

Runout measurements can be made with several different inspection devices. Some of these are

1. A rotary table.
2. A V-block.
3. Centers (if the datum is the center of the part).

   Note. These devices are ANSI-approved "simulated datums."

A rotary table will locate the datum feature and allow it to rotate true so that comparisons can be made. A V-block will do the same thing for a runout on a diameter.

Steps for Circular Runout Measurement

Step 1. Locate the datum feature in the V-block as shown in Figure 5.24.

Step 2. Place the indicator tip on the measured diameter at various points along the diameter while rotating the part.

Step 3. For each point measured (in circular runout) the TIR should be 0.003 in. or less. Each point measured is evaluated separately from the others.

Steps to Total Runout Measurement

Step 1. Locate the datum feature in the V-block as shown in Figure 5.24.

Step 2. Place the indicator tip on the measured diameter at one end, rotate the part, and move the indicator tip along the entire length of the diameter using a uniform sweeping motion.

During the movement of the indicator, watch for the extreme minus and plus readings. These extreme readings will establish the TIR for total runout.

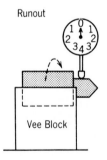

Figure 5.24   Measuring circular runout.

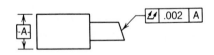

Figure 5.25   End surface total runout.

End Surface Runout

At times, the end surface of a shaft will have a runout tolerance applied to it. This is shown in Figure 5.25. To measure end surface runout, you need a tool that will not allow the shaft to move axially during the measurement. In this case, a V-block will not serve the purpose. A rotary table is suited for this task.

To measure end surface runout, follow the steps outlined below.

Step 1. Locate the datum diameter in a tool (such as a rotary table) that will not allow the part to move axially.

Step 2. Place the indicator in the center of the end surface with plus and minus travel available.

Step 3. Rotate the part and move the indicator tip slowly toward the outer edge of the shaft. Watch for the TIR during this movement. See Figure 5.26 for an illustration of the measurement.

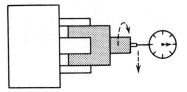

Figure 5.26   Measuring end surface total runout.

Runout tolerances are very easily measured when you understand the datums and the tolerance zones. Runout tolerances can be very helpful in controlling other geometric characteristics, such as

- Roundness
- Taper
- Cylindricity
- Concentricity
- Straightness

In all cases, the datum that is specified must be located properly, and a variable instrument (such as an indicator) must be used. Runout tolerances are regardless of feature size, so there can be no bonus tolerances applied.

PERPENDICULARITY (SQUARENESS) MEASUREMENT

Perpendicularity is another tolerance of orientation that is required for many component parts. However, some industries use the terms *normality* and *squareness,* which mean the same thing. Perpendicularity (Perp.) is always related to a datum plane or feature.

Example

"This surface to be perp. to A dia. within 0.002 TIR." This means that you *must* use A diameter to locate the part for the measurement. The ANSI symbol for perpendicularity is ⊥ . This symbolizes two surfaces at a 90-degree angle to each other.

Let's check some definitions.

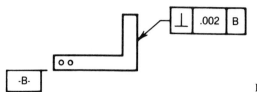

Figure 5.27   Perpendicularity callout.

Feature. Any portion of a part that can be used as a basis for a datum.

Datum. Points, lines, planes, or other geometric shapes that are assumed to be exact for use as a reference for measurement.

In Figure 5.27, the datum feature is B.

Remember. Datum features (or surfaces) are the surfaces you must locate the part on, not measure the part.

The measured surface in Figure 5.27 is the short leg.

In Figure 5.28 the surface can be anywhere in the 0.002-in.-wide tolerance zone. The vertical line shows what perfect perpendicularity would be. The dashed line shows the width of the zone, which is (0.002 in.). The view in Figure 5.28 shows the worst possible condition *within the tolerance zone.*

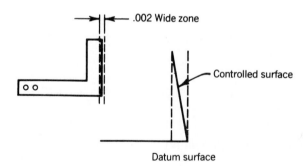

Figure 5.28   Perpendicularity tolerance zone.

To Inspect the Part.   Contact datum B completely on a simulated datum plane (e.g., surface plate); then measure the perpendicularity by using one of the tools shown in Figure 5.29, depending on the tolerance allowed. The machinist's square (Figure 5.29a) and feeler stock can be used when tolerances permit. The cylindrical square (Figure 5.29b) is easy to use. You simply line up any side of the square that matches the measured surface and follow the *topmost* dashed curve to read the out-of-squareness directly. The cylindrical square is accurate to 0.0002 in. The master sequences gage (not shown) has a very accurate vertical member that allows you to measure perpendicularity within 0.0001 in.

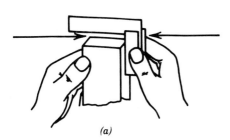

(a)

Figure 5.29*a* Precision square in use.
(L. S. Starett Co. Reprinted by permission.)

(b)

Figure 5.29*b* Cylindrical square.
(Brown & Sharp. Reprinted by permission)

COAXIALITY MEASUREMENT (USING POSITION TOLERANCE)

Coaxiality is a positional tolerance used to control the alignment of two or more holes shown on a common axis when rotation is not involved. In the case below, there is a bolt that must go through the two holes at the final assembly. The largest size of these holes can be a 0.412-in. diameter. If the two holes are not in their true position and/or not at the low limit of size, the bolt will not go through. Therefore, the coaxiality of the two holes is important.

Note. Let's say that in this case, the two holes were drilled at different times in the process. This tells us that their alignment must be inspected in-process to ensure coaxiality.

Coaxiality is the same thing as concentricity because they are both concerned with two features being on the same axis to a certain tolerance. However, using positional tolerance allows you to use MMC and get bonus tolerance.

Example

AS SHOWN ON THE BLUEPRINT (FIGURE 5.30)

This shows two holes with coordinate basic dimensions locating the position (with basic dimensions). The tolerance zone (where the axis of the holes must lie) is a cylindrical zone 0.010 in. in diameter with the true position in the center of the zone.

Remember. Perfect true position here is 2.200 in. and 1.100 in. *exactly.* Note that the holes can vary 0.005 in. from the position in any direction.

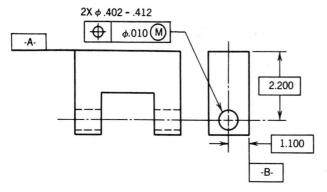

2X φ .402 - .412

Figure 5.30   Position tolerance: coaxial.

THE ABOVE MEANS

The actual centerline of each hole must lie inside the shaded area shown in Figure 5.31.

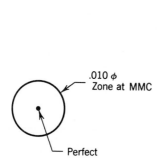

Figure 5.31   Coaxial tolerance zone.

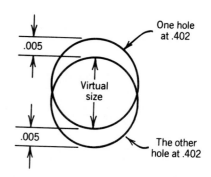

Figure 5.32   Possible location error (virtual size) at MMC.

      The view in Figure 5.32 shows the worst possible condition that the part is allowed to be in (within the tolerance). If one hole centerline is in the extreme of the diametral zone and the other hole centerline is at the opposite extreme *and* they are both at the smallest diameter allowed (0.402 in.), it will cause a condition as shown at the left and contribute to the interference of the bolt going through both holes.

TO INSPECT

Step 1. First inspect the position of each hole from each of the datums A and B. If they are to size (in diameter) and are in the true position zone, there is no problem.

Step 2. To inspect quickly to see if the mating part (the bolt) will go into these parts, simply check the virtual size of the two holes. This can be done by calculating the virtual size.

Now a functional gage can be made. This gage will be a virtual size pin.

$$\text{Virtual size} = \text{MMC (of hole)} - \text{true position tolerance}$$

So

$$0.402 \text{ in.} - 0.010 \text{ in.} = 0.392 \text{ in.}$$

Now, in Figure 5.33, you have a 0.392-in. pin that will go through any good part. Remember, this can be done only if the position tolerance is specified at the MMC condition.

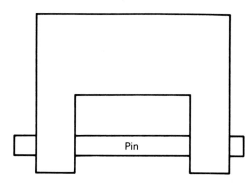

Figure 5.33 Functional gage for coaxiality at MMC.

ANGULARITY MEASUREMENT: ANGULARITY VERSUS SIMPLE ANGLES

Angularity is another important tolerance of "orientation," which is used quite often when the angle of a certain surface *must* be tightly controlled. The ANSI symbol for angularity is $\angle$. This symbolizes a surface at an angle to another surface. There is a big difference between angularity toleranced in linear units and a simple angle having a tolerance in degrees and minutes.

The differences between them are shown below. Example 1 is a simple angle callout, but Example 2 is an angularity callout. Note that they are both always related to a reference surface (or datum).

Example 1: Simple Angle Callout (Figure 5.34).

AS SHOWN ON PRINT

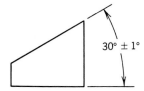

$30° \pm 1°$

Figure 5.34 Simple angle callout.

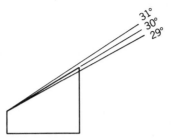

Figure 5.35 Simple angle tolerance zone.

This simply means that the angle of this part must be 30 degrees plus or minus 1 degree. Therefore, the angle could be anywhere from 29 to 31 degrees and still be to print (see Figure 5.35).

TO INSPECT THE PART

This part can be inspected simply with a protractor as shown in Figure 5.36. Protractors can show angles in degrees and minutes.

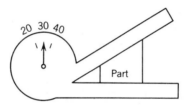

Figure 5.36 Bevel protractor measures simple angles.

Example 2: Angularity

AS SHOWN ON THE PRINT

Shown in Figure 5.37, is an angularity tolerance of 0.015 in. The ANSI symbol for angularity is shown in what is called a *feature control frame*. The main difference between this angularity and a simple angle callout is the fact that angularity is toleranced in linear dimensions and simple angles are toleranced in angular dimensions.

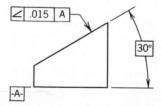

Figure 5.37 Angularity callout.

Angularity within 0.015 in. gives you a 0.015-in.-wide tolerance zone. The surface can lie anywhere within the tolerance zone as long as it is in the zone throughout the entire feature. The view in Figure 5.38 shows that a perfect 30-degree

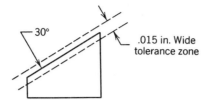

Figure 5.38 Angularity tolerance zone.

angle cuts through the middle of the zone and the zone is equally distributed around the perfect 30-degree plane.

TO INSPECT THE PART

This part requires the use of a sine bar (sine plate) to inspect the 30-degree angle.

The view in Figure 5.39*a* shows the inspection of the part with a sine bar. A sine bar is a simple tool that works using right-triangle trigonometry. You can set up any angle with this tool using gage blocks. An actual sine bar is shown in Figure 5.39*b*.

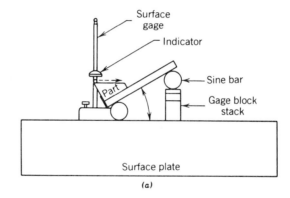

(a)

Figure 5.39*a* Angularity measured using a sine bar.

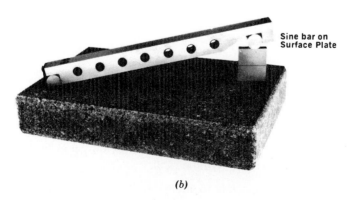

(b)

Figure 5.39*b* Sine bar on a surface plate. (Brown & Sharp. Reprinted by permission.)

In this case we see that the sine bar is set up with gage blocks to form a perfect 30-degree angle. Then the part is set on the sine bar and the only thing to do now is run an indicator across the surface and watch for the TIR. If the TIR is 0.015 in. or less, the part is acceptable. There are other devices used to measure angularity such as angular gage blocks, rotary tables, and indexing heads.

SYMMETRY (RECENTLY CHANGED TO POSITION)

Symmetry is another location tolerance. It is defined as a condition in which a feature (or features) is symmetrically disposed around the center plane of a datum feature. The ANSI symbol for symmetry is ≡. This symbolizes two like lines that are both exactly the same distance from the line in the center, so they are symmetrical.

Figure 5.40 shows a part with a slot in it. This slot, per the drawing, must be symmetrical within 0.010 in. to datum feature B. In short, the actual centerplane of the slot must lie within a 0.010-in.-wide tolerance zone that is equally disposed around the centerplane of the datum feature B, as shown in Figure 5.41.

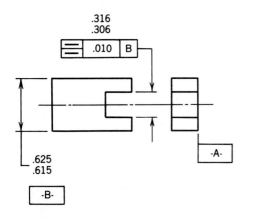

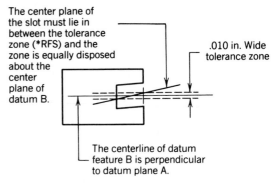

The center plane of the slot must lie in between the tolerance zone (*RFS) and the zone is equally disposed about the center plane of datum B.

.010 in. Wide tolerance zone

The centerline of datum feature B is perpendicular to datum plane A.

*RFS—means regardless of feature size.

Figure 5.40 Symmetry callout (use position symbol).

Figure 5.41 Symmetry tolerance zone (RFS).

To Inspect the Part.   Let's say that we measure the B dimension and it is 0.620 in. and we measure the width of the slot and it is 0.310 in. Therefore, the thickness per side should be 0.620 in. minus 0.310 in. and the answer divided by 2. So each side would be 0.155 in. thick (as shown in Figure 5.42) if the slot and datum B were perfectly symmetrical.

Note. It is important to understand at this point that if the centerplane of the slot were to move by 0.001 in., the thickness on one side would be 0.154 in. and the other side would be 0.156 in. thick in Figure 5.42.

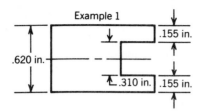

Figure 5.42   Example of perfect symmetry.

To Measure the Part.   You simply measure the thickness of each side, subtract the smallest thickness from the largest, and divide your answer by 2. The answer you get will be the actual distance that the centerplane of the slot has moved away from the centerplane of the datum, and it is allowed to move only 0.005 in. in either direction with a 0.010 total tolerance zone.

Let's say that the part in Figure 5.43 measures as shown. One side is 0.160 in. thick and the other side is 0.150 in. thick. So 0.160 in. minus 0.150 in. equals 0.010, and that divided by 2 equals 0.005 in. that the centerplane has moved. Here the part is acceptable because the centerpiece is allowed to move up to 0.005 in., but it is at the worst allowable condition.

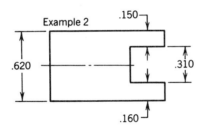

Figure 5.43   Borderline acceptable part.

STRAIGHTNESS (OF AN AXIS)

Straightness is another important tolerance of "form" when it is related to mating parts. The ANSI symbol for straightness is ⎯⎯, symbolizing a straight line (or axis). For example, Figure 5.44 means that the pin can range in diameter from 0.615 in. to 0.605 in. and can be straight up to 0.015 in. when it is at MMC. (MMC is 0.615 in. in this case).

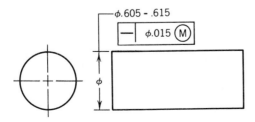

Figure 5.44   Straightness of an axis at MMC.

The drawing in Figure 5.45 shows the worst condition that the part can be and still be acceptable (its virtual condition).

Remember. The virtual size equals the MMC plus the tolerance. So

$$0.615 \text{ in.} + 0.015 \text{ in.} = 0.630 \text{ in.}$$

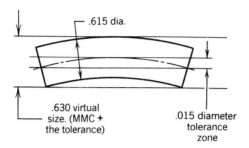

.615 dia.

.630 virtual
size. (MMC +
the tolerance)

.015 diameter
tolerance
zone

Figure 5.45   Diameter versus virtual size.

Note. The drawing says that the part is to be straight within 0.015 in. *at MMC*. Therefore, if the actual part is smaller than MMC, the straightness tolerance (per ANSI) increases by the same amount that the diameter decreases.

Example

If the pin diameter were actually 0.610 in., the diameter would be 0.005 in. smaller than the MMC. So the straightness tolerance would increase to 0.020 in. automatically (0.015 in. + 0.005 in.).

To Inspect the Part.   One way to inspect the part is to have a functional gage made that has a 0.630 inside diameter.

Note. The diameter of the gage must be longer than the pin itself. This gage is simulating the virtual condition (see Figure 5.46).

As you can see in Figure 5.46, a straight pin will fall through the gage, because its virtual size is simply its largest size. Also, in Figure 5.47, you can see that the pin at MMC and 0.015 in. out of straight will still fall through. In Figure 5.48, observe that the pin is at its lowest allowable size and out of straight by 0.025 in., but will still go through because the gage allows for the bonus tolerance you get when the part is under MMC.

Hence, inspecting the part this way simply involves measuring the diameter of the pin and attempting to slip it through the gage. If the diameter is within its

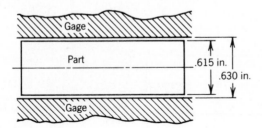

Gage

Part

.615 in.

.630 in.

Gage

Figure 5.46   Part is at MMC size and perfectly straight.

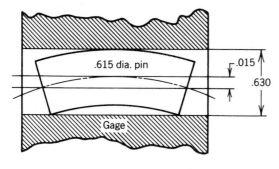

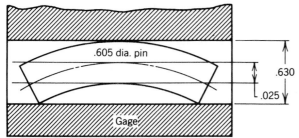

Figure 5.47 Part is at MMC size and bent 0.015 in.

Figure 5.48 Part is at LMC size and bent 0.030 in.

size limits and will go through this gage, this part is straight within tolerances. Always measure the diameter!

STRAIGHTNESS MEASUREMENT: STRAIGHTNESS OF SURFACE ELEMENTS

Another control of straightness is that which is applied to the elements of the surface only (instead of its axis). When straightness is applied to surface elements, neither MMC nor RFS applies because a surface element has no size.

When straightness is applied to surface elements, the feature control frame will have a leader line pointing to the surface, as shown in Figure 5.49. In the figure the tolerance zone is two perfectly parallel imaginary lines 0.002 in. apart. The pin (during measurement) is rotated such that all elements of its surface are passed between these lines and no portion can extend beyond the tolerance zone, as shown in Figure 5.50.

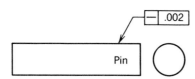

Figure 5.49 Straightness of surface elements callout.

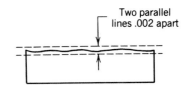

Figure 5.50 Tolerance zone for the straightness of surface elements.

Measuring straightness of surface elements can be done in two basic ways, with the optical comparator and the surface plate setup.

Comparator

With a comparator, the surface of the pin is projected, and an overlay of the tolerance zone is used on the screen. The pin is then rotated 360 degrees and each surface element must lie between the two tolerance zone lines.

Surface Plate Setup

Two jack screws equipped with V-anvils can set up the straightness plane in one element. This setup is shown in Figure 5.51.

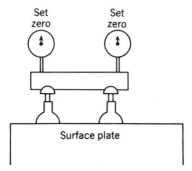

Figure 5.51 Straightness of surface elements measured with jack screws to set up the optimum line element.

Once you establish the straightness plane (in each element), you simply sweep the surface and it should be within the stated tolerance across the surface. This measurement would then be seen using TIR (FIM) of the indicator after establishing 0. Rotate the pin to another element and repeat the procedure. Obviously, use of the optical comparator is a much faster method.

Remember. Straightness of surface elements demands variable measurement. Because MMC cannot apply, no functional gage can be used.

PROFILE TOLERANCES PER ANSI Y14.5M–1982

Profile of a line   Profile of a surface   **Figure 5.52** Profile symbols.

Profile tolerances are used to control the shape of irregular parts. The symbols for profile tolerances are shown in Figure 5.52. Profile tolerances can be used in many ways. Some basic examples are described below.

1. *Profile all around.* This is where a profile tolerance is applied all around a part. Examples of the tolerance callout and tolerance zone are shown in Figure 5.53.

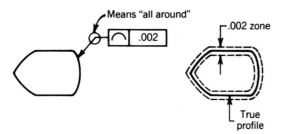

Figure 5.53 Profile (all around) and tolerance zone.

2. *Profile (bilateral tolerance zone).* When a profile tolerance is applied with a leader line pointing directly to a surface, the tolerance zone is the specified width equally disposed around the true profile, as shown in Figure 5.54.

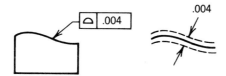

Figure 5.54 Profile of a surface.

3. *Profile tolerance (unilateral zone).* If a profile tolerance is shown using a *phantom* line and two arrows, it means that the tolerance band is totally distributed either inside the true profile or outside it, as shown in Figure 5.55.

Figure 5.55 Profile tolerance: unilateral zones.

4. *Profile tolerance (using datums).* When a profile tolerance is used in reference to a datum plane, it controls shape, location, and size simultaneously. An example of this is shown in Figure 5.56.

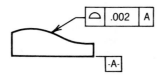

Figure 5.56 Profile tolerance: datums specified.

Measuring Profile Tolerances

Most profile tolerances are typically measured on an optical comparator (Figure 5.57) with the assistance of overlay charts for comparison measurement and translation devices such as tracer units. There are times when direct measurement, such as a sweep gage establishing the true profile, can be employed. The gage follows the template, and the indicator shows high and low areas in comparison to the true profile.

Regardless of the technique or tool used, the important thing is to be able to establish a true profile from which to compare the surface, and some means of measuring the difference. Where it is important for a surface to be in the correct shape and position, profile tolerances should be established from datums.

Profile of a Line

Profile of a line simply means that the profile tolerance must be checked for various elements of the surface, and each line element should be within the stated tolerance. An inspector or machinist would establish the true profile and then measure one line element at a time completely across the surface.

Remember. Each line element must be within the tolerance zone, but they have no relationship with each other collectively.

Profile of a Surface

The profile of a surface simply means that the profile tolerance applies throughout the surface. Several line elements are measured collectively, and all must be within profile tolerance.

USING PROFILE TOLERANCES FOR COPLANARITY

There are times when it is important that two or more interrupted surfaces be in the same plane, or *coplanar*. Coplanarity can easily be accomplished using *profile of a surface tolerances*. Coplanarity is generally applicable when two or more surfaces are interrupted by a feature or features such as slots, holes, pins, tabs, and so on.

Example 1

In this example we have a part that has two surfaces interrupted by a slot. It is important that the two surfaces be coplanar. So we apply profile of a surface tolerance to the part as shown in Figure 5.58.

The tolerance zone is two parallel planes that are 0.005 in. apart; both surfaces must lie between these two planes, as shown in Figure 5.59.

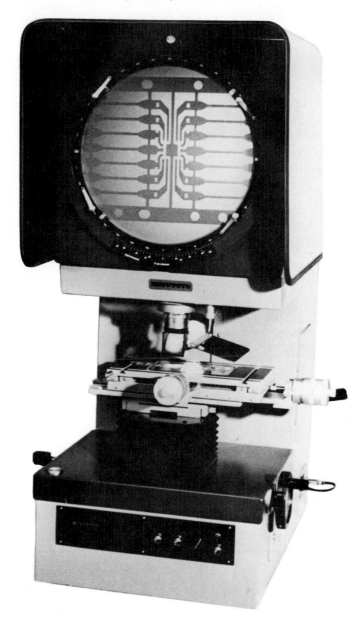

Figure 5.57 Optical comparator. (M.T.I. Corp. Reprinted by permission).

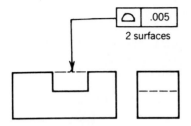

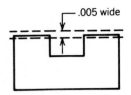

Figure 5.58 Coplanarity control using profile of a surface.

Figure 5.59 Tolerance zone for the part.

Example 2

In this example there is a part that has several slots cut into it and two important surfaces that establish the datum plane for measurement. This part is shown in Figure 5.60.

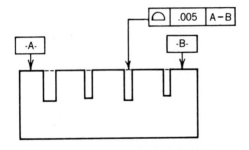

Figure 5.60 Coplanarity: multiple datums.

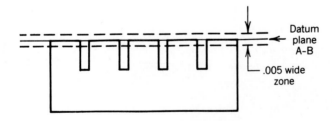

Figure 5.61 Tolerance zone for the part.

In this case, the tolerance zone is two parallel planes that are 0.005 in. apart. The tolerance zone is equally distributed around the datum plane established by datums A and B simultaneously. This is shown in Figure 5.61.

Measurement

In Example 1 we need to establish one plane and compare the other plane to it. This can be done by direct location of one of the surfaces on the surface plate and sweeping the other surface with an indicator.

In Example 2 we must locate datums A and B and compare the other surfaces to them. This can be accomplished by locating datums A and B on a set of parallels (upside down), then sweeping the other surfaces with an indicator.

Coplanarity is well defined when using profile of a surface tolerancing, and the datums and measured surfaces are very clear.

MAKING LINEAR MEASUREMENTS USING THE TRANSFER METHOD OR THE DIRECT READING METHOD

Linear (or length) measurements are most often made by using the transfer method. This method is simply setting the dimension up with gage blocks and transferring the dimension from the gage block stack to the part.

Example

THIS ON THE PRINT (FIGURE 5.62)

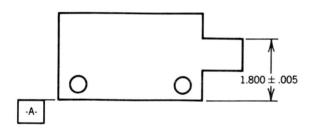

1.800 ± .005

-A-

Figure 5.62 Height requirement.

MEANS THIS

This tab must be 1.800 in. ± 0.0005 in. from datum surface A (no matter what size the tab is).

TO INSPECT THIS PART USING THE TRANSFER METHOD (FIGURE 5.63)

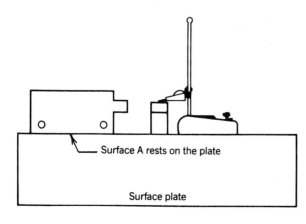

Surface A rests on the plate

Surface plate

Figure 5.63 Transfer measurement with gage blocks.

Step 1. Set up gage blocks to 1.800 in.

Step 2. Set 0 on the indicator (on the blocks).

Step 3. Carefully slide the surface gage to the tab surface and take a reading.

Step 4. The plus or minus amount shown on the indicator tells you how much over or under 1.800 in. the actual dimension is.

Step 5. In this example the indicator should not show more than plus 0.005 in. or minus 0.005 in. (otherwise, the part is not to print).

Measuring the Same Dimension Using the Direct-Reading Method

This method of linear measurement is also often used. Here you use the surface plate as a starting point (since surface A is where the dimension begins, and surface A is resting on the plate).

The first step is to rest the tip of the indicator on the surface plate and (using the fine adjustment knob) set 0 on the indicator (see Figure 5.64). Next, read the vernier scale on the height gage and write the reading down. This reading will be used later. For this example, let's say that the reading was 1.200 in.

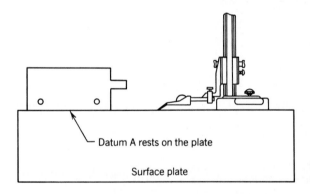

Datum A rests on the plate

Surface plate

Figure 5.64 Dual-reading technique: zero on the surface plate.

The next step is to raise the indicator making it rest on the tab surface (as shown in Figure 5.65). Take care in doing this and do not bump the indicator against anything. If the indicator does get bumped, start over. Now, using the fine adjustment knob, set 0 again on the indicator. At this point, take a second reading on the vernier scale and write it down. Let's say that this reading was 3.002 in.

Now, the actual dimension of the part is simply the difference between the highest and lowest readings you have taken. This is true because you set the indicator to 0 on the plate and 0 on the tab surface. Hence

$$
\begin{array}{r}
3.002 \\
-1.200 \\
\hline
1.802
\end{array}
$$

1.802   Since the tolerance is 1.800 in. ± 0.005 in., the part is to print.

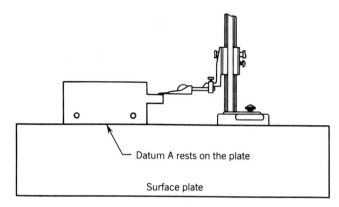

Datum A rests on the plate

Surface plate

Figure 5.65 Dual-reading technique: zero on the part surface.

There are other ways to make the type of measurements shown above, such as transferring measurements from height masters or using digital height gages for direct reading.

CENTERLINE MEASUREMENT OF FEATURES

Many measurements require that the centerline of a feature (a feature being a hole, slot, tab, etc.), must be a specified distance from a "datum" or reference surface. To obtain an accurate measurement, the centerline must first be found. For example, the actual centerline of a hole depends on the actual size of the hole. Linear measurements to the centerline of a feature cannot be accurately made unless the feature size is first measured. Below are a few basic steps in making measurements to centerlines of features.

Step 1. Locate the datum surface first on the surface plate.

Step 2. Measure the actual size of the feature. (In this example, it is a hole.)

Step 3. Write down the actual feature size.

Step 4. Measure the distance from the reference surface to your choice of either side of the feature.

Step 5. Write down this distance.

Step 6. Now, depending on which side of the feature you chose to measure, you simply add or subtract one-half the actual hole size from your measurement.

An Example of Centerline Measurement of a Hole

The drawing in Figure 5.66 means that the actual centerline of the $0.410 - 0.400$ in. hole must be within 1.500 ± 0.010 in. from datum A no matter what the actual hole size is.

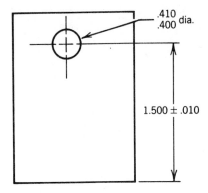

Figure 5.66   Coordinate dimension.

Therefore, you can see the importance of the need to measure the hole first, because it has 0.010 in. tolerance on it and could be any actual size within that tolerance. You can never assume that the hole is a specific size.

Let's say that we measured the actual hole size here and that it was 0.402 in. The preferred way to measure this part, in most places, is to get a gage pin that fits the hole and insert it into the hole. After doing this, you set the part on the surface place with datum A resting on the plate (since it is the locating surface) (see Figure 5.67). In this case we choose to measure to the top of the hole. So you calculate the distance to the top of the pin assuming that the hole is exactly 1.500 in. from surface A to its centerline. This is simply 1.500 in. plus one-half the pin size. So it is 1.500 in. + 0.201 in. = 1.701 in. Next, you set up a gage block stack of exactly 1.701 in. and zero the indicator on this gage block stack. Now, the zero on your indicator is set at 1.701 from the surface plate. Transfer this measurement by sliding your surface gage to where the indicator tip is at the very top of the pin. If the part is within blueprint tolerance, your indicator will not show a reading lower than minus 0.010 in. or higher than plus 0.010 in.

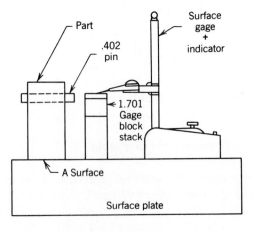

Figure 5.67   Coordinates measured using transfer technique.

Remember. Never allow the indicator to be bumped after you have set zero. If you bump the indicator, *start all over again*.

Make certain that you contact the very top of the pin during the measurement.

REVIEW QUESTIONS

1. (True or False) Measurement of parallelism must be with respect to a datum.

2. A geometric characteristic of a reference plane is _____.
3. Which geometric tolerance is often measured using three jack screws?

4. Optical flats measure flatness within millionths using _____ light bands.
5. Which of the runout tolerances is evaluated using one composite TIR?
6. Name two of the three coaxiality controls in ANSI Y14.5.
7. Angularity differs from simple angle tolerances in that the results are not in degrees but in _____ units.

8. What tolerance replaced the symmetry symbol in the 1982 ANSI Y14.5 standard?
9. Which of the straightness tolerances (on a shaft, for example) has a cylindrical tolerance zone: straightness of an axis, or straightness of surface elements?

10. Profile tolerance zones are always _____ from true profile unless otherwise specified.
11. Which of the profile tolerances is used to control coplanarity?
12. The tolerance zone for perpendicularity of a surface to a datum surface is _____ _____ planes.

13. All runout measurements are evaluated in terms of the resultant FIM. Give another abbreviation that means the same thing as FIM.
14. (True or False) Concentricity measurement is not a TIR measurement.
15. Name two other geometric characteristics that are controlled by total runout applications.

6

Graphical Inspection Analysis*

Graphical inspection analysis (GIA), also referred to as *layout gaging,* is an inspection verification technique employed to ensure functional part conformance to engineering drawing requirements without high-cost metal functional gages. With the increased use of dimensioning and tolerancing methods, where we are concerned with concepts such as maximum material condition, cylindrical-shaped tolerance zones, and *bonus* tolerancies, we must use inspection methods that will evaluate parts conformance on a functional basis. Graphical inspection analysis provides the benefit of functional gaging without the expense and time required to design and manufacture a close-toleranced, hardened metal functional gage. Graphical inspection analysis is well suited when small lots of parts are being manufactured; however, the inspection concept applies equally well to large quantity production. Also, inspection can be performed on parts too large and parts too small for conventional, functional receiver gages.

Inspection with functional receiver gages is the most desirable method to verify part conformance for interchangeability. Functional gages are really three-dimensional, worst-case condition, mating parts. If the gage assembles to the part being inspected, we are assured assembly with all conforming mating parts. The drawbacks of functional gages are the expense and time required for design and manufacturing. Graphical inspection analysis provides a technique that will provide immediate, accurate inspection results without high cost and time delays.

---

* This chapter was written and figures reprinted by permission of George O. Pruitt, Technical Documentation Consultants.

GEOMETRIC TOLERANCING PRINCIPLES

Graphical inspection analysis is very dependent on the methods and principles used in geometric dimensioning and tolerancing. It is assumed that readers of this chapter are knowledgeable in the concepts and rules of the geometric dimensioning and tolerancing standard as specified in American National Standards Institute (ANSI), standard Y14.5. The ANSI principles will be reviewed to ensure common interpretation for an effective understanding and use of GIA.

Principle 1: Authority Standard

It is imperative that the parts inspector knows which revision of the ANSI Y14.5 standard is used as the authority document in the preparation of the drawing. In this chapter we point out various differences existing between the three revisions of the standard (USASI Y14.5–1966, ANSI Y14.5–1973, and ANSI Y14.5M–1982).

Principle 2: Basic Dimensions

A basic dimension is a theoretical, exact dimension without tolerance. When used in conjunction with position tolerance specification, the basic dimension locates the exact center of the position tolerance zone. The center plane, or axis, of an acceptable feature is allowed to vary within the basic, located, position tolerance zone, as shown in Figure 6.1. Basic dimensions on drawings are identified by the word BASIC, or the abbreviation BSC, near the dimension. Symbolically, the dimensions can be enclosed in a box to signify the basic notation. When features are located by basic chain dimensioning on the drawing, there is no accumulation of tolerance between features (Figure 6.2). Basic dimensions are absolute values and, when added together, they equal an absolute value. The

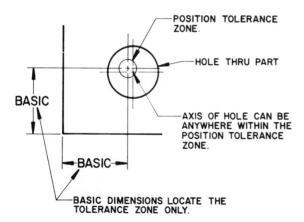

Figure 6.1   Positional tolerance zone.

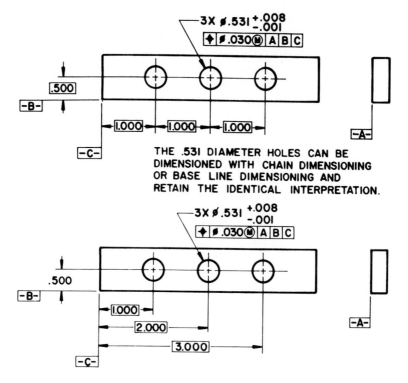

THE .531 DIAMETER HOLES CAN BE
DIMENSIONED WITH CHAIN DIMENSIONING
OR BASE LINE DIMENSIONING AND
RETAIN THE IDENTICAL INTERPRETATION.

Figure 6.2 Chain dimensioning versus baseline dimensioning.

basic dimensioning locates the position of the tolerance zone, not the manufactured feature locations.

Principle 3: Condition Modifiers

There are three material condition modifiers used with position tolerancing: maximum material condition (MMC), regardless of feature size (RFS), and least material condition (LMC). The symbol for maximum material condition is Ⓜ. When the MMC symbol is associated with the tolerance or a datum reference letter in the feature control frame, the specified tolerance applies to the feature *only* if the feature is manufactured at its maximum material condition size.

Let's review the maximum material condition as it applies to position tolerance. The maximum material condition is the condition of a feature when it contains the maximum amount of material or weight within its allowable size limits. For example, a 0.500 ± 0.010 diameter pin will be at maximum material condition only if the machinist manufactures the pin at 0.510 diameter—a 0.509 diameter pin would not be at maximum material condition. Another example is: a 0.525 ± 0.010 diameter hole in a part would be at maximum material condition, or its greater weight, if drilled at 0.515 diameter. In other words, maximum material condition is maximum allowable shaft size or minimum allowable hole size.

As shown in the feature control frame (Figure 6.3), the 0.028 tolerance value applies *only* if the feature with which the frame is associated is produced at MMC. What happens if the associated feature or datum depart from MMC? Principle 5 will discuss additional tolerance allowances.

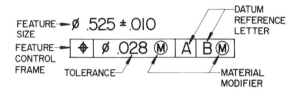

Figure 6.3   MMC feature control frame.

The symbol for regardless of feature size is ⓢ. When the RFS symbol is selected to modify the tolerance or datum reference in the feature control frame, the specified tolerance applies to the location of the feature regardless of the feature's size. In Figure 6.4, the 0.028 tolerance applies if the hole is manufactured at 0.515, 0.535, or any other value in between. The 0.028 is the total location tolerance for the feature regardless of its size.

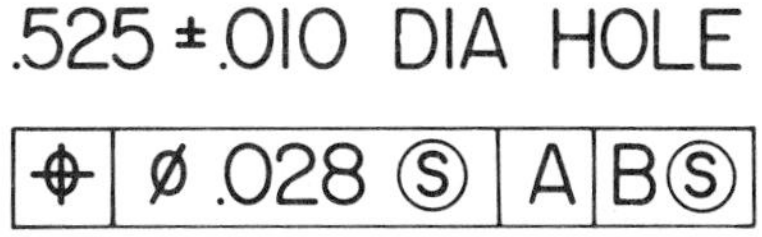

.525 ±.010 DIA HOLE

⟦ ⌖ | ⌀ .028 ⓢ | A |Bⓢ ⟧

Figure 6.4   RFS feature control frame.

The symbol for least material condition is Ⓛ. The least material condition concept and symbol were not introduced until the approval of ANSI Y14.5M–1982. Least material condition is the condition of a feature when it contains the least amount of material or weight; for example, smallest shaft size and largest hole size—just opposite of the MMC concept. When a feature control frame references LMC to the tolerance or datum reference letter, the specified tolerance *only applies* when the feature is produced at the LMC size. Additional discussion of what happens when the feature or datum size departs from LMC will be discussed in Principle 5.

Principle 4: Feature Control Frame

The *feature control frame,* formally called *feature control symbol,* tells the drawing user many things. The shape and size of the tolerance zone is specified, the surfaces and order of precedence for setup of the part is dictated, and the material condition modifiers are assigned to the tolerance and datum reference letters.

As shown in Figure 6.5, the inspector is aware that the 0.014 tolerance value applies only when the feature being verified for locations is produced at its maxi-

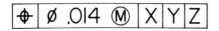

Figure 6.5 Feature control frame.

mum material condition size. Also, in Figure 6.5, the diameter symbol ($\emptyset$)
preceding the tolerance indicates to the inspector that the 0.014 tolerance zone
will have a cylindrical shape. Absence of the diameter symbol would indicate that
the tolerance zone shape would be the area between two parallel planes or two
parallel lines 0.014 apart. No symbol is specified to indicate the latter tolerance
zone shape. The 1966 standard required for the abbreviation DIA for diameter
tolerance zone and R for radius tolerance zone to be included in the feature
control frame. The 1966 standard also required the specifications of TOTAL for
total wide tolerance zone and R for one-half of total wide tolerance zone.

The order of specified datum references in the feature control frame is very
important. The primary datum is shown at the left; the least important datum is
shown at the right. This datum precedence allows manufacturing and inspection
to determine part orientation for their respective functions. The part is fixtured to
allow a minimum of three points of contact on the primary datum surface, minimum two points of contact for the secondary datum feature and minimum one
point of contact for the tertiary datum feature.

The 1966 revision of the ANSI Y14.5 standard required the datum reference
letters to precede the tolerance compartment of the feature control frame—-while
the 1982 revision requires the datum reference letters to follow the tolerance
compartment. Either method was accepted in the 1973 revision of the standard.
The left-to-right datum procedure concept is identical in both methods (Figure
6.6).

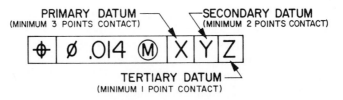

Figure 6.6 Datum precedence.

Principle 5: Bonus Tolerance and Additional Tolerance

As discussed in Principle 3, when a tolerance or datum reference letter is modified
with the MMC symbol, the specified tolerance in the feature control frame *only
applies* to the feature location when the feature is manufactured at its MMC size.
As the feature departs from MMC size, the position tolerance is increased. The
amount that the feature deviates from MMC size is added to the position tolerance
specified in the feature control frame. This extra tolerance is called a *bonus*
tolerance. As shown in Figure 6.7, the 0.014 diameter position tolerance applies

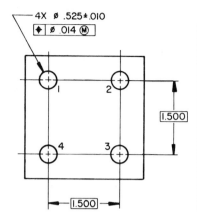

FEATURE SIZE	BONUS TOLERANCE	TOTAL POSITIONAL TOLERANCE
.515 (MMC)	.000	.014
.516	.001	.015
.517	.002	.016
.525	.010	.024
.533	.018	.032
.534	.019	.033
.535 (LMC)	.020	.034

Figure 6.7 Bonus tolerances.

when the hole is drilled at 0.515 diameter, the MMC size. If hole 2 were drilled at 0.525, a departure of 0.010 from the MMC size of the hole, we would gain a bonus tolerance of 0.010. The 0.010 bonus tolerance is added directly to the original 0.014 position tolerance to give an allowable positional tolerance of 0.024 (see the table in Figure 6.7). The axis of hole 2 must be within the 0.024 diameter toler-ance. The allowable positional tolerance zone size for each hole must be deter-mined in conjunction with the actual manufactured hole size.

We can also gain extra locational tolerance as a datum feature of size, referenced in the feature control frame as MMC, departs from MMC size. This added tolerance is not called a bonus tolerance and is treated somewhat differ-ently than a bonus tolerance. In Figure 6.8, we could gain up to 0.005 bonus tolerance as the 0.260 diameter hole departs from MMC. Additional tolerance could also be gained as the datum B hole departs from MMC. The added toler-ance as datum B departs from MMC does not add directly to the original position tolerance as a bonus tolerance would. The datum additional tolerance must be

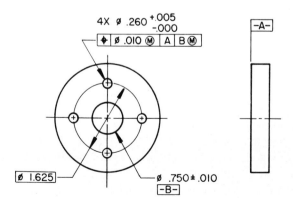

Figure 6.8 Additional tolerance from a size datum.

applied to the hole pattern as a *group*. For example, the added tolerance could allow the four-hole pattern to shift to the right as a group. No additional hole-to-hole tolerance gain is realized within the four hole pattern.

The rules and principles of MMC bonus and additional tolerance also apply to tolerances and datum references that are modified at LMC. The bonus and additional tolerances are determined from the feature's departure from LMC size.

The 1966 and 1973 revisions of ANSI Y14.5 allowed the MMC symbol to be implied by the drawing user when the material condition modifier had been omitted from the position feature control frame. It should be noted that this rule is contrary to international practice, which implies RFS in the same situations. The 1982 revision of ANSI Y14.5 requires the feature control symbol to be completed with the appropriate modifier (no implied modifier).

Principle 6: Datums

Datum specification in the feature control frame is of great importance to the drawing user. Proper functional datum selection allows the part to be fixtured for manufacturing and inspection, as it will be assembled as a finished product. Consistency with quality control will be assured when manufacturing uses the specified datum features in the proper order of precedence for machining operations.

The 1966 and 1973 revisions of the ANSI Y14.5 standard allow the use of implied datums by not specifying datum reference letters in the feature control frame. This required the print users to make certain assumptions in the course of manufacturing and inspection, such as which features to fixture upon and in what order of precedence. The use of implied datums can lead to problems if everyone concerned does not make the same assumptions. Designers and drafters should be encouraged to specify all necessary datums to provide uniform drawing interpretation. The 1982 revision of the standard has discontinued the use of implied datums.

The three datum plane reference frame is ideal for ensurance of common drawing interpretation. For noncylindrical parts the manufacturing/inspection fixture shown in Figure 6.9 can be constructed to ensure uniformity during manufacturing and inspection operations. All related measurements of the part originate from the fixture datum planes. For cylindrical parts the three plane reference frame is more difficult to visualize. The primary datum is often described as a flat surface perpendicular to the axis of the cylindrical datum feature, as shown in Figure 6.10. This axis can be defined as the intersection of two planes 90 degrees to each other, at the midpoint of the cylindrical feature. The tertiary datum is used if rotational orientation of the cylindrical feature is required because of the interrelationship of radially located features. The tertiary datum is often a locating hole, slot, or pin. If no angular located features are involved in the hardware requirements, the tertiary datum is omitted.

There are two types of datum features—datums of size and nonsize datums. Nonsize datums are established from surfaces. A datum surface has no

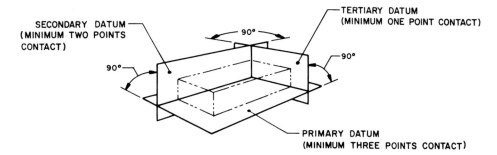

Figure 6.9 Three-plane reference frame for noncylindrical features.

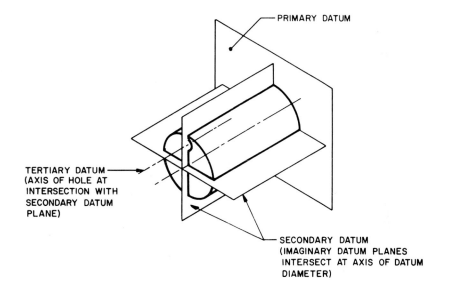

Figure 6.10 Three-plane reference frame for cylindrical features.

size tolerance because it is a plane from which dimensions or relationships originate. Datums of size are established from features that have size tolerance, such as holes, outside diameter, and slot widths. The centerplane or axis of datum feature of size is the actual datum. For example, a 0.250 ± 0.005 hole specified as a datum would be a datum feature of size (± 0.005 tolerance on feature size). The centerline or axis of the manufactured hole is the actual datum.

Principle 7: Position Tolerance

A position tolerance is the total permissible variation in the location of a feature about its exact position. For cylindrical features, such as holes and outside diameters, the position tolerance is the diameter of the tolerance zone within which the axis of the feature must lie. The center of the tolerance zone is located at the exact position. For other than round features, such as slots and tabs, the position tolerance is the total width of the tolerance zone within which the center plane of the feature must lie. The center of the tolerance zone is located at the exact location. Position tolerance zones are three-dimensional and apply to the thickness of the part. As shown in Figure 6.11, a diameter position tolerance zone is really a cylindrical tolerance zone through the part within which the feature axis must lie. This cylindrical tolerance zone is 90 degrees basic from the specified datum reference surface. Note that the tolerance zone will also control perpendicularity of the feature within the position tolerance requirement.

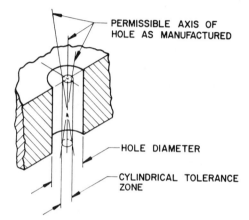

PERMISSIBLE AXIS OF
HOLE AS MANUFACTURED

HOLE DIAMETER

CYLINDRICAL TOLERANCE
ZONE

Figure 6.11   Position: cylindrical tolerance zone.

Composite position tolerancing provides a method of allowing a more liberal locational tolerance for the pattern of feature as a group, and then controlling the feature-to-feature interrelationship to a finer requirement. Each horizontal entry of the feature control frame shown in Figure 6.12 constitutes a separate inspection operation. The upper entry specifies the positional requirements for the pattern as a group, and the lower entry specifies the position tolerance within the pattern.

⊕	Ø .060 Ⓜ	A	B	C
	Ø .028 Ⓜ	A		

Figure 6.12   Composite feature control frame.

Principle 8: Screw Thread Specification

Where geometric tolerancing is expressed for the control of a screw thread or where a screw thread is specified as a datum reference, the application shall be applied to the pitch diameter. If design requirements necessitate an exception to this rule, the notation MINOR DIA or MAJOR DIA shall be shown beneath the feature control frame or datum reference as applicable.

Principle 9: Virtual Condition and Size

Virtual condition is the worst possible assembly condition of mating parts resulting from the collective effects of size and the geometric tolerancing specified to control the feature (see Figure 6.13). Virtual condition is primarily a tool used by product and tool/gage designers to calculate basic gage element size or to perform tolerance analysis to ensure assembly of mating parts. The following formulas can be used to determine virtual conditions:

External features = MMC size + tolerance of form attitude
or location

Internal features = MMC size − tolerance of form, attitude
or location

    The virtual condition is calculated from information specified on the engineering drawing (such as size and geometric tolerancing), but virtual size is determined by the parts inspector from the actual measured size and location, or attitude configuration of each feature. The virtual size of features must be considered for accurate GIA verification (see page 332 for additional discussions).

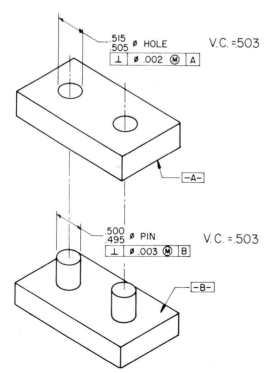

Figure 6.13 Virtual condition of mating parts.

GRAPHICAL INSPECTION ANALYSIS PROCEDURE

Example 1a

The performance of graphical inspection analysis (GIA) is a relatively simple, four-step procedure. The first step of the process requires plotting the basic location of the feature to be inspected in relationship to the datum features as described by the engineering drawing. This plot is called the *data graph* and is drawn on 10 × 10 graph paper at any desirable scale; $8\frac{1}{2}$ × 11 paper can be used for simple parts, whereas larger sheets will be required for complex parts. We will use the part shown in Figure 6.14 as an introductory explanation of a GIA procedure.

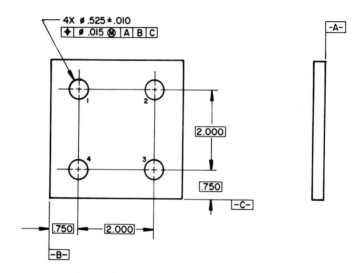

Figure 6.14 Example 1*a* part.

Figure 6.15 shows the *data graph* for Example 1*a*. The basic feature locations are plotted on the graph in relationship to the datum features at any convenient scale. We shall refer to this scale as the *configuration scale*.

Step 2 of the analysis requires plotting the locations of the actual inspected features on the data graph in relationship to the basic plot. These data are established by the parts inspector through surface plate inspection techniques as shown in Figure 6.16, or with the coordinate measurement machine as shown in Figure 6.17. The data are recorded on the appropriate inspection report form (Table 6.1) to become part of the inspection report.

To establish accurate data, the part shown in Figure 6.14 must be fixtured for surface plate inspection as specified in the feature control frame. When setup for inspection, the part must contact the angle plate at a minimum of three points for

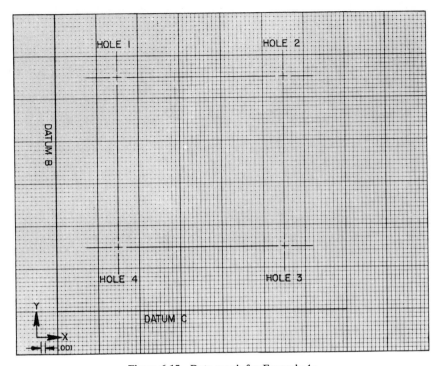

Figure 6.15 Data graph for Example 1*a*.

Figure 6.16 Surface plate inspection setup.

Figure 6.17   Coordinate measuring machine inspection setup.

the primary datum (datum A) and a minimum of two points of contact on the surface plate for the secondary datum (datum B) as shown in Figure 6.18.

The tertiary datum surface (datum C) is established by placing the second angle plate against the original plate as shown in Figure 6.19. Since the part being inspected has a symmetrical configuration, the machinist must mark the surfaces that he used to establish datums A, B, and C for the placement of the holes. This will ensure consistency between manufacturing and quality control. Notice that the orientation of the two angle plates and the surface plate has created the three-plane reference frame shown in Figure 6.19. The part will be clamped to the angle plate to restrict its movement when performing inspection measurements. Snug-fitting gage pins are placed in each hole, and a measurement is established from the surface plate to the top of the pin as shown in Figure 6.20. One-half of the gage pin diameter is subtracted from the measured value to establish the dimension from datum surface B (surface plate) to the center of the hole. Measurements are recorded for each hole in this setup. The angle plate is rotated 90 degrees

TABLE 6.1   INSPECTION DATA FOR EXAMPLE 1*A*

FEATURE NUMBER	FEATURE LOCATION						FEATURE SIZE				POSITION TOLERANCE		
	X AXIS			Y AXIS						MATERIAL CONDITION			
	SPECIFIED	ACTUAL	DEVIATION	SPECIFIED	ACTUAL	DEVIATION	SPECIFIED	ACTUAL	DEVIATION		SPECIFIED	BONUS	TOTAL
1	.750	.754	.004	2.750	2.751	.001	.515	.531	.016	MMC	.015	.016	.031
2	2.750	2.760	.010	2.750	2.750	.000	.515	.533	.018	MMC	.015	.018	.033
3	2.750	2.762	.012	.750	.748	.002	.515	.535	.020	MMC	.015	.020	.035
4	.750	.753	.003	.750	.749	.001	.515	.530	.015	MMC	.015	.015	.030

Figure 6.18   Surface plate setup (primary, secondary datums).

(without loosening clamps) to establish similar measurements from datum C as shown in Figure 6.16.

A three-plane reference frame must also be employed when using the coordinate measurement machine to verify hole locations of the part shown in Figure 6.14. The tabletop establishes the primary datum (Figure 6.21a) and the two

Figure 6.19   Surface plate inspection setup.

Figure 6.20   Surface plate inspection from datum B.

parallels provide contact surfaces for the secondary (Figure 6.21*b*) and tertiary datums (Figure 6.21*c*).

The appropriate measurement probe is placed in the coordinate measurement machine and calibrated to a zero reading at the intersection of the parallels that establish datums B and C. The probe is placed in each hole of the part to verify the actual measurements from datums B and C as shown in Figure 6.17. These data are recorded on the inspection report form.

Step 2 of the GIA process, for Example 1*a*, can be completed with the data tabulated in Table 6.1. The feature location deviations from the basic specified location will be in the order of magnitude of thousandths of an inch. We will let each grid square of the graph equal 0.001 in. to allow sufficient accuracy for the gaging operations. We will refer to this scale as the *deviation scale*. The location deviations are plotted for each feature as shown in Figure 6.22.

In step 3, a transparent *tolerance zone overlay gage* will be generated (see Figure 6.23) by placing a transparent material over the data graph. With a compass, the allowable position tolerance zone will be drawn at each of the basic locations for the specified tolerance zone plus any bonus tolerance. The bonus tolerance data have been recorded in the feature size deviation block of the inspection report shown in Table 6.1. The tolerance zones will be generated at the same scale (deviation scale) used to plot hole locational deviations in step 2.

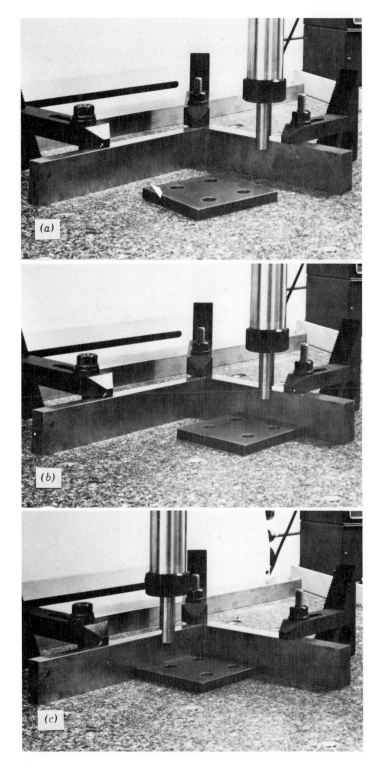

Figure 6.21 Setting up the three-plane reference frame on a coordinate measuring machine.

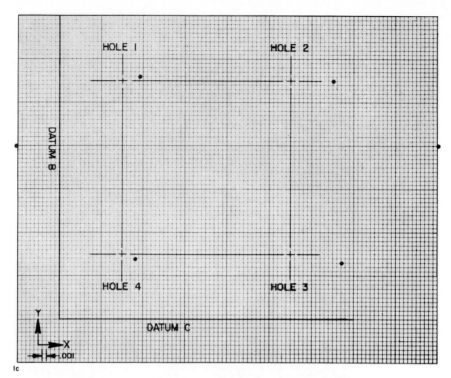

Figure 6.22 Data graph for Example 1a. Locations are plotted.

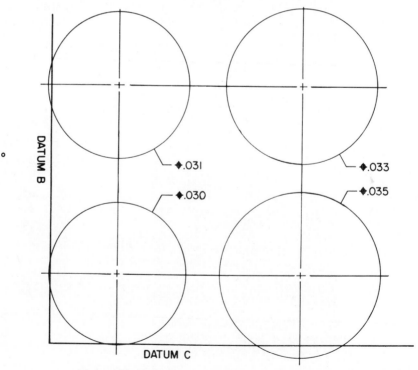

Figure 6.23 Example 1a tolerance zone overlay gage.

320

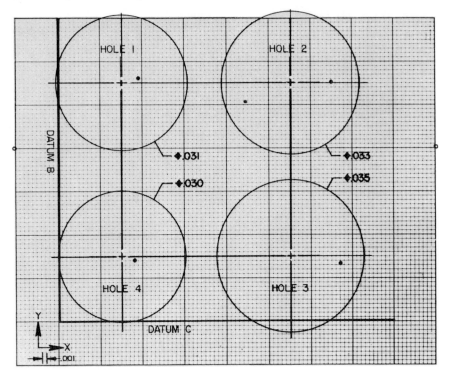

Figure 6.24 GIA of Example 1a.

In step 4, the transparent tolerance zone overlay gage is superimposed over the data graph and rotated and/or translated to provide alignment with the datum feature. We have an acceptable part if the actual feature axis of the four holes falls within the allowable tolerance zones at proper setting of the overlay (shown in Figure 6.24).

The part drawing for Example 1a is well done, in that datum surfaces have been specified, and the feature control frame has indicated a datum setup precedence. Both manufacturing and inspection will locate the part with minimum three-point contact of surface A, minimum two-point contact of surface B, and minimum one-point contact of surface C. The manufacturer marked the surface he used for datums A, B, and C in his drilling operations to ensure consistency with inspection. The MMC modifier associated with the 0.015 position tolerance zone in the feature control frame has alerted manufacturing and inspection of the bonus tolerance possibilities as the hole size departs from 0.515 diameter. In this example the manufacturer used the largest standard drill available to take advantage of this additional production tolerance.

Inspection data for a number of parts can be plotted on the same data graph as shown in Figure 6.25. Notice that the tolerance zone overlay gage includes

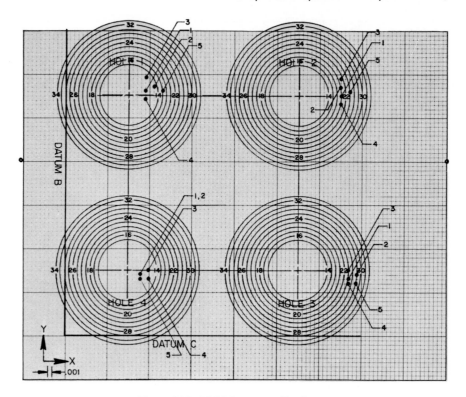

Figure 6.25 Multiple-part verification.

numerous tolerance zones to allow evaluation of the features as they depart from MMC. Each feature would be checked for conformance at its allowable position tolerance.

Another technique of graphical inspection which employs common axis data plotting and a tolerance zone overlay of prepared concentric circles can be used for feature location verification of simple parts as shown in Example 1a. However, this technique does not lend itself to the evaluation of parts with secondary and tertiary datums of size or where angular orientation is concerned. Therefore, we will only demonstrate the graphical inspection technique where data plotting resembles the geometry of the part.

Example 1b

In Example 1b, Figure 6.26 is the composite position tolerancing method of dimensioning. Example 1b does not combine the coordinate dimensioning system with the position tolerancing system. The 0.750 dimensions are specified basic.

Each entry of the composite feature control frame must be verified separately. This is an expensive gaging operation when using functional receiver

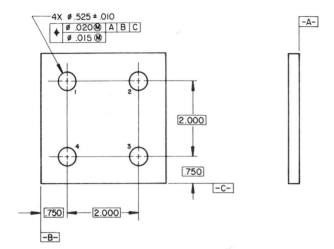

Figure 6.26 Example 1*b* print.

gages, since separate gages must be designed and manufactured for each entry. The upper entry has specified selected datum features and order of precedence for common interpretation. The lower entry is independent of any datum location requirement. The four-step GIA verification process can be performed with the inspection data shown in Table 6.2.

Steps 1 and 2 require plotting the actual locations of the feature in relationship to their basic locations (Figure 6.27). In step 3, tolerance zone overlay gages are prepared for both the upper and lower entry of the composite position feature control frame.

Notice the presence of datum feature lines in the upper tolerance zone overlay gage (Figure 6.28), which allow overlay alignment with the specified datum surfaces on the *data graph* as specified by the feature control frame. The lower tolerance zone overlay gage (Figure 6.29) is not datum related and is, therefore, allowed to float at will to establish hole-to-hole locational requirements. Step 4, Figure 6.30, requires the plotted hole located to be within the

TABLE 6.2 INSPECTION DATA FOR EXAMPLE 1*B*

FEATURE NUMBER	FEATURE LOCATION						FEATURE SIZE			POSITION TOLERANCE			
	X AXIS			Y AXIS						MATERIAL CONDITION	SPECIFIED	BONUS	TOTAL
	SPECIFIED	ACTUAL	DEVIATION	SPECIFIED	ACTUAL	DEVIATION	SPECIFIED	ACTUAL	DEVIATION				
1	.750	.754	.004	2.750	2.757	.007	.515	.515	.000	MMC	.020	.000	.020
1										MMC	.015	.000	.015
2	2.750	2.755	.005	2.750	2.743	.007	.515	.516	.001	MMC	.020	.001	.021
2										MMC	.015	.001	.016
3	2.750	2.745	.005	.750	.743	.007	.515	.516	.001	MMC	.020	.001	.021
3										MMC	.015	.001	.016
4	.750	.746	.004	.750	.757	.007	.515	.519	.004	MMC	.020	.004	.024
4										MMC	.015	.004	.019

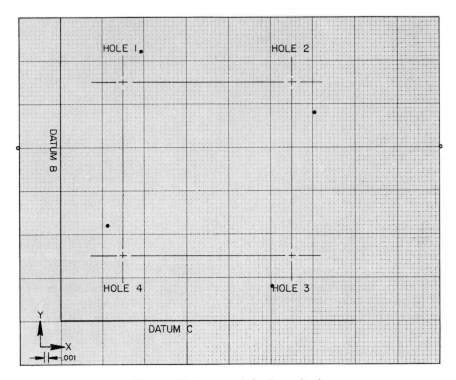

Figure 6.27   Data graph for Example 1*b*.

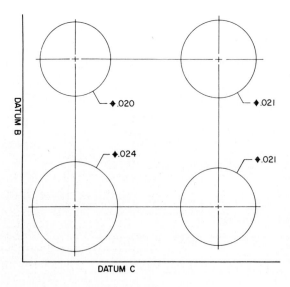

Figure 6.28   Upper tolerance zone overlay gage.

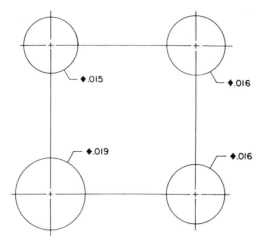

Figure 6.29 Lower tolerance zone overlay gage.

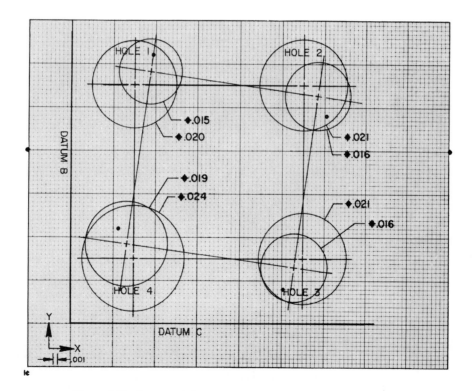

Figure 6.30 Data graph for Example 1*b*/tolerance zone overlay gage.

tolerance zones of both the upper and lower tolerance zone overlay gages in order to meet drawing requirements. This part is acceptable.

WORK PROBLEMS

Example 2 is provided so that you can experience in the concepts of graphical inspection analysis. Refer to the part drawing, inspection data, and pictures of the inspection process for the information required to generate data graphs and tolerance zone overlay gages (a completed example is provided).

Example 2

Example 2 provides a part with radially located holes related to a datum of size as shown in Figure 6.31. The inspection procedure shown in Figure 6.32 was used to generate the inspection data shown in Table 6.3.

Steps 1 and 2 of the GIA procedure require the generation of the data graph and the incorporation of the inspection data (see Figure 6.33). Notice from the inspection data that hole 1 has been used by the inspector to establish a starting point. Step 3 requires generation of the tolerance zone overlay gage for the holes (Figure 6.34) and a tolerance zone overlay gage (not shown) for any tolerance gained as the datum feature of size departs from MMC size.

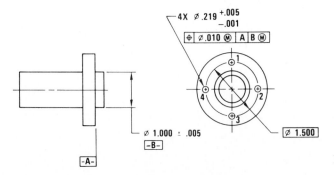

Figure 6.31   Example 2 print.

TABLE 6.3   EXAMPLE 2 INSPECTION DATA

FEATURE NUMBER	FEATURE LOCATION						FEATURE SIZE			POSITION TOLERANCE			
	X AXIS			Y AXIS						MATERIAL CONDITION			
	SPECIFIED	ACTUAL	DEVIATION	SPECIFIED	ACTUAL	DEVIATION	SPECIFIED	ACTUAL	DEVIATION		SPECIFIED	BONUS	TOTAL
1	.000	.000	.000	.750	.754	.004	.218	.219	.001	MMC	.010	.001	.011
2	.750	.743	.007	.000	+.007	.007	.218	.220	.002	MMC	.010	.002	.012
3	.000	-.006	.006	.750	.745	.005	.218	.219	.001	MMC	.010	.001	.011
4	.750	.757	.007	.000	+.007	.007	.218	.219	.001	MMC	.010	.001	.011
DATUM B	—	—	—	—	—	1.005	.995	.010	MMC	—	—	—

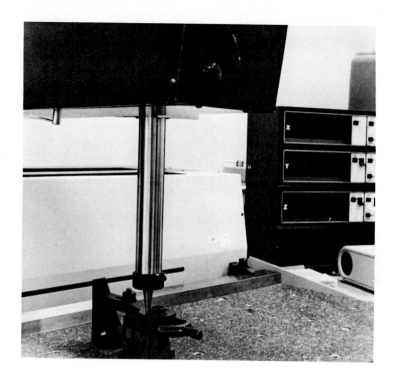

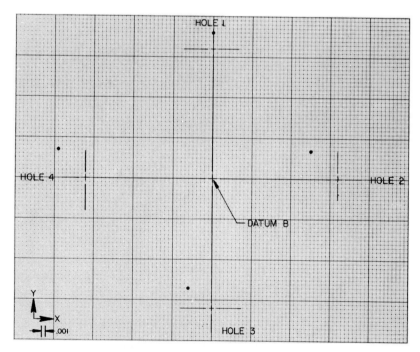

Figure 6.33 Data graph for Example 2.

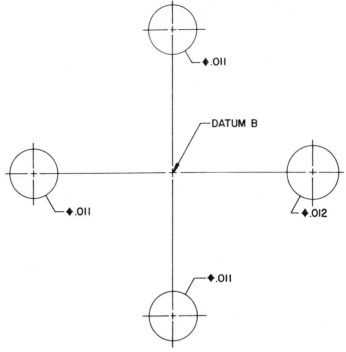

Figure 6.34 Example 2 tolerance zone overlay gage.

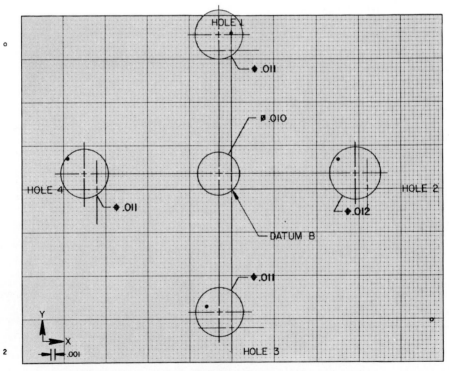

Figure 6.35 Example 2 data graph/tolerance zone overlay gage.

Step 4 shows the hole tolerance zone overlay gage rotated and translated within the added tolerance, resulting from datum B departure from MMC in an attempt to accept the hole locations (Figure 6.35). This part would not be accepted if verified by a technique that was incapable of evaluating the additional tolerance allowed as datum B departs from MMC.

GRAPHICAL INSPECTION ANALYSIS ADVANTAGES

There are many advantages to the graphical inspection analysis concept for hardware verification. A partial list includes the following.

1. *Functional acceptance.* Most hardware is designed to provide interchangeability of parts. As machined features depart from their MMC size, location tolerance of the features can be increased while maintaining functional interchangeability. Graphical inspection analysis techniques evaluate these added functional tolerances in the acceptance process.

2. *Reduced cost and time.* The high cost and lead time required for the design and manufacture of a functional gage can be eliminated with the GIA technique. An immediate, inexpensive, functional inspection can be conducted by the parts inspector at his or her workstation.

3. *No gage tolerance or wear allowance.* Functional gage design allows 10 percent of the part tolerance to be used for gage manufacturing tolerance. Often, an additional wear allowance (of up to 5 percent) will be designed into the functional gage. This could allow up to 15 percent of the parts tolerance to be assigned to the functional gage. GIA techniques do not require any portion of the product tolerance to be assigned to the verification process. The GIA concept does not require a wear allowance since there is no wear.

4. *Allows functional verification of MMC, RFS, and LMC.* Functional gages are designed primarily to verify hardware designed with the MMC concept applied. In most instances, it is not practical to design functional gages to verify hardware designed with RFS or LMC tolerance requirements. With GIA techniques all three material modifiers can be verified with equal ease.

5. *Allows verification of any shape tolerance zone.* Virtually any shape tolerance zone (round, square, rectangular, etc.) can be easily constructed with graphical verification methods. On the other hand, hardened steel functional gaging elements of nonconventional configurations are difficult and expensive to produce.

6. *Visual record of tooling capabilities and problems.* Capabilities of machines can be evaluated through the GIA inspection data. For example, the data graph for the five hole patterns shown in Figure 6.25 would indicate a machine problem in the X axis. There appears to be a 0.004-per-inch offset in the one axis. Tooling wear and misalignment can also be detected during the production operation with periodic GIA verification of parts.

7. *Visual record for material review board.* Material review board (MRB) meetings are postmortems that examine rejected parts. Decisions on the disposition of nonconforming hardware usually are influenced by engineering rank and perseverance rather than engineering information and evaluation techniques. GIA can provide a visual record of verification methods that can be evaluated with tolerance zone overlay to ensure part function.

8. *Minimum storage required.* The inventory and storage of functional gages can be a problem. Functional gages can also rust and corrode if not properly stored. The GIA graphs and overlays can be stored in drawers similar to drawings.

Accuracy

The overall accuracy of graphical inspection analysis is affected by such factors as the accuracy of the graphs and layouts, the accuracy of the inspection data, the completeness of the inspection process, and the ability of the drawing to provide common drawing interpretations.

An error equal to the difference in the coefficient of expansion of the materials used to generate the data graph and tolerance zone overlay gages may be encountered if the same materials are not used throughout. Papers will also expand with an increase in humidity and should be avoided. Mylar is a relatively stable material and, when used for both the data graph and tolerance zone overlay gage, the expansion/contraction error will be nullified.

The GIA procedure outlined in this chapter requires the generation of the tolerance zone overlay gage using the grid of the data graph to determine the proper zone size. Error in the grid printing accuracy will be canceled with this procedure.

Layout of the data graph and tolerance zone overlay gages will allow approximately a 0.010 error in the positioning of lines. This error is minimized by the scaling factor selected for the data graph. If a 10×10 to the inch grid is used, with each grid representing 0.001 in., a scale factor of 100-to-1 will be provided. With the following formula we can calculate the actual error from line positioning:

$$\frac{\text{line position error}}{\text{scale factor}} = \text{actual error}$$

The 0.010 assumed line position error will equal an actual error of 0.001 in. This error can be minimized by consistently working to one side of the line or by increasing the scale factor.

A certain amount of error is inherent in all inspection measurements. The error factor is dependent on the quality of the inspection equipment, facility, and inspection personnel. Most inspection operations will produce an error of less than 5 percent.

Accuracy of graphical inspection analysis operations is dependent on the quality of the inspection report. Inspection reports must contain adequate infor-

Figure 6.36 Inspecting perpendicularity.

mation to ensure common understanding and provide complete variables data. Drawings that specify properly selected datums and provide datum precedence in the feature control frame, will allow common drawing interpretations. When datum features and datum precedence must be selected by the inspector because of the incompleteness of the drawing, the datums and setup procedure must be included in the inspection report.

Complete inspection data must also be provided to ensure GIA accuracy. For example, the inspection data included in this chapter have disregarded the perpendicularity condition of the inspected features. As discussed in Principle 7, position tolerance controls the location of feature through the part's thickness. Figure 6.36 shows the inspector inspecting both sides of the feature axis. This information would allow the establishment of both ends of the axis of the data graph for each hole axis and joined by a line to indicate that they are the axis for a single hole (Figure 6.37). This procedure creates the effect of a three-dimensional gage at virtual condition (see Principle 9).

Figure 6.37 shows a data graph, for the part shown in Figure 6.14, where the location of the features are plotted at their virtual size (see the inspection data in Table 6.4). The complete axis of each hole must be within the tolerance zone overlay gage to be an acceptable part, as shown in Figure 6.38. This method of

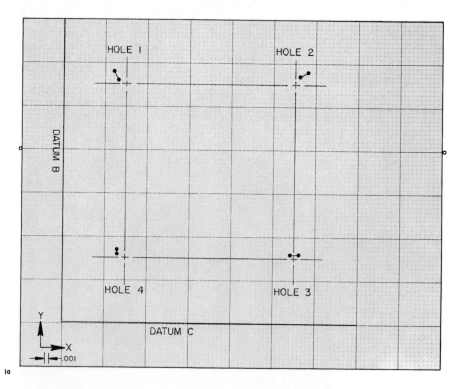

Figure 6.37   Three-dimensional hole verification (two coordinates and perpendicularity).

TABLE 6.4   VIRTUAL SIZE DATA FOR EXAMPLE 1*A*

Inspection Data

Feature Number	Feature Location						Feature Size			Position Tolerance			
	X Axis			Y Axis						Material Condition	Specified	Bores	Total
	Specified	Actual	Deviation	Specified	Actual	Deviation	Specified	Actual	Deviation				
1	.750	.748	.002	2.750	2.751	.001	.515	.516	.001	MMC	.015	.001	.016
		.747	.003		2.753	.003							
2	2.750	2.751	.001	2.750	2.752	.002	.515	.516	.001	MMC	.015	.001	.016
		2.753	.003		2.753	.003							
3	2.750	2.749	.001	.750	.751	.001	.515	.517	.002	MMC	.015	.002	.017
		2.751	.001		.751	.001							
4	.750	.748	.002	.750	.751	.001	.515	.516	.001	MMC	.015	.001	.016
		.748	.002		.752	.002							

verification is imperative if accurate, meaningful results are to be achieved with graphical inspection analysis.

    A datum feature of size that is interrelated by a separate tolerance of attitude or form and is referenced within the same feature control frame, must also be evaluated at its virtual size (often referred to as General Rule 5). The examples

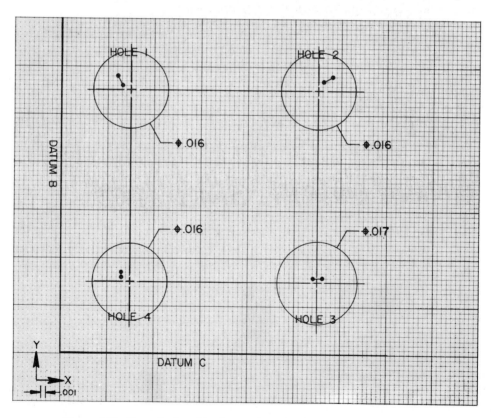

Figure 6.38   Example 1*a* with virtual size data graph/tolerance zone overlay gage.

shown previously for parts that referenced datums of size did not depict complete part definition, since the attitude requirements of the datums feature of size have been omitted. When we added this additional control required for drawing completeness, we must consider the virtual size of the datum feature when evaluating the allowable additional tolerance that can be gained as the datum feature departs from MMC. The parts inspector will determine the datum feature virtual size (by measurement) and subtract this value from the datum feature virtual condition (calculated from the drawing specifications) to determine the additional tolerance allowance.

Computer Inspection Analysis

Computer programs can be written to process the X and Y coordinate measurements and tolerance from inspection data and provide the required rotation/translation of the data to determine functional acceptance of parts. The computer can print out data indicating acceptance or rejection, or produce the data in graphical form, resembling manually prepared graphical inspection analysis. Recommended procedures can be provided for rejected parts to allow rework and ultimate part acceptance.

REVIEW QUESTIONS

1. A basic dimension is a theoretically exact dimension with no _____.
2. A tolerance stated at MMC applies only when the feature controlled is at _____ size.
3. (True or False) Bonus tolerance comes from controlled features, and additional tolerance comes from size datums.
4. The plot of basic locations of features related to the datum features in GIA is called a _____ graph.
5. The scale used for the GIA technique is referred to as the _____ scale.
6. (True or False) The determination of whether features are out of position tolerance is based on the amount the location departs from basic dimensions.
7. The GIA technique superimposes a tolerance zone _____ gage for measurement.
8. Name two advantages of GIA over hard functional gages.
9. The recommended material to use for the overlay gage is _____.
10. (True or False) An inspection report (variables) is needed in order to perform the GIA technique.
11. Another name for GIA is _____ gaging.
12. The authority standard for position tolerancing is ANSI _____.

7

Quality Costs

One of the methods used to improve quality and provide effective management is an accurate system of reporting quality costs. If a company can accurately measure the costs of quality, it opens several doors to improving quality and productivity. A quality cost system defines the areas of high costs, which, in turn, define areas of concentration for corrective action to reduce those costs.

CATEGORIES OF QUALITY COSTS

There are four categories in which quality costs are reported:

1. Prevention.
2. Appraisal.
3. Internal failure.
4. External failure.

Prevention

These costs are related to any function in a company that attempts to *prevent* poor quality from being produced. Some examples are:

- Quality engineering tasks (planning).
- Manufacturing engineering tasks (work instructions).
- Design engineering tasks (design reviews).

Appraisal

These costs are related to functions that appraise (or evaluate). Inspection and testing are two examples.

Internal Failure

These costs are related to failures (nonconformance) that occur *in-house*. Two examples are scrap and rework.

External Failure

These costs are related to failures (nonconformance) *in the field* (or at the customer's facility). Some examples are:

- Returned products from the customer.
- Handling customer complaints.
- Returned goods that are scrap.
- Returned goods that must be reworked.

The previous examples are only a few quality cost examples. Most costs will fit one of the four categories; however, some of them may be difficult to categorize because they could fit either one or another. The important thing to remember is that these costs should be reported and corrective action taken to reduce them.

Examples of Quality Costs

Appraisal costs include the costs of hiring full-time inspectors, inspection supervisors, inspection clerks, and people performing tests. They also include the costs of maintaining materials and equipment used for testing the product, preparing and handling test or inspection records, and performing product audits and field tests.

Internal failure costs include the direct and associated costs incurred due to scrap, rework, and repairs. The direct cost of scrap, rework, and repair is the loss of the product. The associated costs of scrap, rework, and repair (often called the "hidden factory") are the costs of handling, paperwork, replacing, material review board functions, and loss in profits due to reduced prices for substandard products.

External failure costs include the cost of repairing or replacing products returned by the customer, handling customer complaints, recalls for repair or replacement parts during warranty periods, and the costs of testing, legal services, settlements, and other costs associated with product liability problems. External failures also come with "hidden factory" costs, such as processing paperwork, corrective action teams, and other actions that were originated directly because of returned materials. Other costs are not directly measurable, such as customer dissatisfaction (or the impact on the company's business due to customer dissatisfaction).

Prevention costs include the costs of quality planning, process control, quality training, designing equipment and processes to measure and control quality, special studies, vendor relations, variability analyses, design reviews, manufacturing planning, and other activities that are performed by anyone in the company for the purpose of preventing defects.

QUALITY COST RELATIONSHIPS

When a company first establishes a quality cost monitoring system, they are often startled by the following relationships and proportions of quality cost expenditures:

1. The total quality cost expenditures are usually a large proportion of sales, profits, or other selected cost denominators.
2. The amounts of internal failure costs (such as scrap, rework, and repair) and appraisal costs (such as inspection, test, and auditing) are the largest portion of the total (and sometimes equal to each other).
3. The amount of external failure costs (due to products delivered to the customer which are not to specifications) are fairly low because of the mass amount of inspection and testing being performed. In general, the more you inspect, the more defects you find. The more defects you find, the more you inspect.
4. The amount of prevention cost expenditures is also a small portion of the total.

More on Prevention

Prevention is a vital part of quality costs, mainly because it should be the major area of investment by a company. The reason for this is that prevention activities help stop poor quality *before* it occurs. Any action taken after poor quality occurs is corrective action. Corrective action is necessary to improve poor quality, but preventive action saves money by preventing poor quality. Prevention is the one category where increased investments will reduce the other categories and the total costs of quality.

Preventive action

This action costs money because extra time and effort are needed to plan for good quality.

Corrective action

This action costs more than prevention because you not only have the costs of investigating and correcting, but you also have the costs of the defective products (if they were scrap, they must be replaced, and if they were able to be reworked, it costs you to rework them).

Remember. Quality costs are not only the costs of the quality assurance department, they are the costs of all departments in a company that have an influence on product quality.

BASIC UNDERSTANDING OF PARETO'S LAW

One important subject to understand when attempting to solve problems is Pareto's law. The basic concept of Pareto's law is that in almost all cases, 80 percent of the problems are caused by 20 percent of the people, machines, and other factors. Pareto's law says that you should concentrate on the major problems immediately and work on the minor problems later. This is often referred to as "The vital few over the trivial many." When setting up a quality costs reporting system, one of the most important parts is being able to apply Pareto's law to the reports in such a way that you are concentrating on the vital few.

Example

You collect several scrap costs (internal failure) and these are shown below.

Department	Quantity Scrapped	Cost per Unit
Welding	12	$ 3.00
Grinding	3	$ 25.00
Stamping	103	$ .30
Forging	1	$ 10.00
Heat treat	3	$100.00

Since you are reporting quality costs, one of the first things you must do is find the total cost for each department (the quantity scrapped times the cost per unit). They are shown here:

$$\text{Welding} = \$\ 36.00 \text{ scrap cost}$$

$$\text{Grinding} = \$\ 75.00 \text{ scrap cost}$$

$$\text{Stamping} = \$\ 30.90 \text{ scrap cost}$$

$$\text{Forging} = \$\ 10.00 \text{ scrap cost}$$

$$\text{Heat treat} = \$300.00 \text{ scrap cost}$$

Next, you report these costs in Pareto's form (the highest cost first):

Heat treat = $300.00 scrap cost

Grinding = $ 75.00 scrap cost

Welding = $ 36.00 scrap cost

Stamping = $ 30.90 scrap cost

Forging = $ 10.00 scrap cost

The Pareto list you just made shows that the highest quality cost area to the company is in the heat treat department. This scrap should be investigated first, and attempts made to correct the problem. Next, the grinding department should be investigated to correct the $75.00 scrap problem, and so on.

It would be a great mistake for a company to investigate the problem in the stamping department (where the scrap quantity is high, but the cost is only 30 cents each) while letting the high-cost problems in the heat treat department continue to occur.

If you are not on a quality cost reporting system, it would be a good idea to start one. Until then, you may use Pareto's law in other ways. One approach is considering quantity instead of costs.

Example

A company makes one product where overall scrap costs remain approximately the same. Here, you can use Pareto's law by the "quantities" of scrap. This can be done on companywide, departmentwide, or by operation.

Department	Quantity Scrapped	Quantity Produced
Grind	123	1,000
Heat treat	21	100,000
Stamping	1100	3,000

Pareto Analysis by Quantity	Pareto Analysis by Percent of Product[a]
Stamping = 1100	Stamping = 37%
Grind = 123	Grind = 12.3%
Heat treat = 21	Heat treat = 0.021%

[a] Quantity scrapped/quantity produced × 100.

In the above examples, no matter how you report it, the stamping department is at the top of the Pareto analysis and requires immediate attention to correct the problem(s).

Now, you see that understanding Pareto's law can be an asset to any company, or even at home to find out where most of your money is spent.

GRAPHS, CHARTS, AND QUALITY COSTS

There are many ways to report quality costs in meaningful relationships to each other or over time. Most of them have a lot to do with Pareto's law (the vital few versus the trivial many). On almost all kinds of reports, there are charts or graphs of some kind. These charts and graphs are used mainly to pictorially represent trends (the magnitude of the problem). Several different types of charts and graphs are used, but we will only look at three: line graphs, bar charts, and pie charts. These are the most widely used.

Line Graphs

Line graphs are usually drawn on square grid paper. They are used for many things, but for our purposes, we will use them for scrap reports. The graph in Figure 7.1, for example, shows that there was $1100 worth of scrap in week number 1, $1200 in week number 2, and so on.

Some line graphs serve a dual purpose. For example, Figure 7.2 shows a line graph of scrap cost and quantity of scrap (% of production) for each week. The graph in this figure shows that in week 2 there was $1200 scrap and it was 4 percent of production.

Bar Graphs

Bar graphs are also used in reports of various kinds. Two main styles are used in bar charts. One shows the magnitude of a single problem, and the other makes comparisons.

Bar graph (magnitude)

The graph in Figure 7.3 shows, for example, that there was $1300 worth of scrap in week 4.

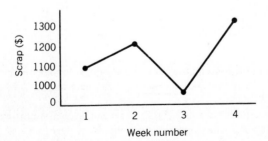

Figure 7.1 Line graph.

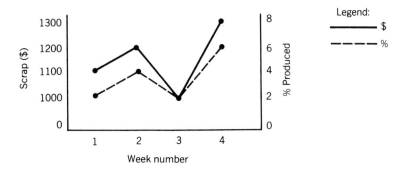

Figure 7.2 Dual-purpose line graph, including legend.

Bar graph (comparison)

The graph in Figure 7.4 shows the total quantity of scrap for each week and the machine that contributed most to that scrap (machine 201). The length of the bar surrounding the machine number indicates the magnitude (in this chart, the quantity of scrap).

Pie Charts

Pie charts get their name because they are shaped like a pie. They are very effective in showing Pareto problems. Two examples of the use of pie charts are shown in Figures 7.5 and 7.6.

Pie chart examples

Figure 7.5 is a pie chart that reflects the total company scrap costs by department. This shows that departments 21 and 19 contribute most of the scrap. (*D*

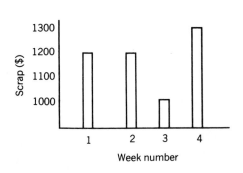

Figure 7.3 Bar graph.

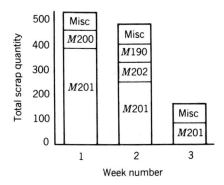

Figure 7.4 Comparison bar graph.

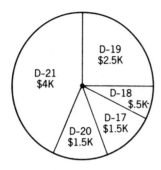

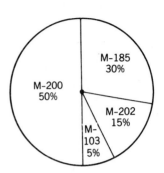

Figure 7.5 Pie chart.   **Figure 7.6** Pie chart.

means department and *K* means thousands of dollars.) Figure 7.6 is a pie chart of the percentages of total scrap by machine. This chart shows that if you attempt to solve problems at machines 200 and 185, most of your scrap will have been reduced.

Graphs and charts are very effective in showing trends and relationships in whatever you are reporting or comparing. They show you when things are getting worse or better, and they also help you recognize the most important thing—the *specific areas that need improvement.*

PREVENTION: THE RETURN ON QUALITY INVESTMENT (ROQI)

A primary consideration in any business has to do with the *return on investment* (ROI). Management must consider that if they invest in necessary quality programs, there will be a return on their investment and more. This return is not always immediate which is usually frowned upon.

One investment that management often discards is in the area of prevention. They do this because the returns are seldom immediate and because it is difficult to measure them. I call prevention techniques a *return on quality investment* (ROQI). This is where management makes a certain investment in the people, tools, systems, training, and other factors that will help prevent problems downstream. As stated earlier, an ROQI is not directly measurable because you do not know the problems that may have occurred if you did not take preventive measures. You can measure the effects of prevention, providing you have measured quality costs (scrap, rework, warranty, etc.) prior to taking preventive action. All you need to do is continue measuring the progress you are making in these areas, and you can say that some of this progress is due to prevention and some would have happened anyway.

It is just good business to use preventive action. Common sense tells you that if you prevent the occurrence of one problem one time, then you may have prevented its occurrence several times. So, what are some areas of prevention that could use more investment?

1. *Training*. Training is one of the most important areas in which to invest. People who are highly trained will still make some mistakes because it is human nature; however, people who are not trained will make even more frequent and costly mistakes.

2. *Process control*. Statistical process control (SPC) is another prevention technique. It is costly to implement but will be a major factor in quality improvement. With SPC, costly bad parts (and scrap and rework) can be prevented.

3. *Quality engineering*. This group of people shares a sole purpose that is to *plan* for *quality* and *analyze* the results. Planning always pays off because if quality is well planned for, things are usually done right the first time around.

4. *Calibration*. Equipment calibration (performed properly) is a major investment in people and tools that can be a turning point in quality improvement. Without proper calibration, inspection and test personnel cannot be certain that their measurements are accurate.

5. *Effective reporting of quality costs*. Reporting the various quality costs incurred during given time periods can detect areas in need of immediate improvement. Correcting these problem areas can be a factor in preventing future costs in these areas.

6. *Communication*. The improvement of communications between customers and vendors can be a major factor in quality improvement and also in the improvement of internal communications. Many problems are born simply because people cannot or do not communicate.

There are other prevention areas that exceed the purpose of this chapter, but some of the major areas have been discussed. Management must be aware of the options they have. These options are basically only two:

1. Take no action and leave quality to chance.
2. Take sound preventive action and have something to say about quality and productivity levels.

Preventive action is costly, but corrective action costs more in the final analysis.

REVIEW QUESTIONS

1. Name the one quality cost category in which more investment will reduce the other three categories and total costs of quality.

2. Handling nonconforming material in house is a _____ _____ quality cost.

3. Which of the following is not a quality cost?
   a. Inspection
   b. Process control
   c. Shipping costs
   d. Audits

4. (True or False) Quality costs are only dependent on the costs of the quality assurance department.

5. Quality cost reporting deals with all activities associated with finding or preventing

   _____.

6. Pareto analysis helps identify the vital few problems versus the _____ many.

7. The primary purpose for collecting and analyzing quality costs is the reduction of

   _____ quality costs.

8. Give one example of external failure costs.

9. Give one example of internal failure costs.

10. Give one example of prevention costs.

11. Give an example of appraisal costs.

8

Inspection Planning

Inspection planning is simply planning what you are going to inspect and how you are going to inspect it. Inspection plans can be written by anyone who understands the quality requirements of the product, depending on how many parts are produced and how complex they are. Below are guidelines for choosing the person best suited to do the planning.

TYPE OF JOB	PLANNING PREPARED BY:
A small job shop making five pieces at a time of relatively simple products.	The inspection foreman or the inspector himself.
A small job shop making five pieces at a time of several complex parts.	Usually, the inspection foreman (or supervisor).
A large plant mass-producing simple products.	Inspection foreman or quality engineer.
A large plant mass-producing complex component parts.	Inspection foreman or quality engineer.

There are a few things to be considered when planning for inspection. Some of these are

1. *What characteristics to inspect.* Not all characteristics need to be inspected every time.
2. *Sampling plans.* How many should you inspect?
3. *Tools to use.* It is better to plan to make sure that there is consistency from inspector to inspector in tools and methods used.

4. *The cost of inspection.* Too much inspection is costly in dollars, and too little inspection can be costly in poor outgoing quality.

5. *Work area design.* Inspectors can be more effective when the work area is designed to assist them. They need good lighting, comfortable temperatures, even material flow, well-planned workbenches, readily available tools, and specifications.

6. *Records.* How will inspectors report their findings in a manner that will be easy to review and track quality problems?

Inspection Activities

There are several key activities for inspection. Some of these are

1. *Receiving inspection.* When purchasing finished parts or castings from a vendor (or source).

2. *First piece inspection.* At the beginning of a production run. Defects could be very costly if the setup is bad.

3. *Process inspection.* Inspecting at certain key points during the process before the part goes to the next operation.

4. *Final inspection.* Inspection of parts just before they are sent to the customer, storage, or the assembly line.

5. *Assembly inspection.* Inspecting the assembly just before shipment to the customer, or the next level of assembly.

6. *Patrol inspection.* Where one inspector covers a group of stations each of which do not warrant a full-time inspector.

7. *Source inspection.* Where the inspector inspects parts at the vendor's facility.

Inspection Instructions

At any of the inspection activities above there must be instructions for the inspector on *what* or *how* to inspect. These instructions state clearly:

1. What to inspect.
2. Tools or method to use.
3. Sampling plan (how many).
4. Standards to use.
5. Records to keep.
6. What to do with acceptable and nonacceptable parts.

FLOWCHART USE

When planning for inspection, a good idea is to produce a flowchart for the product. Flowcharts are very useful in inspection planning because they can reveal places where inspection points should be placed and can also present the entire picture of what the part goes through during the process. You can thus evaluate areas that require inspection before going to the next operation. You might note turning operations that must leave stock for the grinding operation, for example.

An example of a flowchart is shown in Figure 8.1. Flowcharts should follow the part through every stage and can be easily made up by asking questions, following the part through the process, or looking at the sequence of operations on the operation sheets.

There are three inspection points already listed:

1. Receiving inspection.
2. Final inspection.
3. Assembly inspection.

Others (such as process inspection) can be added at important points in the process from *weld* through *deburring*.

Example

Improper heat treating should be detected before grinding is done (Inspection ④: heat treat inspection).

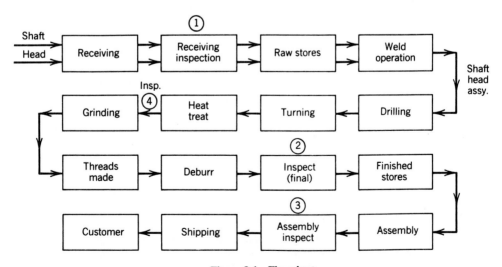

Figure 8.1 Flowchart.

HOW TO PLAN FOR INSPECTION

The principles of inspection planning apply regardless of the type of product or the quantity to be inspected. They are just as applicable to the truck manufacturer making 30 units per day as they are to a toy manufacturer making hundreds per day. The most important question is *why* inspection should be planned, and the answer is simple. It is more economical than not planning. The quality professional's job is to keep costs at a minimum. That is what the concept of quality costs is all about.

The expense of inspection planning is almost always more than offset by a decrease in failure costs (scrap, rework, etc.). In addition, good planning will significantly reduce the time spent in inspection and will allow this operation to be performed by a less skilled individual. The higher the volume, or the more times that a part is inspected, the greater the savings will be.

One example of such savings is a complex housing project that is molded by an outside supplier. Before the inspection was planned on this part, seven man-hours per receipt were spent inspecting it. As a result of 32 hours of quality engineering time spent planning for this item, the company:

1. Reduced the inspection time to three hours per receipt.
2. Could use a less skilled and, therefore, lower-paid inspector.
3. Reduced the number of rejections by providing the supplier with a copy of the inspection plan.
4. Virtually eliminated the acceptance of defective product.

The improvements saved the company thousands of dollars each year on this one product.

Industrial engineers have been planning manufacturing operations since the turn of the century when F. W. Taylor and the Gilbreths (Frank and Dr. Lillian) began developing methods engineering. We in the quality profession have been slow learners. We have been content to let each shop inspector decide what to inspect and how to inspect it. This has caused inconsistency in the characteristics checked and the methods used, not to mention gross inefficiencies and errors.

Once, some years ago, at one company an inspector worked for several hours to set up a large, complex, machined housing. He then began his inspection while marking on the blueprint those dimensions that he verified. Then he broke down the setup, turned the part over, made his new setup, and began checking dimensions on the alternate side. Then he went over the drawing to determine what dimensions he had missed.

That is when it got interesting! For if he found a characteristic he had forgotten and that required an additional two-hour setup to check, he had to make an important decision. Did he spend the time to do the inspection, or did he assume that the dimension was correct?

This dilemma was eliminated at this company by inspection planning. The planner decided not only what characteristics required inspection, but also how to sequence the inspection so that situations like the previous one could be avoided.

Another company did something similar for all in-process inspections as well as receiving inspection. Taking a page from the industrial engineer, they actually wrote methods sheets for the inspectors. Some of these were extremely detailed. One such set of instructions was a series of 10 photographs showing the method used to inspect assembled components. Under each photo were instructions detailing the characteristics to be inspected. This is an excellent example of documentation accomplished at a minimum of expense.

A side benefit of this type of inspection planning is using it as a training aid for new employees. They can readily see what is required and how to do the job. Not all inspection planning needs to be this detailed, but it *must* document the proper method of inspecting the product. In every case, these instructions must be available for use by the inspectors.

Another reason why it is important to plan inspection is the need for consistency. Suppose that a complaint has been received from the field about a specific defect that has been shipped. The accepted approach in industry is, of course, to take corrective action to assure that the mistake is not made again. Frequently, a portion of that corrective action is to change the inspection procedure. That information can be communicated in a variety of ways ranging from verbal direction to a written change (written changes are always preferred). At one company, the inspection procedure is revised, and the revised procedure is followed from that point forward.

Intelligent planning of inspection operations eliminated double inspections. If a characteristic is to be checked, it should be done as soon in the manufacturing sequence as possible. Waiting until the product is complete to detect an error that could have been caught earlier and cheaply corrected is an economic disaster. Proper planning allows a selection of the most economical point of inspection.

How to Plan

The first step in effective inspection planning is examining how the item is produced and used to establish proper quality levels. A typical method of doing this is to make a simplified block flow diagram of the primary steps in manufacturing the product, as shown in Figure 8.2. The example in the figure was done by a toy manufacturer for a toy car. It is excellent because of its simplicity.

Each block in Figure 8.2 represents a major manufacturing operation. The numbers indicate the acceptable quality level (AQL) of the product (left) for functional characteristics and (right) for cosmetic characteristics. As can be seen, the finished product has a looser AQL than its components because of the influence the components have on the quality of the finished assembly.

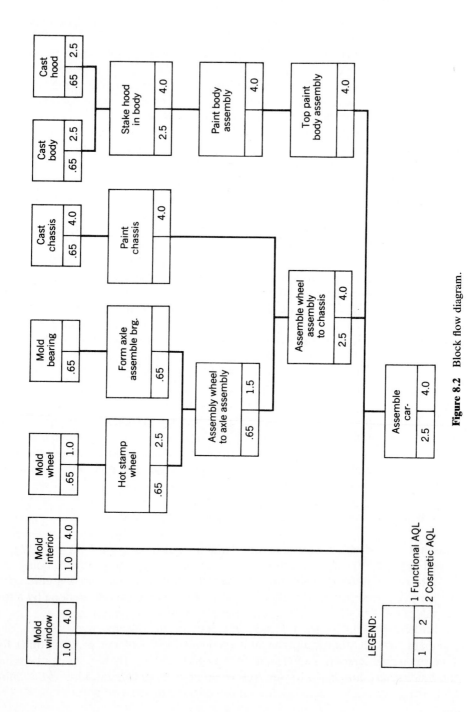

Figure 8.2 Block flow diagram.

In establishing the AQL, start with the finished item and determine from experience and knowledge what AQL is acceptable. Then work backward, keeping in mind what can be economically expected from purchased parts and manufacturing processes. Where conflict exits and it becomes obvious that the process cannot yield the quality required, manufacturing engineering must be advised so that screening or sorting activities can be included in the basic manufacturing process. (Sorting good from bad is a production responsibility, not a quality control responsibility.) The decision that sorting is required must be made early to assure that the expense is calculated in the cost of the product.

It is essential that separate AQLs be established for functional and cosmetic characteristics. Customers normally are less critical of defects of appearance than those relating to function. The extent to which this is true, however, is dependent upon the type of item being sold. Although both engineering and marketing must be consulted regarding the proper AQL, quality control is the group that must make the final decision.

Once AQLs have been established, it is necessary to determine what characteristics require inspection. This is best done by classifying the characteristics, normally using no more than four classifications:

1. Critical
2. Major
3. Minor
4. Incidental

The designation *critical* should be reserved for characteristics that, if discrepant, are likely to result in hazardous or unsafe conditions for the individuals using, maintaining, or depending on the product. This restrictive definition assures that the term *critical* is used only when mandatory controls are required. No matter how important a characteristic is to the operation of the end item, if it does not affect life or limb, it should not be called critical.

At most companies, major characteristics are, by definition, those that can cause defects likely to prevent operation of the product or to cause the customer to return the product before the next scheduled maintenance period (materially reduce the usability of the product).

Minor characteristics can cause defects that are likely to result in a customer complaint or a request for service during routine maintenance periods rather than a request for immediate service. The defect is not likely to reduce the usability of the product. Such defects may cause additional labor during manufacturing but will not be severe enough to make the parts unusable.

Incidental characteristics are all other characteristics. Other companies use similar logic to establish their definitions; the important thing is that the definitions be clearly written.

Classification of Characteristics

Once characteristics are defined, the easiest method of classifying them is again to start at the end item and work back to the individual parts. As this is done, actual classifications can be marked adjacent to characteristics on the engineering drawings. An obvious but frequently overlooked necessity is that the person doing the classifying understands the application of the product and the definitions of classification of characteristics. A closely toleranced dimension is not automatically a major characteristic, nor does a wide tolerance always mean an incidental classification. In other words, *tolerances do not equal severity*.

Some companies have the design engineer do the classification. A good quality control engineer can review the designer's decisions to see if they are reasonable. With the classification completed, the inspection plan can be written. Generally, criticals are inspected 100 percent, majors are inspected to a stringent sampling plan (e.g., 1 percent AQL), minors and incidentals are inspected using a less stringent plan (e.g., 4 percent AQL).

Planning for Purchased Parts

Since buyers have little control over the production of items they purchase, inspection for these parts is designed as an after-the-fact process. An example is a company that uses substantial amounts of molded plastic parts which, for the most part, are made by outside sources using buyer-owned tooling.

In these instances the dimensions to be inspected are selected to show that the tool has been properly operated. Included is at least one major length and width dimension to determine shrinkage after molding. An across-the-parting-line dimension is selected to assure that the tool is properly closed and injection pressure is maintained. Any major characteristics not listed above is also verified. Parts are also inspected for flash, short shot, excessive gate, and sink.

Any of these characteristics can change over relatively short periods of time because of the lack of control at the supplier. This concept of inspection is predicated upon inspecting the tooling first (known as a first article inspection). From then on, it is assumed that if the tool is run properly (and this is verified by checking the characteristics previously mentioned) the parts will conform.

Other inspections are also planned into the system. Each month during the time that the tool is operated, parts are subject to more thorough inspection. If the tool is pulled for any reason, an inspection is done. This is done not only to verify quality but also to allow corrections to be made while the tool is not needed for production. After tool rework, the tool is inspected for the reworked area. The same philosophy is used in planning parts produced on punch presses and on numerically controlled tools.

In the case of complex, nontooling-produced parts, the quality control engineer is faced with a trade-off between the expense of checking all characteristics and the risk of accepting a nonconforming condition. In general, the more important the characteristic, the more logical the inspection is. Other important considerations include how likely it is that the defect will be detected during assembly and, if found, what the rework/repair or replacement cost would be. If the operator can tell quickly during the assembly that the part is defective, there is less of a problem than if the assembly would have to be completed and tested before the defect is found. The quality control engineer is responsible for making this decision.

Inspection plans should be written for all purchased items regardless of complexity. Often, a single plan can be written for a family of products and then modified to fit each product in that family. This type of plan relies heavily on *approved samples* to display the cosmetic limits of acceptance. These samples often have to be developed as the situations develop. Each time a decision is made regarding the acceptance or rejection of product due to a cosmetic characteristic, a properly identified example of the condition should be added to that part's sample file.

Special sampling plans are frequently a part of the inspection plan for purchased parts. This is essential when a common lot size can be as large as 600,000 parts. Using single sampling, a sample size of 1250 pieces is required. If the part has many dimensions or has a low unit price, obviously some reductions must be made in the sampling.

One company's first choice was to go to double sampling; for this, it has a combined sampling plan that started with single sampling and switched to double sampling when the lot size reached a given quantity (see Figure 8.3). This type of plan technically violated some sampling principles, but since Mil-Std-105D provides the operating characteristic curve of single and double sampling, it is possible to determine the risks.

The company's plan saved money and was far superior to allowing the inspector to select his or her own sample size. It was also statistically more valid than using a fixed lot size regardless of sample size. Sampling plans should be detailed in the inspection plan. Quality engineering needs to analyze the economics of the sampling plan (the cost of inspection versus the risk of acceptance of defective product) and specify the best plan to use.

Planning for Manufactured Parts

To plan for manufactured parts, a page can be taken from the industrial engineer's handbook. A flowchart (shown in Figure 8.4) is prepared which shows the basic manufacturing steps involved to produce the product and includes every storage function but ignores all present inspection points.

To start planning the first important rule is to do each inspection as early in the manufacturing as possible. Rule 2 is to do any given inspection only once. Rule 3 is to have the inspection done the same way, regardless of the inspector, by

Lot size	Sample size	Cumulative sample size	Acceptable quality level—AQL					
			0.65 (Ac Re)	1.0 (Ac Re)	1.5 (Ac Re)	2.5 (Ac Re)	4.0 (Ac Re)	6.5 (Ac Re)
2-8	2		↓	↓	↓	↓	↓	0 1
9-15	3						0 1	↑
16-25	5					0 1	↑	↓
26-50	8				0 1	↑	↓	1 2
51-90	13			0 1	↑	↓	1 2	2 3
91-150	20		0 1	↑	↓	1 2	2 3	3 4
151-280	20 20	20 40	↑	↓	0 2 1 2	0 3 3 4	1 4 4 5	2 5 6 7
281-500	32 32	32 64	↓	0 2 1 2	0 3 3 4	1 4 4 5	2 5 6 7	3 7 8 9
501-1200	50 50	50 100	0 2 1 2	0 3 3 4	1 4 4 5	2 5 6 7	3 7 8 9	5 9 12 13
1201-3200	80 80	80 160	0 3 3 4	1 4 4 5	2 5 6 7	3 7 8 9	5 9 12 13	7 11 18 19
3201-10000	125 125	125 250	1 4 4 5	2 5 6 7	3 7 8 9	5 9 12 13	7 11 18 19	11 16 26 27
10001-35000	200 200	200 400	2 5 6 7	3 7 8 9	5 9 12 13	7 11 18 19	11 16 26 27	
35001-150000	315 315	315 630	3 7 8 9	5 9 12 13	7 11 18 19	11 16 26 27		
150001-500000	500 500	500 1000	5 9 12 13	7 11 18 19	11 16 26 27			
500001 and over	800 800	800 1600	7 11 18 19	11 16 26 27	↑	↑	↑	↑

Figure 8.3 Combined sampling plan.

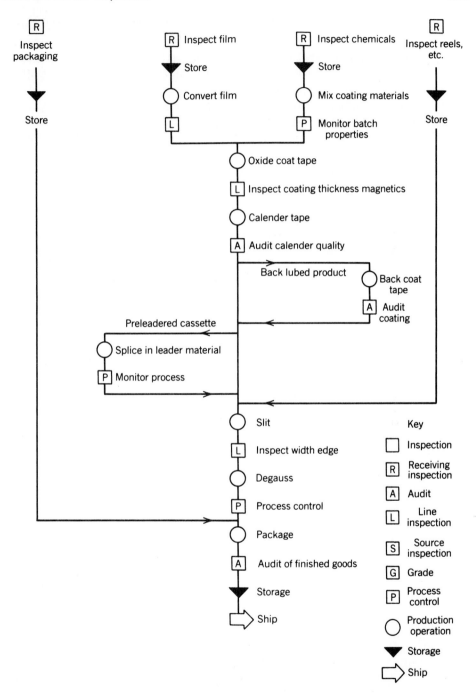

Figure 8.4 Process/product flowchart.

having an inspection plan including careful selection of the measuring instrument. It is far better to take more care doing the first inspection correctly than to have to reinspect that characteristic later in the manufacturing cycle.

Begin at the end! Use the flowchart in Figure 8.4 to identify each characteristic in the finished item that requires inspection. Then decide at what point during manufacturing that characteristic could first be inspected. Mark this information on the flowchart. Continue to do this for every characteristic requiring inspection. Then work backward during the manufacturing cycle doing the same thing at each point along the way. After you are finished, a clear pattern of inspection points will be defined on the flowchart.

Now, starting from the beginning of the manufacturing cycle, detail the inspection points and decide whether it is more economical to move the inspection to a later point in the process. Can more be saved in inspection time than can be risked should the defect be found later in the process? That is the only decision.

After the decision on when to inspect is made, the individual inspection plans can be written. At each inspection point, the sampling plan must be defined and the inspection operations sequenced. The sequence must be planned with the same care that the industrial engineer uses in planning manufacturing methods. In fact, advice from the industrial engineering department may be helpful to the inspection planner. Remember, the inspection is probably going to be performed every day, day in and day out, for as long as the product is made. Any savings in labor is one that the company is going to enjoy from now on.

Inspection Checklists

Checklists offer excellent inspection planning for progressive assembly-line operations (as an example). One company's final assembly checklist system has been instrumental in reducing the average number of defects per assembly by 40 percent while improving the outgoing quality of the product. The assembly line is divided into 11 zones, each with a checklist covering the characteristics requiring inspection (see Figure 8.5).

Characteristics on each checklist are listed in the sequence for inspection. A separate set of checklists is completed for each product built. The classification of each characteristic is shown (*M* for major and *m* for minor). If the item is critical it is shown in boldface type.

The inspector is to inspect the characteristics as soon as the production operation is complete. In other words, the inspector works up and down the assembly line within the area, inspecting the assembly as it is built instead of waiting at the end of the line. He or she is encouraged to become a part of the team composed of the assemblers, the repair person, and the inspector. It is best if the inspector can detect a mistake as it occurs and point it out at that time.

The production repairperson is encouraged to assist in the inspection activity (and usually does). It is to their advantage to reduce the amount of mistakes since that directly reduces the amount of repairs to be performed. The most

Characteristics and Acceptance Criteria	Defect. Class.	Okay	Def.	Comments	Repair	Insp.	Supv'r
1. Radiator: Coolant level and inhibitor added	M						
2. Fuel tank steps: Positioned and secured	M						
3. Sanders: Positioned and secured	m						
4. Lower grille: Positioned and secured	m						
5. Access steps: Positioned and secured	M						
6. Steel hoods: Paint damage, color and cutouts when applicable	M						
7. Fenders: Paint damage, color, positioned and secured	m						
8. Road lamps: Positioned and secured, when required	m						
9. Grille guards: Positioned and secured, when required	m						
10. Exhaust system: Positioned and secured	M						
11. Air cleaner: Positioned and loose assembled	M						
12. Mud flaps: Positioned and secured	m						
13. Swing hoods: Paint damage, color, loose assembled	M						

Figure 8.5 Checklist for assembly inspection.

efficient areas are those where the repair people spend the majority of their time training personnel and checking, rather than correcting mistakes.

If an end item is rejected, the defect column on the checklist is stamped "D" and the particular defect is noted in the remarks column. Production makes the repair within the area. Inspection verifies that the defect has been corrected. The inspector records, using tally marks, the number of defects occurring each hour (Figure 8.6).

The information is posted in a particular area of the assembly line. The production supervisor reviews this information and initiates corrective action as required. At final assembly inspection, the inspector reviews all checklists to see that all characteristics have been accepted. The inspector may also perform additional inspections, using a checklist, to assure that other standards are met (an example of other factors could be federal safety standards).

Other Factors in Inspection Planning

Certain human factors should be considered during the inspection planning. For instance, the inspector's workstation must be designed with provisions for holding the product and displaying it during the inspection. Lighting and good air circulation must be provided, as well as a proper place for keeping the inspection equipment, paper work, inspection tags, and other essentials.

Another consideration is the form used to collect the data generated during the inspection. It is disastrous to inspect without documenting the findings. How else can corrective action be taken? The best form is one that requires minimum time for the inspector to complete, sees that the data can be readily tabulated and summarized and that the data are all meaningful. Whenever possible, expected defect categories should be prerecorded. The inspector can then record the frequency of the defects in each lot inspected. When the columns are totaled, the summary data are ready to publish.

Another consideration during planning is what will be done when defects are found. The major reason for continuous sampling being unsuccessful is that no one plans for rejections. Typically, a continuous sampling plan requires more inspector time when excessive defects are detected than when the line is in control with few defects. The problem is where to get those *extra* people. Most textbooks say to borrow them from production. However, production has its own problems when defects occur and is least likely to have extra people. In short, unless an adequate supply of qualified personnel can be found to handle the line when rejects occur, do not use continuous sampling.

Another consideration is the inspection and handling of rejects on a *lot* basis. It is easier to handle rejects when using lot inspection, but facilities must be available to hold the rejected lot until production has the time to sort the product. An orderly method must be provided to get this sorted material back to the inspector for verification that the sorting has been done properly.

Characteristics and Acceptance Criteria	8:30	9:30	10:30	11:30	1:00	2:00	3:00	4:00	Total
1. Radiator: Coolant level and inhibitor added									
2. Fuel tank steps: Positioned and secured									
3. Sanders: Positioned and secured									
4. Lower grille: Positioned and secured									
5. Access steps: Positioned and secured									
6. Steel hoods: Paint damage, color and cutouts when applicable									
7. Fenders: Point damage, color, positioned and secured									
8. Road lamps: Positioned and secured, when required									
9. Grille guards: Positioned and secured, when required									
10. Exhaust system: Positioned and secured									
11. Air cleaner: Positioned and loose assembled									
12. Mud flaps: Positioned and secured									
13. Swing hoods: Paint damage, color, loose assembled									
14. Quarter fenders: Paint damage, color, positioned secured									
15. Battery box covers: Positioned and secured									
16. Trailer electrical cable: (In cab) positioned and secured									
17. Seats: Positioned and secured									

Figure 8.6 Tally sheet.

If the inspection process requires comparison "to approved samples," such samples must be easily accessible to the inspector. If those samples deteriorate easily by handling or exposure, they must be carefully protected. Such is the case with samples showing maximum allowable scratches, or with color chips that may fade.

Who Should Plan and When

The person best able to determine the proper method of inspection and the characteristics to check is usually the trained quality planner or engineer. As mentioned earlier, because of the complexity of the product and other factors, the planner must be part industrial engineer to specify methods, part inspector to determine procedure, part design engineer to determine what to inspect, part statistician to determine sampling plans and, most important, part prophet to select the most economical method based upon what will happen during the manufacturing process. A good inspection planner is right most of the time. However, if the inspection planner is right all of the time he or she is probably in the wrong profession.

In a large firm a separate department may do all the planning. In a small firm, the responsibility may fall to the chief inspector. In every firm where inspection is performed, someone must plan. The greatest economic gains come when planning is done before production begins. It takes the same amount of time then as it does later and will save money in both reduced inspection costs and a better product. If changes are required in the manufacturing process, they can be planned into it rather than having to alter the process to accommodate them later.

REVIEW QUESTIONS

1. A fundamental tool, often used in inspection planning, is a _____ chart.

2. One reason for an inspection plan is to achieve _____ among all inspectors.

3. Which of the following is not part of an inspection instruction?
   a. The sampling plan
   b. The measuring equipment
   c. Characteristics to inspect
   d. Control charts
   e. The applicable inspection record

4. When classifying characteristics for inspection, one must understand the application of the product. Tight _____ does not necessarily mean that the characteristic is functionally important.

5. The classification _____ means that the characteristic is important to the fit, form, and function of the product.

6. The classification _____ means that the characteristic has no direct relationship to fit, form, or function of the product.

7. Name the tool that is beneficial for providing assembly-related inspection planning.

8. Flowcharts are used in inspection planning to identify appropriate _____ points in the process.

9

Lot-by-Lot Acceptance Sampling Plans

Lot-by-lot acceptance sampling plans are often used to judge whether a particular lot submitted for inspection is acceptable or not, depending on the quality levels agreed upon by the producer and the consumer. These plans are often used on completed lots at one stage of processing or another to judge the quality of the lot based on statistical samples taken at random. There are two types of lot-by-lot sampling plans. There are a variety of lot-by-lot sampling plans from which to choose whether one is looking at attributes or variables of the output of the process (refer to the glossary for definitions). Attribute sampling plans covered in this chapter includes Mil-Std-105D and Dodge–Romig sampling plans. Variables sampling plans covered in this chapter include Mil-Std-414, and Chapter 10 covers the most effective variables sampling plan, *statistical process control*.

The basis for lot-by-lot acceptance sampling is a percentage called an acceptable quality level (AQL). An AQL (or the maximum percent defectives in a given lot which are considered acceptable as a process average) must be agreed upon by the producer and the consumer before lot-by-lot acceptance sampling can be used. Although acceptance sampling can be, and has been, more effective than 100% inspection at judging lot-by-lot quality, neither method is generally acceptable today as a process control tool. One of the main reasons for this is that neither is effective at *preventing* defects. They are both designed to *find* defects from the process.

Acceptance sampling is often used in conjunction with process control to judge the results of the process and to feed back information to the processors regarding defective output at various stages of operations. The sampling tables covered in the rest of this chapter are designed based on probabilities, and when used properly, will help reduce inspection and still detect poor lot-by-lot quality levels. It is recommended that lot-by-lot acceptance sampling should not be the only method for attempting to control outgoing quality.

RELATIONSHIPS BETWEEN SAMPLING AND 100 PERCENT INSPECTION

When deciding whether to 100 percent inspect items or use the various sampling schemes, you must first understand the main advantages and disadvantages of both. Listed below is a comparison of sampling inspection to 100 percent inspection.

100 PERCENT INSPECTION	SAMPLING INSPECTION
Definition. This is the inspection of every part produced.	*Definition.* This is inspecting a *random* sample of those parts produced.
Costs. 100 percent inspection is *costly* due to the time involved to inspect every part.	*Costs.* Sampling inspection takes less time and is, therefore, less expensive.
Effectiveness. Due to boredom, fatigue, and other human problems, 100 percent inspection is not 100 percent effective. In fact, it is usually only 70 to 90 percent effective.	*Effectiveness.* With sampling inspection, there are also risks involved, but these risks are calculated.
Application. In certain situations (such as destructive tests) 100 percent inspection cannot be done.	*Application.* Sampling inspection is a must for destructive tests.
Handling. In 100 percent inspection you have a higher probability of damaging a part because you must handle every part.	*Handling.* In sampling inspection, you only handle a few of the parts and, therefore, handling damage is less likely.
Bias. 100 percent inspection sometimes encourages personal bias on the part of the inspector.	*Bias.* Sampling inspection and its decision making helps the inspector to be more objective. Bias is not often a problem here.

The previous information given on sampling and 100 percent inspection could lead a person to believe that there is no place for 100 percent inspection in industry. This is not true. There are some positive points to be made about 100 percent inspection that should be considered. They are

1. 100 percent inspection is more effective than sampling and is not prohibitive with regard to cost when it is *automated*. Machines do not get tired, bored, or biased.

2. 100 percent inspection (when done manually) can be made more effective by rotating the people who are doing the 100 percent inspection. Rotating (or having someone else take over the job for a while) offers less chance that the person will get tired or bored.

3. 100 percent inspection (done properly) gives you more information about quality levels than sampling does. This is because you had the opportunity to evaluate the quality of the entire lot being inspected.

All things considered, sampling inspection still comes out ahead so far as economic choices are concerned. For all of the various plans for sampling (Mil-Std-105D, Mil-Std-414, Dodge–Romig, etc.) there are certain situations where they should be used, and if you choose the right one for the job and stick to it, the quality levels can be maintained.

Caution should be observed, however, when you make the decision to sample or 100 percent inspect; certain end items (especially those performing major or critical functions) should always be inspected and tested 100 percent to assure the customer maximum product quality and reliability.

RELATIONSHIPS BETWEEN LOT-BY-LOT ACCEPTANCE SAMPLING AND PROCESS CONTROL SAMPLING

There are important differences between lot-by-lot acceptance sampling and process control sampling that should be considered when making the choice to sample. First is the effectiveness of the sampling plans. The overall effectiveness of sampling has already been discussed. Now the question is primarily where to use sampling, or at what stages of the process. Process control sampling (refer to Chapter 10) involves the use of samples taken during the process and data are graphed (such as control charts and frequency distributions) in a manner that will measure and help monitor the variation of the process. Lot-by-lot acceptance sampling plans are designed to sample semifinished or finished lots for acceptance purposes. This method of sampling is reactive, in that the quality has already been built into the product by the time that acceptance sampling is used.

This is one of the primary reasons why process control sampling is more effective. Samples are taken during the processing of the product such that the identification of defects is closer to real time, and the process is controlled to prevent defects from being made. Lot-by-lot sampling plans are after the fact, and when defects are found, it is usually too late (at this stage) to identify the true causes. When one considers the effect of finding problems at late stages of the process, as opposed to finding problems at the time they happen, it is not too difficult to identify the most effective sampling plan.

The best decision is to find a balance between process control sampling during the process for preventive action, and lot-by-lot sampling, which will help track the results of process control. Process control will usually be the primary choice as a tool for preventing defects from reaching the customer. The balance of this chapter is dedicated to how lot-by-lot sampling plans work, and is intended to give the reader a basis for decisions regarding using the plans properly, or deciding which plan to use. One of the most fundamental things in lot-by-lot sampling is being able to ensure that the quality of a sample is representative of the quality of the lot. One aspect of representative samples is that samples are taken at random from the lot.

RANDOM SAMPLING

Many companies use sampling plans to inspect lots because of the advantages in sampling over 100 percent inspection. The goal of sampling is to be able to accept good lots and reject bad lots a high percentage of the time. Besides being able to read sampling tables and understand them, the inspector must have the ability (and the desire) to take samples at *random*. Randomness is important when you are sampling a lot of parts and making a decision on the entire lot based on a small sample. That sample must be representative of the lot (not just the top layer).

Definition

Random sampling is giving every part in the lot an equal chance of being selected for the sample, regardless of its quality.

Drawbacks

Too often, it has been noted that when a company sample inspects, emphasis is put on how to read the sampling tables and very little on the importance of random samples and how to take them. This is only one problem. Another is that some know how to take random samples but do not because it is too much work, or sometimes they cannot because of the way lots are packaged. It must be understood that when a company chooses to save money by sampling, they must also provide the means by which random samples can be taken. This may be in the form of different packaging, or simply giving the inspector time to do so. Either way, it must be done. Sampling without randomness ruins the effectiveness of any sampling plan.

How Lots Are Packaged

This book will explain four main ways in which lots are delivered to inspection:

1. Single layer on a pallet.
2. Boxes stacked on top of each other.

3. One box where parts are in layers.

4. Individual parts on a conveyor.

Single layer on a pallet.   The easiest of all to random sample are those lots presented on a single layer on a pallet, as shown in Figure 9.1. In this case, every single box on the pallet should be given an equal chance of being selected (not just those nearest you). Then every part in those boxes chosen are given an equal chance of being selected for the sample.

Boxes stacked on top of each other.   This situation is a little harder to sample randomly. To do so, you must unstack the boxes and draw your sample from the parts when all of them are exposed (giving them an equal chance) (see Figure 9.2). Often, this calls for a little muscle, but it is worth the effort.

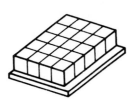

Figure 9.1   Single layer on a pallet.

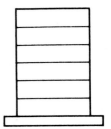

Figure 9.2   Stacked layers in boxes.

One box where parts are in layers.   This one is the most difficult of all (especially if the parts are either very heavy, or very small and numerous) (see Figure 9.3). This requires that the packaged layers are removable or that the inspector "dig" to obtain a random sample. It is very hard to dig down and get randomness because of the parts in the center and those at the very bottom. Sometimes "thief" sampling doors can be put into the box so that the inspector can look at the bottom layer.

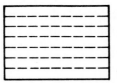

Figure 9.3   Multilayers in a container.

Individual parts on a conveyor.   This method of delivery makes it easy to get random samples. All of the parts are exposed, and your only decision is what sequence of parts to look at (every third one, or number 3, 9, 16, 24, etc.). There are many ways of obtaining random samples if this is the method of delivery.

1. *Numbers in a hat.* A hat with various numbers in it can be used. If your sample size for the entire lot is to be 13, draw 13 numbers from the hat and select the parts based on this sequence of numbers.

2. *Numbered balls in a jar.* This works the same way as the hat. For example, your sample size may be 5 out of 50 parts total. You draw 5 numbered balls from the jar. They are 3, 17, 21, 34, 48. Now, you simply inspect the 3rd, 17th, 21st, 34th, and 48th part as they come down the conveyor.

3. *Table of random numbers.* The sample table in Figure 9.4 is an example of a portion of a table of random numbers. These numbers have no predetermined sequence and are completely random. The tables allow you to select at random to avoid bias in your sample.

Line \ Column	(1)	(2)	(3)	(4)	(5)
1	10480	15011	01536	02011	81647
2	23368	46573	25595	85393	30995
3	24130	48360	22527	97265	76393
4	42168	93093	06243	61680	07856
5	37570	39975	81837	16656	06121

Figure 9.4   Example table of random numbers.

Using the Table of Random Numbers

You can find these tables in any book of math tables or a statistical text. If you do get the chance to use it, the steps below basically explain how.

Step 1. Determine your sample size. (*Example:* an 80-piece lot and a five-piece sample.)

Step 2. Choose any line and column combination you want. (For this example, I chose line 1, column 1. The first number here is 10480.)

Step 3. Since 80 (the lot size) is two digits, you can use any combination from the five digits shown.
For example, in 10480 you can use 10, 04, 48, or 80 as your first number selected. I chose 10.

Step 4. Now go down the column using the first two numbers in each one. They are (line 1, column 1):

10	These are the numbered parts I will inspect.
22	*On a conveyor,* I will inspect the 10th, 22nd,
24	24th, 37th, and 42nd pieces.
42	*Parts in a box* must be set out and numbered prior to reading the
37	table.

Note. If you reach a number that is repeated, skip that number and use the next one. *If you reach a number that is too large* (over 80), skip that number also, and select the next one.

Summary. There are numerous other ways to get random numbers (such as a deck of cards or rolling dice), but the main thing to remember is that *without randomness the risks of sampling error are greater.*

USING MIL-STD-105D FOR SAMPLING INSPECTION

Military Standard 105D when first developed was called the "ABC" (America, Britain, and Canada) standard. Generally, Mil-Std-105D is used as a basis for sampling to effectively sample-inspect products. Other sampling tables (such as Dodge–Romig tables) can also be used; these tables work well when the process average is known.

Notes

1. Before reading further, refer to the Glossary for sampling terms and definitions.
2. See the Appendix for the actual tables taken from Mil-Std-105D.

With any sampling plans, however, there are certain pieces of information you (as the inspector) must be supplied ahead of time. These are

1. *Lot size.* This is usually known on lots submitted for inspection.
2. *Inspection level.* This is usually given in inspection plans or procedures.
3. *AQL.* This is also usually given in inspection plans or procedures.
4. *Type of sampling to use.* This is single, double, or multiple.

When these things are known, it is simply a matter of reading the appropriate tables and understanding what to do with them.

Mil-Std-105D plans are primarily intended to be used to sample-inspect a continuing series of lots (or batches), as opposed to isolated lots.

Single Sampling Tables*

Shown in Figure 9.5 is an example of Mil-Std-105D *single sampling tables.* This type of sampling is the main area of discussion here. The actual code letter table appears on page 501.

---

* Refer to the table on page 502.

The thing you are looking for is called the *sampling plan*. This is simply the sample size (*n*), the accept number (Ac), and the reject number (Re). The steps to finding the sampling plan for a lot presented for inspection are as follows.

Example

Lot size = 510 pieces, AQL = 4.0 percent, general inspection level II.

Step 1. Starting with Figure 9.6, find the range of numbers under "lot size" where the lot size you are sampling (510) falls within. This range is 501 to 1200.

Step 2. Look straight across from the lot size range you found earlier until you are directly under the inspection level you are working with (which in this example is general level II). There, you will find the code letter J.

Step 3. Turn to the sampling table, Figure 9.5, and find code letter J again.

Step 4. Look straight across the page from your code letter J and stop when you are under the AQL you've been given (4.0 percent). This is where you find the accept number (Ac) of 7 and the reject number (Re) of 8.

Step 5. Look in the same row J under "sample size" to find what size your sample should be. In this example, it is 80 pieces.

In practice, remember to use the actual tables such as those on page 502. **Now you have found the sampling plan:**

n	Ac	Re
80	7	8

This tells you to take a *random* sample of 80 pieces from the lot and if you find seven defectives (or less) in your sample, you set the defectives aside and accept the rest of the lot. But if you find eight (or more) defectives in the sample, this is cause for rejection of the entire lot and returning it for screening (or sorting).

The Arrows

In the sampling table in Figure 9.5, you will see that there are arrows with some pointing up and some down. These arrows are used as follows. First, you follow the previous steps to find your sampling plan. If you come to a sampling plan that shows an arrow, this means that you use the sampling plan just above or just below (depending on which way the arrow is pointing).

Example

Lot size = 200, general level II, AQL = 1.0 percent
So the code letter is G, but under 1.0 percent you find an arrow pointing down. This means that you go to code letter H, and the sampling plan is

N	Ac	Re
50	1	2

Note. It is important to use the entire plan, including the new sample size, that coincides with letter H.

Acceptable Quality Levels

Sample size code letter	0.25	0.40	0.65	1.0	1.5	2.5	4.0	6.5	10	15	25	40	65	Sample Size
	Ac Re	Ac Re	Ac Re	Ac Re	Ac Re	Ac Re	Ac Re	Ac Re	Ac Re	Ac Re	Ac Re	Ac Re	Ac Re	
A	↓	↓	↓	↓	↓	↓	↓	↓	↓	0 1	1 2	2 3	3 4	2
B	↓	↓	↓	↓	↓	↓	↓	↓	0 1	1 2	2 3	3 4	5 6	3
C	↓	↓	↓	↓	↓	↓	↓	0 1	1 2	2 3	3 4	5 6	7 8	5
D	↓	↓	↓	↓	↓	↓	0 1	1 2	2 3	3 4	5 6	7 8	10 11	8
E	↓	↓	↓	↓	↓	0 1	1 2	2 3	3 4	5 6	7 8	10 11	14 15	13
F	↓	↓	↓	↓	0 1	1 2	2 3	3 4	5 6	7 8	10 11	14 15	21 22	20
G	↓	↓	↓	0 1	1 2	2 3	3 4	5 6	7 8	10 11	14 15	21 22	↑	32
H	↓	↓	0 1	1 2	2 3	3 4	5 6	7 8	10 11	14 15	21 22	↑	↑	50
J	↓	0 1	1 2	2 3	3 4	5 6	7 8	10 11	14 15	21 22	↑	↑	↑	80
K	0 1	1 2	2 3	3 4	5 6	7 8	10 11	14 15	21 22	↑	↑	↑	↑	125
L	1 2	2 3	3 4	5 6	7 8	10 11	14 15	21 22	↑	↑	↑	↑	↑	200
M	2 3	3 4	5 6	7 8	10 11	14 15	21 22	↑	↑	↑	↑	↑	↑	315
N	3 4	5 6	7 8	10 11	14 15	21 22	↑	↑	↑	↑	↑	↑	↑	500
P	5 6	7 8	10 11	14 15	21 22	↑	↑	↑	↑	↑	↑	↑	↑	800
Q	7 8	10 11	14 15	21 22	↑	↑	↑	↑	↑	↑	↑	↑	↑	1250
R	10 11	14 15	21 22	↑	↑	↑	↑	↑	↑	↑	↑	↑	↑	2000

↓ = Use the first sampling plan, and sample size *below*. If the sample size equals, or exceeds the lot size, inspect the *whole* lot.

↑ = Use the first sampling plan, and sample size above the arrow.

Ac = The acceptance number. If the number of defectives is equal to, or less than this number, remove the defectives, and accept the balance of the lot.

Re = The rejection number. If the number of defectives is equal to, or greater than this number, reject the *whole* lot.

Figure 9.5 Mil-Std-105D single sampling table: normal inspection.

Mil-Std-105D: Code Letters for Sample Sizes

Lot Size			Special Inspection Levels				General Inspection Levels		
			S-1	S-2	S-3	S-4	I	II	III
2	to	8	A	A	A	A	A	A	B
9	to	15	A	A	A	A	A	B	C
16	to	25	A	A	B	B	B	C	D
26	to	50	A	B	B	C	C	D	E
51	to	90	B	B	C	C	C	E	F
91	to	150	B	B	C	D	D	F	G
151	to	280	B	C	D	E	E	G	H
281	to	500	B	C	D	E	F	H	J
501	to	1200	C	C	E	F	G	J	K
1201	to	3200	C	D	E	G	H	K	L
3201	to	10000	C	D	F	G	J	L	M
10001	to	35000	C	D	F	H	K	M	N
35001	to	150000	D	E	G	J	L	N	P
150001	to	500000	D	E	G	J	M	P	Q
500001 and over			D	E	H	K	N	Q	R

Figure 9.6 Table of code letters for sample sizes.

Remember that with any sampling plan, taking a *random sample* is important for the plan to be effective in accepting good quality and rejecting poor quality.

Double Sampling per Mil-Std-105D*

Double sampling (per Military Standard 105D) is used from time to time by inspection. Generally, double sampling plans are used in inspection when product quality levels are "real good" or "real bad." Depending on the actual quality levels, double sampling can reduce the amount of inspection you do and still give you confidence that you will stop poor quality and accept good quality (per the AQL). Double sampling works similarly to the single sampling plans discussed previously. Double sampling plans, however, usually result in the least amount of inspection over time.

You still need to know:

- AQL
- Reduced, normal, or tightened
- Lot size
- Level I, II, III, or others

You still use the table of code letters to obtain a code letter for your plan. However, *you must use the double sampling tables* provided in Mil-Std-105D.

The actual tables are given on page 503. Only an example of what you will see is shown in Figure 9.7. When you have selected your code letter and followed it over to just below the AQL you are using, you will see two sample sizes, two numbers, and two reject numbers.

Letter	Sample	Sample size	Cum. Sample size	Ac	Re
M	First	200	200	3	7
	Second	200	400	8	9

Figure 9.7 Section of a double sampling table.

The example in Figure 9.7 uses code letter M and 1.0 percent AQL. This simply tells you that you sample 200 pieces on the first sample. If you find three (or less) defectives, accept the lot and set the defectives aside. If you find four, five, or six defectives, take the second sample of 200 pieces. If you find seven or more defectives in the first sample, reject the entire lot for screening inspection.

If you do take the second sample, the accept numbers and reject numbers include the number of defects found in the first sample.

* Refer to page 503 in the Appendix for the actual table.

Example

You found six defectives in the first sample. This tells you to take the second sample of 200 pieces. If you find two more defectives, this will total eight defectives found; you still accept the lot. If you find three more defectives, reject the entire lot because the reject number is nine.

There is another type of sampling in Mil-Std-105D, called *multiple sampling*. This does exactly the same thing as double sampling, except it gives the lot seven chances to be accepted or rejected. The technique for multiple sampling is the same as the technique for double sampling except you are given seven chances to accept or reject the lot, instead of only two. Multiple sampling tables are rarely used; therefore, they are not covered in this chapter.

INTRODUCTION TO DODGE–ROMIG SAMPLING TABLES*

Dodge–Romig sampling tables are another method of sampling by attributes. These tables are most effective when you know the process average (or fraction defective) of the lot. This process average is called ($\bar{p}$). If Dodge–Romig tables are to be used with an unknown process average ($\bar{p}$), you should use the largest $\bar{p}$ in the tables until you gather enough data to establish a process average.

Dodge–Romig sampling tables are used with the understanding that rejected lots will be screened (or 100 percent inspected). When used properly, Dodge–Romig sampling plans reduce the amount of inspection without increasing the risk of poor outgoing quality. This reduced amount of inspection saves time and money and allows you to inspect lots quicker and still be effective in stopping poor quality.

Dodge–Romig tables come in four types.

1. Single stamping (LTPD) lot tolerance percent defective (see page 504).
2. Double sampling (LTPD) (see page 505).
3. Single sampling (AOQL) (see page 506).
4. Double sampling (AOQL) (see page 509).

Percentages at the top of each table reflect the percent defective lot that will be rejected using any of the plans within that table. Refer to Figure 9.8.

When double sampling (with Dodge/Romig tables) instead of single sampling, fewer units need to be inspected, provided that the decision to accept or reject the lot can be made on the first sample. See the example in Figure 9.9 (the actual table is in the Appendix).

The table in Figure 9.9, another example of a Dodge/Romig Inspection table, shows that with a lot size ranging from 801 pieces to 1000 pieces, your first sample

---

* Refer to pages 504 to 508 in the Appendix for the actual tables.

Lot Size 401–500	Process Average 0 to 0.04%			Process Average 0.05 to 0.40%		
	n	c	$P_T\%$	n	c	$P_T\%$
	18	0	11.9	18	0	11.9

n = sample size.
c = accept number.
P_T = LTPD corresponding to this AOQL plan.

Figure 9.8 Section of a Dodge–Romig sampling table. Single sampling plans (AOQL = 2.0%).

Lot Size 801–1000	Process Average						Process Average					
	Trial 1		Trial 2				Trial 1		Trial 2			
	n_1	c_1	n_2	n_1+n_2	c_2	$P_T\%$	n_1	c_1	n_2	n_1+n_2	c_2	$P_T\%$
	55	0	30	85	1	5.2	60	0	75	135	2	4.4

n_1 = first sample size.
c_1 = first accept number.
n_2 = second sample size.
c_2 = second accept number.
$n_1 + n_2$ = cumulative sample size.
$P_T\%$ = the corresponding LTPD for this AOQL plan [with a consumer's risk (P_c)] of 0.10.

Figure 9.9 Section of a Dodge–Romig double sampling table. Double sampling plans (AOQL = 1.0%).

will be 55 pieces and that you can accept the lot if there are no defectives in the first sample. If there is one defective in the first sample, you must take another sample. If there is more than one defective, you reject the lot. If you take a second sample, your cumulative sample size $(n_1 + n_2)$ is 85 pieces, and you can accept the lot if there is only one defective, but you must reject the lot if there is more than one defective part.

MIL-STD-414 VARIABLES SAMPLING TABLES

Note. Actual Mil-Std-414 tables are included in the Appendix.

Sampling by variables is different from sampling by attributes (as in Mil-Std-105D or Dodge/Romig plans) because here you are measuring actual sizes and variation on a continuous scale (such as inches or pounds) and are not simply concerned with defective or not defective.

Mil-Std-414 has four sections: A, B, C, and D.

Section A. Introduction.

Section B. Consists of sampling plans that are used when the variability is *unknown,* and the *standard deviation method* is used.

Section C. Consists of sampling plans that are used when the variability is *unknown,* and the *range method* is used.

Section D. Consists of sampling plans that are used when the variability *is known.*

There are five inspection levels to Mil-Std-414:

I   II   III   IV   V

If there is no level specified, *use level IV.*

The use of Mil-Std-414 requires that you

1. Choose the level.
2. Choose the method.
3. Know the AQL.
4. Know the lot size.

There is a table in Mil-Std-414 that shows the ranges of AQLs, and you must use this table to establish the AQL needed in the plans. See page 486 in the appendix for these tables. For Mil-Std-414 definitions, see Appendix pages 499 to 500.

Example

SPECIFIED AQL RANGE	USE THIS AQL
0.110 to 0.164	0.15

So, if you are working with an AQL of 0.13 percent, you must use 0.15 percent in the sampling tables.

There are two methods to use in finding the variability of a lot when sampling per Mil-Std-414. These are the standard deviation method and the range method.

Standard Deviation Method

Q_u is the number of standard deviations that the mean $(\overline{X})$ is away from the upper tolerance limit.

$Q1$ is the number of standard deviations that the mean $(\overline{X})$ is away from the lower tolerance limit.

s is the estimated standard deviation. Refer to pages 393–394 for details.

$$s = \sqrt{\frac{\Sigma(x - \bar{x})^2}{n - 1}}$$

$$Q = \frac{\text{tolerance limit} - \text{mean}}{\text{standard deviation}}$$

Range Method

You must find $\bar{R}$, which is the average range of the subgroups. (Sometimes there is only one subgroup; if so, R is equal to $\bar{R}$.)

$$Q = \frac{\text{tolerance limit} - \text{mean} \ (\bar{X})}{\text{average range} \ (\bar{R})}$$

or

$$Q = \frac{\text{tolerance limit} - \bar{X}}{\bar{R}}$$

Note. The mean ($\bar{X}$) is the average of the data.

These methods will be expanded on in the following pages.

There are three different severities for inspection (as in Mil-Std-105D): *normal, tightened,* and *reduced.* Each of these severities has rules about when to switch from one to the other. The severity must also be known for the sampling plan to be found. Refer to Mil-Std-414 for further information.

As with most plans, when the sample size is equal to or larger than the lot size, every part must be inspected.

Mil-Std-414 Standard Deviation Method: Single Specification Limit

Example

A manufacturer produces 28 parts where datum A must have a maximum surface finish of AA32. The severity is normal inspection, level IV, and the AQL is 1.5 percent.

Step 1. Find the AQL to be used in the conversation table. It is, in this case, 1.5 percent, since 1.5 percent falls between 1.10 and 1.64 range (see page 486).

Step 2. Find the sample size code letter according to the lot size of 28 pieces and level IV. This is letter *D* (see page 486).

Step 3. Go to the table for standard deviation method: single specification limit. Read across from code letter D and find that the sample size is 5. Read also under the 1.5 percent AQL and find the *K* constant to be 1.40. This *K* value must be remembered for future steps (see page 487).

Step 4. Randomly select five units from the lot and measure the actual surface finish. Record each measurement. Let's say that they were 5, 9, 10, 7, and 8.

Step 5. Compute $\bar{X}$ and the standard deviation(s). $s = 1.924$ $\bar{X} = 7.8$.

Step 6. Find Q_u (since it is only the upper limit you are concerned with). This upper limit is AA32.

$$Q_u = \frac{\text{upper tol.} - \bar{X}}{s} = \frac{32 - 7.8}{1.924} = 12.578$$

Step 7. Compare Q_u (which is 12.578) to the K constant you found in step 3. If Q_u is larger than K, the lot is acceptable. If it is smaller, the lot must be rejected.

This lot is acceptable because Q_u (12.578) is larger than K (1.40).

Mil-Std-414 Standard Deviation Method: Double Specification Limit

Example

A lot size of 35 manufactured parts must be within a surface finish tolerance of AA20 to AA10 on the datum C. The severity is normal inspection, level III, and an AQL of 2 percent.

Step 1. Find where the 2 percent AQL falls in the conversion table of AQLs. Here the AQL of 2 percent is within the range of 1.65 to 2.79, so you would use 2.5 percent (see page 486).

Step 2. Find the sample size code letter. Here the lot size is 35 and level III applies, so the code letter is B (see page 486).

Step 3. Go to the table for standard deviation method: double specification limit. Read across from code letter B and find the sample size to be 3 and the M constant number to be 7.59 (reading under the AQL of 2.5 percent) (see page 487).

Step 4. Draw a random sample of three units and measure the surface finish of datum C. Record the actual measurements. Let's say that they were 15, 16, and 18.

Step 5. Find $\overline{X}$ and the s (standard deviation). $\overline{X} = 16.33$; $s = 1.53$ using previously given formulas.

Step 6. Find Q_u and Q_L (since it is a double specification limit). This upper limit is AA20.

$$Q_u = \frac{U - \overline{X}}{s} = \frac{20 - 16.33}{1.53} = 2.40$$

$$Q_L = \frac{L - \overline{X}}{s} = \frac{10 - 16.33}{1.53} = 4.14$$

(Do not be concerned here with minus numbers.)

Step 7. Now you must find the *estimated lot percent defective* from the appropriate tables in Mil-Std-414. In these tables, a Q_u of 2.40 is estimated as 0.000 percent defective, and a Q_L of 4.14 is estimated also as 0.000 percent defective according to the sample size of 3. Samples of these tables are given on pages 489 to 492. Therefore, the total estimated percent defective here is

$$P = P_u = P_L = 0$$

Step 8. The last step is to compare P with M. The lot is acceptable if P is equal to or less than M. M was 7.59 (in step 3 above) and P is 0.

Accept the lot.

Mil-Std-414 Range Method: Single Specification Limit

Example

A manufacturer produces 30 parts where datum B must have a surface finish of AA32 maximum. The inspection level is not specified (so use IV), normal inspection, and AQL of 1.5 percent.

Step 1. Select the AQL to use from the AQL conversion table. Here the specified 1.5 percent falls between 1.10 and 1.64, and 1.5 percent is used (see page 486).

Step 2. Find the sample size code letter in the table according to lot size of 30 and level IV. It is code letter D (see page 486).

Step 3. Go to the table for range method: single specification limit and read across from code letter D and under 1.5% AQL to find that the sample size is 5 and the K constant number is 0.565. Remember this K number for later use (see page 493).

Step 4. Select (at random) five units from the lot, measure the actual surface finish of each, and record the results. Let's say that they are 5, 9, 10, 7, and 8.

Step 5. Find $\overline{X}$ and $\overline{R}$.

$$\overline{X} = \frac{5 + 9 + 10 + 7 + 8}{5} = 7.8 \qquad \text{(let's say 8)}$$

$$\overline{R} = \text{average of the ranges}$$

But here there is only one range. Highest − lowest = range, so $10 - 5 = 5$.

Step 6. Find Q_u.

$$Q_u = \frac{U - \overline{X}}{\overline{R}} = \frac{32 - 8}{5} = 4.8$$

Step 7. Compare Q_u to K. If Q_u is larger than K, the lot is acceptable. Q_u is 4.8, K is 0.565.

Accept the lot.

Mid-Std-414 Range Method: Double Specification Limits

Example

A lot of 28 parts is to be inspected. The tolerance for surface finish on a surface is AA12 to AA8. We have for this example normal inspection, level IV, AQL = 1.68 percent.

Step 1. Select the AQL to use from the table. The specified AQL of 1.68 percent falls between 1.65 and 2.79; therefore, use 2.5 percent (per the table) (see page 486).

Step 2. Find the sample size code letter in the table according to the lot size (28) and level IV. It is D (see page 486).

Step 3. Refer to the table for range method: double specification limit. Read across from code letter D and under 2.5 percent AQL to find the sample size, which is 5, and the M number, which is 9.90. Remember this M number for later use (see page 494).

Step 4. Randomly select five units from the lot, measure the actual surface finish, and record the measurements. Let's say they are 9, 10, 10, 12, and 12.

Note. This lot could be rejected due to inherent variation.

Step 5. Find $\overline{X}$ and R of the measurements.

$$\overline{X} = 10.6 \text{ (say 11 here)} \qquad R = 3$$

Step 6. Find the c factor in the table. This is 2.474; the M number, remember, is 9.90 (see page 494).

Step 7. Find Q_u and Q_L.

$$Q_u = \frac{(U - \overline{X})c}{R} = \frac{(12 - 11)2.474}{3} = 0.825$$

$$Q_L = \frac{(L - \overline{X})c}{R} = \frac{(8 - 11)2.474}{3} = 2.47$$

Step 8. Find the estimated lot percent defective from the tables in Mil-Std-414 called "Estimated percent Defective Using the Range Method."
Where Q_u is 0.825 (and sample size is 5), the estimated percent defective P_u is 21.56 (see page 496).
Where Q_L is 2.47 (and sample size is 5), the estimated percent defective P_L is 0 (see page 498).

$$P_u - P_L = P$$

So

$$21.56 - 0 = 21.56$$

Step 9. Compare P to M. If P is equal to, or larger than M, reject the lot.

$$P = 21.56 \qquad M = 9.90$$

Reject the lot.

Note. All of the measured values of this lot were within specification limits, but the lot was rejected. It was rejected due to its inherent variation. Notice that the measured values were 9, 10, 10, 12, and 12, and that 12 is the upper tolerance limit. The spread, and the fact that the measurements were too near the high limit, caused the lot to be rejected.

OPERATING CHARACTERISTIC (OC) CURVES

All sampling plans (sample size, accept number, and reject number) have their own operating characteristic curves. These are called OC curves for short. The inspector is usually not concerned with using OC curves, but should know that

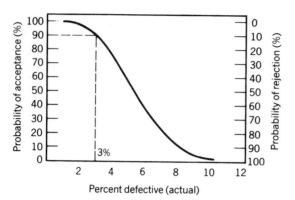

Figure 9.10   Sample operating characteristic (OC) curve.

they exist, and understand what they are for. Figure 9.10 shows an example of what an OC curve looks like.

Definition of an OC Curve (per Mil-Std-105D)

An OC curve indicates the *percentage* of *lots* or batches that may be expected to be *accepted* under the various sampling *plans* for a given *process quality*.

An OC curves gives you an advantage for choosing the sampling plan that you can expect good results from. For example, let's say that using the sampling plan connected with the OC curve in Figure 9.10, you inspect a lot that is actually 3 percent defective. The curve shows that when a lot that is 3 percent defective is inspected with this plan, there is a 90 percent probability that it will be accepted and a 10 percent probability that it will get rejected.

It is simply a matter of reading the OC curves and choosing the sampling plan that will accept (or reject).

Example

You want a plan that will reject any lot worse than 2 percent defective. The OC curve above is not the one that will do this because it will accept lots that are 2 percent defective for 100 percent of the time. Therefore, you must look for another plan. OC curves are constructed primarily according to the sampling plan (sample size, accept number, reject number). The lot size is least important when working with OC curves; however the lot size should be considered, since once the OC curve and plan are chosen, the lot size will be used to determine the sample size.

As stated earlier, OC curves are not often used by the inspector, but it is good for the inspector to know that they exist and understand how they help in the selection of sampling plans that are effective in maintaining a maximum allowable percent defective.

Another thing about OC curves is the fact that when the curve is steeper (nearly vertical), its ability to discriminate between good lots and bad lots is better.

OC curves are widely used to help the quality engineer or manager select sampling plans that are effective in reducing the risks of sampling (accepting poor quality or rejecting good quality), and that also help to keep the high cost of inspection down.

REVIEW QUESTIONS

1. Which of the following is more effective, sampling or 100% inspection?
2. For samples to be representative of the lot, it is important to draw samples from the lot at _____.
3. When using Mil-Std-105D double sampling plans, what should be done if the reject number is found in the first sample?
4. Double sampling Mil-Std-105D is generally better than single sampling when the process average (percent defectives) is very _____ or very _____.
5. Using Mil-Std-105D, single sampling, normal inspection, level II, AQL 1.0, find the sampling plan for a lot of 2000 pieces.
6. Find the sampling plan for Question 5 using the double sampling tables.
7. If lot quality were very good, how many fewer products would have to be sampled in Question 6 as opposed to Question 5 plans?
8. Find the range of the following values: 0.500, 0.505, 0.509, 0.501, 0.505.
9. Find the sample standard deviation for the values in Question 8.
10. When using Dodge–Romig sampling tables, what process average ($\bar{p}$) should be used when the actual process average is unknown?
11. It is understood with Dodge–Romig sampling tables that when using them, rejected lots will be _____.
12. Using Dodge/Romig single sampling tables for AOQL = 2.0%, lot size = 700 parts, and the actual $\bar{p} = 0.9\%$, find the single sampling plan.
13. Using the following information and Mil-Std-414 range method for single specification limit, is the lot acceptable? Information: range method—single specification limit, normal inspection, level is not specified. Lot size = 50, AQL = 1.0, tolerance is 0.500 maximum, $\bar{R} = 0.002$, $\bar{x} = 0.492$.

10

Statistical Process Control

Statistical process control (SPC) is a funnel for quality and productivity (Figure 10.1). It uses very basic statistics to control processes.

Process. Any collection of things that gets you from one point to another. A process could be anything from a washline, to a machine, or even driving to work in the morning.

All things are bugged with *variation*. No two things are ever alike. Even though the difference between the sizes of two machined parts is very small, they are different!

There are two kinds of variation with regard to process control.

Random variation. This is variation that has several common causes. *Common* causes of variation are things such as vibration, heat, and humidity that either individually or collectively cause variation. These common causes are very difficult to identify and correct.

Nonrandom variation. This is variation that is not built into the process. Causes of variation like this are called *special* causes. Special causes of variation are those that can be directly identified and corrected (particularly when you know exactly *when* they are occurring).

Variation is the reason we must have tolerances on the engineering drawing.

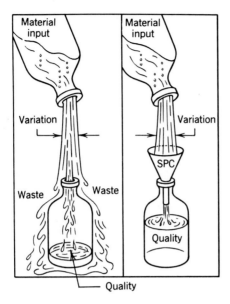

Figure 10.1 Statistical process control—a funnel for quality and productivity.

Drawing tolerance. These are the tolerance limits on a blueprint (or other specification). These limits define whether or not the product is acceptable. It is assumed that most parts produced will be near the nominal size.

Natural (process) tolerance. This is the measure of the ability of the process itself according to its own inherent variation. Keep in mind that a process will produce parts according to its own variation regardless of what the specifications say. The important thing is to make the process do what it can do consistently (control), then make sure that what it can do matches the blueprint requirements (capability).

OBJECTIVES OF SPC

The first objective of SPC is to get the process control, which means the identification and elimination of *special* causes of variation.

A process in control. A process that is in control will consistently produce parts within its own *natural tolerance limits*. This is done by eliminating all of the special causes of variation that may exist. Some examples of special causes are tool wear, loose workholding devices, and power surges. There are many more.

After a process is in control (stable) and producing consistently within its natural tolerance, it can then (and only then) be compared to the engineering tolerance limits to see if it is *capable* of meeting those limits.

A capable process. The capability of a process is directly related to the *ability* of the process to produce parts consistently within the drawing tolerance limits. Capability cannot be studied until the process is *in control,* because it is not consistent enough to trust the results of the study. We discuss capability in more detail later.

TOOLS FOR PROCESS CONTROL

The best tools for process control are *control charts.* They are very simple statistical charts that are powerful tools for:

1. Detecting special causes of variation in the process *at the time* they exist.
2. Measuring the natural tolerance of the process that is due to *normal variation* (or common causes).
3. Assisting you in getting the process in control and reaching a capability of meeting drawing tolerance limits consistently.

Normal variation. When the variation in the process (due to common causes only) is exhibited by a bell-shaped curve (called the *normal distribution).*

Frequency distribution. A graphical technique for studying measured data to see if they are distributed normally is called a frequency distribution.

THE IMPORTANCE OF DISTRIBUTION

One example of distribution is when a group of measured values are put on a scale and an × is placed for each time a value occurs (see Figure 10.2). If you draw a line around the outside of the plotted ×'s you will see that this is a normal distribution because it is shaped like the bell curve.

There are times when you can plot measured values on a distribution and the shape of the curve will not be normal. Some examples are shown in Figure 10.3.

One of the most important things to remember is that if you have a normal distribution, you can *accurately predict the output* of the process. This is the reason why control charts are such powerful tools for process control. Even though the individual measurement may not be normally distributed, the *averages* plotted on a control chart will tend to take the shape of the normal distribution. Therefore, if the control chart is out of control, you know that it is abnormal variation (special causes) that are affecting the process.

Definitions and Symbols

There are a few definitions and symbols that are important to understand at this point.

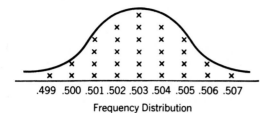

.499 .500 .501 .502 .503 .504 .505 .506 .507

Frequency Distribution

Actual Diameters	Number Of Pieces
0.499	1
0.500	2
0.501	4
0.502	5
0.503	6
0.504	5
0.505	4
0.506	2
0.507	1
Total	30

Figure 10.2 Frequency distribution and data.

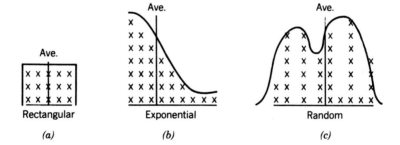

Ave.   Ave.   Ave.

Rectangular   Exponential   Random

(a)   *(b)*   *(c)*

Figure 10.3 (*a*) Rectangular distribution. (*b*) Exponential distribution. (*c*) Random distribution.

Definitions

Variables. Variables are measurable characteristics of the part, such as inches of length, weight in pounds, or diameters in millimeters.

Attributes. Attributes are characteristics that must be (or are chosen to be) studied in terms of whether they are good or bad, on or off, Go or NoGo, and so forth.

Variable control charts. These charts require variable measurements. The X-bar ($\overline{X}$) or range (R) charts are examples. The X-bar and R chart are two charts in one. (This is the prime subject of this chapter.) Other variables charts include X-bar sigma, median charts, charts for individuals, charts for moving ranges, and regression charts.

$\overline{X}$ **chart.** A control chart for averages (average length, weight, etc.) A measure of how the process is centered.

R **chart.** A control chart for ranges (the difference between the largest and the smallest measurement). A measure of the variation (spread) of the process.

X-bar (the average). X-bar $(\overline{X})$ is the average of measured values. It is found by adding the values then dividing by the number of values added.

Example: Shaft lengths 5 in., 6 in., 9 in., 8 in., and 5 in.

$$X\text{-bar } (\overline{X}) = \frac{5 + 6 + 9 + 8 + 5}{5} = \frac{33}{5} = 6.6$$

Range (R). The range is simply the difference between the largest value and the smallest.

Example: 5 in., 6 in., 9 in., 8 in., and 5 in.

$$\begin{aligned}\text{Range } (R) &= \text{max.} - \text{min.} \\ &= 9 - 5 \\ &= 4\end{aligned}$$

Subgroup (sample) (n). A few parts measured consecutively at certain times during the process. X-bar and R charts are usually five pieces.

Symbols

Σ means sum (or add).

$\hat{\sigma}$ is the approximate standard deviation using $\overline{R}$.

k, means the total number of values added.

n, means the sample size.

X_i, means one "individual" measurement.

X-bar $(\overline{X})$ is the average.

X-double-bar $(\overline{\overline{X}})$ is the grand average of all of the $\overline{X}$ values.

s is the sample standard deviation.

R is the range.

$\overline{R}$ (R-bar) is the average range.

$A_2\, D_4\, D_3\, d_2$ are certain constant numbers depending on the sample size. See the table in Figure 10.4.

R_m is the moving range.

$\tilde{X}$ is the median or midpoint.

		Table of Factors		
n	A_2	D_4	D_3	d_2
2	1.880	3.268	0	1.128
3	1.023	2.574	0	1.693
4	0.729	2.282	0	2.059
5	0.577	2.114	0	2.326

Figure 10.4 Table of X–R chart factors.

Normal Distribution Practice Problems and Solutions

1. The following surface finish values were obtained and are shown with the frequency at which each value occurred. Is the distribution of these values a normal distribution?

Surface Finish	Frequency
2	1
3	2
4	3
5	4
6	5
7	6
8	6
9	4
10	3
11	2
12	1

Solution: Arrange the values in a distribution (as shown below).

```
 2×
 3× ×
 4× × ×
 5× × × ×
 6× × × × ×
 7× × × × × ×
 8× × × × × ×
 9× × × ×
10× × ×
11× ×
12×
```

2. What is the percent of area between minus 1 sigma and plus 2 sigma?
   Solution: a. From the mean to minus 1 sigma is ½ of 68.26 percent, which is 34.13 percent.

b. From the mean to plus 2 sigma is ½ of 95.46 percent, which is about 47.73 percent.

c. These two percentages added will give you the area between minus 1 sigma and plus 2 sigma.

$$
\begin{array}{r}
34.13\% \\
+\ 47.73\% \\
\hline
81.86\% \quad \textit{Answer}
\end{array}
$$

3. The following shaft diameters were obtained. Is the distribution normal?

Shaft Dia.	Frequency	Solution
0.500	2	0.500 ××
0.501	3	0.501 ×××
0.502	4	0.502 ××××
0.503	5	0.503 ×××××
0.504	7	0.504 ×××××××
0.505	8	0.505 ××××××××
0.506	9	0.506 ×××××××××
0.507	10	0.507 ××××××××××
0.508	11	0.508 ×××××××××××
0.509	5	0.509 ×××××
0.510	3	0.510 ×××

No, the distribution is not normal.

4. The following data have been collected. Is the distribution normal or not? (Answer Yes or No.)

Shaft Dia.	No. of Pieces
0.499	2
0.506	12
0.500	3
0.507	10
0.501	4
0.505	16
0.502	3
0.504	14
0.503	13

Solution: Arrange the data in order (either ascending, or descending order). I chose descending (or from the largest number down). Then put an × beside the data point for each number of times it occurred.

Descending order:

0.507	× × × × × × × × × ×	(10 times)
0.506	× × × × × × × × × × × ×	(12 times)
0.505	× × × × × × × × × × × × × × × ×	(16 times)
0.504	× × × × × × × × × × × × × ×	(14 times)
0.503	× × × × × × × × × × × × ×	(13 times)
0.502	× × ×	(3 times)
0.501	× × × ×	(4 times)
0.500	× × ×	(3 times)
0.499	× ×	(2 times)

Answer: No, the distribution is not normal (or equally distributed about the mean.)

MEASURES OF CENTRAL TENDENCY

In statistical quality control, there are various measures of *central tendency*. Only three of these will be discussed here:

Mean   Mode   Median

Measures of central tendency are simply one number (or value) that represents the center of a group of data. These measures of *averages* are vital to SPC in many ways. They are often used in conjunction with measures of *dispersion* (or variation), which will be discussed later.

The Mean [Called *X*-Bar ($\overline{X}$)]

The mean is simply the arithmetic average of a group of numbers. For example, if there are five numbers in all, you simply add them up and divide the sum by 5.

$$\text{Formula: } \overline{X} = \frac{\Sigma X}{n}$$

Note

$\overline{X}$ is the mean (or average).
X represents each number.
Σ means summation (or add).
n is the sample size (or the amount of numbers).

Example

(10 Numbers)

$$4 \quad 3 \quad 7 \quad 9 \quad 8 \quad 3 \quad 2 \quad 4 \quad 6 \quad 8$$

Find $\overline{X}$.

Solution

$$\overline{X} = \frac{4 + 3 + 7 + 9 + 8 + 3 + 2 + 4 + 6 + 8}{10}$$

Answer: $\overline{X} = 5.4$.

The Mode

The mode of a group of numbers is the easiest of all to find. It is simply the one single number that occurs most often in a group of numbers.

Example

(10 Numbers)

$$8 \quad 2 \quad 3 \quad 6 \quad 7 \quad 9 \quad 4 \quad 2 \quad 3 \quad 2$$

Solution   The mode here is 2 because the number 2 occurs *three* times above, which is more times than any of the other numbers occur.

Note. Groups of data can have more than one mode (or can be multimodal).

Example

Bimodal (2 Modes)

$$2 \quad 2 \quad 3 \quad 5 \quad 3 \quad 6 \quad 8$$

Answer: The modes are 2 and 3.

Example

Trimodal (3 Modes)

$$3 \quad 2 \quad 5 \quad 2 \quad 5 \quad 3 \quad 6$$

Answer: The modes are 2, 3 and 5.

The Median (or Midpoint)

The median is the value in the middle where one-half of the data are above it and the other half below it, when the data have been *ordered*.

Ordering data. This means rearranging the data and listing them in ascending or descending order.

Example

Data are 7, 3, 2, 4, 5.

ASCENDING ORDER	DESCENDING ORDER
2	7
3	5
4	4
5	3
7	2

There are two formulas for the median, depending on whether the sample size (n) is an *even* number or an *odd* number.

MEDIAN
ODD NUMBER OF DATA ITEMS

$$\text{median} = \left(\frac{n+1}{2}\right)\text{th value}$$

(when the data are ordered)

Example: 9, 2, 5, 13, 6, 1, 11

Ordered: 13

11   $\text{Median} = \dfrac{n+1}{2}$

9

6   $= \dfrac{7+1}{2}$

5

2   $= \dfrac{8}{2}$

1

$= 4$ (4th value)

So, the median is the fourth number from either end.
Answer: 6

MEDIAN
EVEN NUMBER OF DATA ITEMS

$$\text{Median value} = \frac{M1 + M2}{2}$$

(when the data are ordered)

$$M1 = \left(\frac{n}{2}\right)\text{th value}$$

$$M2 = \left(\frac{n+2}{2}\right)\text{th value}$$

Example: 9, 3, 7, 5, 2, 1, 5, 3, 5, 8
Ordered: 9

8   $M1$ here is $= 5$

7   $M2$ here is $= 6$

6

6   $\text{Median} = \dfrac{M\ 1\text{st} + M\ 2\text{nd}}{2}$

5

3   $= \dfrac{11}{2}$

3

2   $= 5.5$

1

So, the median is 5.5.

Central Tendency Practice Problems and Solutions

1. Find the mean of the following data: 8, 6, 5, 9, 4, 7, 8, 10, 2, 4.
Solution:

$$\text{Mean} = \frac{8 + 6 + 5 + 9 + 4 + 7 + 8 + 10 + 2 + 4}{10} = \frac{63}{10} = 6.3$$

2. Arrange these numbers in descending order: 5, 4, 8, 1, 6, 4, 3, 5, 6.
   Solution: Descending order is from the largest down to the smallest number. The numbers are 8, 6, 6, 5, 5, 4, 4, 3, 1.
3. Arrange these numbers in ascending order: 3, 9, 3, 5, 6, 1, 5, 6, 8, 8, 2.
   Solution: Ascending order is from the smallest to the largest number. The numbers are 1, 2, 3, 3, 5, 5, 6, 6, 8, 8, 9.
4. Find the mode of these numbers: 1, 6, 3, 8, 3, 6, 8, 5, 3, 6, 5, 3, 4, 3.
   Solution: The mode is the number that occurs more often than any other number.
   The number above that occurs most often is 3.
5. Find the median of the following numbers: 3, 8, 6, 5, 1, 7, 9.
   Solution: The median of an *odd* set of numbers is found by arranging them in ascending (or descending) order, and it is the value where there are the same amount of numbers above it as there are below it.
   The numbers are 1, 3, 5, 6, 7, 8, 9
            6 is the median here.
6. Find the median of the following even set of numbers: 5, 3, 7, 2, 5, 9, 4, 6, 8, 7.
   Solution: There are 10 numbers ($n = 10$).

$$\text{Median} = \frac{M1 + M2}{2} = \text{median of ordered data} \qquad M1 = \frac{n}{2} = \frac{10}{2} = 5$$

$$\text{Median} = \frac{5 + 6}{2} = \frac{11}{2} = 5.5\text{th number (or 5.5)} \qquad M2 = \frac{n + 2}{2} = \frac{12}{2} = 6$$

7. What is the mode of these numbers (or data)? 4, 6, 7, 2, 4, 2, 6, 8, 9, 10, 4, 2, 6.
   Solution: The numbers 2, 4 and 6 are each listed three times; therefore, the mode of this group of data is trimodal (or has three modes).

MEASURES OF DISPERSION

In statistical process control there are three main measures of dispersion to understand: *range, variance,* and *standard deviation.*

Range, *R*

The range, R, is easy to find. It is simply the highest value minus the lowest value in the data.

Example

You measure the Rockwell hardness (C scale) of five shafts, and get 58, 61, 60, 64, 55.

TO FIND THE RANGE

The highest value = 64
The lowest value = 55
Range $(R) = 64 - 55$

Answer: R = 9.

Variance, V

The variance, V, is equal to the sum of the deviations from the mean, squared, and divided by the sample size. The formula for variance is

$$\text{variance } (V) = \frac{\Sigma(X - \overline{X})^2}{n}$$

where Σ means the sum of.

Note. Use n for population variance and $n - 1$ for sample variance.

X is each value.
$\overline{X}$ is the average (mean).
n is the sample size.

Calculating the variance requires a few steps.

Step 1. Find X-bar $(\overline{X})$.
Step 2. List the X's (or the value of each measurement) in a column.
Step 3. From each X, subtract X-bar.
Step 4. Square each answer in step 3.
Step 5. Add all of the answers in step 4.
Step 6. Divide the sum in step 5 by the sample size.

Example

(using the Rockwell hardness readings)

Step 1 (Finding the Mean, $\overline{X}$	Step 2 X Values	Step 3 $(X - \overline{X})$ $(\overline{X})$ Subtracted from X	Step 4 (Square Each Answer in Step 3)
58	58	−1.6	2.56
61	61	1.4	1.96
60	60	0.4	0.16
64	64	4.4	19.36
55	55	−4.6	21.16
$298 \div 5 = 59.6\ (\overline{X})$			$\Sigma = 45.20$

Step 6. $45.20 \div 5 = 9.04$ (variance)   Add the answers in step 4

Standard Deviation, σ

The population standard deviation is the square root of the population variance.

$$\sigma = \sqrt{V} = \sqrt{\frac{\Sigma(X - \overline{X})^2}{n}}$$

Notes

1. For sample standard deviation, use $n - 1$ in the equation.
2. The variance is equal to the standard deviation squared.

$$V = \sigma^2$$

COMPUTING THE AVERAGE, RANGE, AND STANDARD DEVIATION WITH NEGATIVE NUMBERS IN THE DATA

Process control charts are often structured in such a way that negative numbers appear in the data. If this occurs, computations must be carefully made to avoid error. The following suggestions can be used to make those computations.

Computing the Average ($\overline{X}$)

The average is the sum of all of the values divided by the total number of values. Consider the following data.

$$5, 3, 4, -2, 0$$

The addition of these numbers can be confusing if the correct rules are not followed. A quick rule to use is: add all of the positive numbers and subtract the negative number from the total. Therefore, the sequence of the problem above is

$$5 + 3 + 4 + 0 - 2 = 10$$

To compute the average, it must be remembered that the 0 is a value. Therefore, there are five numbers in all. The average is $\frac{10}{5}$ or 2.

Computing the Range (R)

The range is the largest value minus the smallest value in the data. To compute the range with negative numbers in the data, simply follow this rule: To subtract a negative number from a positive number, change its sign and *add* it. For an example we will use the same values previously mentioned (5, 3, 4, -2, 0).

$$
\begin{aligned}
R &= 5 - (-2) \\
&= 5 + 2 \\
&= 7
\end{aligned}
$$

For further explanation of this rule, refer to the discussion of the real number line in Chapter 2.

Computing the Standard Deviation (σ)

The steps to compute the standard deviation (σ) are covered earlier in this chapter (using positive values). Using negative values, the problem surfaces in the step where $X - \overline{X}$ is required. In this step, positive numbers must often be subtracted from negative numbers. In this case, the following rule applies: To subtract a positive value from a negative value, add the two values and the answer will still be negative.

Example

$$(-2) - (+3) = -5$$

After this obstacle is overcome, the calculation data can continue as was shown earlier in this chapter.

Dispersion Practice Problems and Solutions

1. Find the range of the following numbers: 3, 6, 1, 6, 8, 2, 5, 8, 9.
   Solution: Range is the largest number minus the smallest. The largest above is 9, and the smallest is 1.

$$\text{Range } (R) = 9 - 1 = 8 \text{ answer}$$

2. Find the variance of the following numbers: 4, 6, 8, 2, 4, 5.
   Solution: The variance is the sum of X minus $\overline{X}$ squared, then divided by the sample size.

X	$X - \overline{X}$	$(X - \overline{X})^2$
4	−0.8	0.64
6	1.2	1.44
8	3.2	10.24
2	−2.8	7.84
4	−0.8	0.64
5	0.2	0.04

$$\overline{X} = \frac{4 + 6 + 8 + 2 + 4 + 5}{6} = \frac{29}{6} = 4.8$$

Sum is 20.84; then $20.84 \div 6 = 3.5$, the variance.

3. Find the standard deviation of the numbers above. The standard deviation is the square root of the variance. The variance above is 3.5 and the square root of 3.5 is 1.87.

4. Find the standard deviation of these numbers: 5, 2, 8, 6, 4.

Solution: The standard deviation is the square root of the variance; *or the* square root of the sum of $X - \overline{X}$ squared, then divided by the sample size.

X	$X - \overline{X}$	$(X - \overline{X})^2$
5	0	0
2	−3	9
8	3	9
6	1	1
4	−1	1

$$\overline{X} = \frac{5 + 2 + 8 + 6 + 4}{5} = \frac{25}{5} = 5$$

Sum is $20 \div 5 = 4$, the variance.
Four is the variance, so the square root of 4 is the standard deviation.
Answer: 2.

5. Process A has a 6 sigma spread of 0.0003 in. and process B has a 6 sigma spread of 0.0009 in. Process A costs $50 to produce a part, and process B costs $10 to produce a part. If the total tolerance of the dimension is 0.006 in., what process would you choose to produce the part?
Solution: Both processes will produce the part easily, simply by comparing their 6 sigma spread to the tolerance (total). Therefore, use process B and save $40 per piece produced.

DEFINITION OF A PROCESS IN CONTROL

A process is said to be in a state of control when special causes of variation have been eliminated to the extent that the points plotted on a control chart remain within the control limits, and exhibit random (normal) variation between control limits. (A state of control does not indicate a capable process.)

Natural Patterns of Variation (Figure 10.5)

When points are plotted on a control chart, the following guidelines can be used to check for control.

1. No points are outside the control limits.
2. Points are almost equally disposed on both sides of the central line.
3. About two-thirds of all points are near the central line.
4. Only a few points are near either control limit.
5. There are no continuous runs of about four or more points on one side of the centerline.

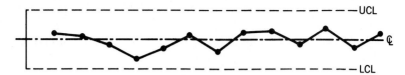

Figure 10.5 A process in control.

6. There are no groups of continuous points heading in a trend toward either control limit.
7. There are no hugging patterns (e.g., all points lie within the plus or minus 1σ zones).
8. There are no straight-line patterns.

Some Advantages of a Process in Control

Some advantages of having a process in control are given in the list that follows.

1. There is more uniformity (less variation) between units, which decreases the chances of producing a defective product.
2. Fewer samples are necessary to judge product quality, since they are more uniform.
3. Inspection costs are reduced.
4. A more accurate definition of capability and sound business decisions can be made such as:
   a. Selection of specification limits.
   b. Knowledge of the yield of the process.
   c. Selection of the appropriate process.
5. The percentage of product that will be produced within specification limits can more accurately be calculated (% yield).
6. Considerably less scrap, rework, and other wastes of productive time and money.

CONTROL LIMITS VERSUS SPECIFICATION LIMITS

CONTROL LIMITS	SPECIFICATION LIMITS
1. Must be calculated from data gathered during the process. Calculate grand average ($\overline{X}$) and average range (R) *Example:* Subgroup size $= 5$	Are given on the print, operation sheet, or other specifications. *Example:* 4.000 ± 0.010 Specification limits are 4.010 and 3.990

CONTROL LIMITS

Number of subgroups = 10

$$\overline{\overline{X}} = 4.001$$
$$\overline{R} = 0.007$$

UCL (upper control limit)
LCL (lower control limit)
$$UCL = \overline{\overline{X}} + A_2(\overline{R})$$
(when the sample size is 5, A_2 = 0.577)

$$= 4.001 + (0.577)(0.007)$$
$$= 4.001 + 0.004$$
$$= \underline{4.005}$$

$$LCL = \overline{\overline{X}} - A_2(\overline{R})$$
$$= 4.001 - (0.577)(0.007)$$
$$= 4.001 - 0.004$$
$$= \underline{3.997}$$

SPECIFICATION LIMITS

UTL (upper tolerance limit) = 4.010
LTL (lower tolerance limit) = 3.990

2. Are used to determine whether or not the *process* is in *statistical control*.

3. For process control and capability, the control limits must be within the specification limits.

4. If a plotted point is outside the control limits, it indicates that the process is *not* in statistical control, and assignable causes exist at the time the sample was drawn for inspection, as shown in Figure 10.6.

Are used to determine whether or not the parts inspected are per the specification.

To produce parts at minimum costs, the specification limits should be outside the control limits. This indicates that the process has the capability to produce parts per specifications.

If a plotted point is within the control limits, but outside the specification limits (see Figure 10.7).

This indicates that the process is in statistical control, but *does not have* the capability of producing the parts to specifications. So either the process should be improved, or the speciation should be changed.

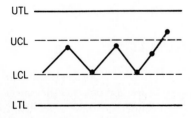

Figure 10.6 Control limits are inside tolerance limits.

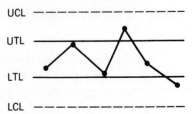

Figure 10.7 Control limits are outside tolerance limits.

TABLE FOR THE INITIAL SELECTION OF SUBGROUPS SIZES AND INTERVALS

There is no absolute rule for how often we should sample. The realities of the factory layout and the cost of sampling must be balanced with the value of the data obtained. In general, it is best to sample quite often at the beginning, and reduce the sampling frequency when the data permit. Table 10.1 can be used for estimating the amount of initial sampling required.

TABLE 10.1

Production Rate per Shift	Total Number of Pieces to Be Sampled (per Shift)
1 to 65	5
66 to 110	10
111 to 180	15
181 to 300	25
301 to 500	30
501 to 800	35
801 to 1,300	40
1,301 to 3,200	50
3,201 to 8,000	60
8,001 to 22,000	85

For example, if the process is expected to produce 3000 pieces per shift, then from the table we should sample 50 pieces per shift. If we use the typical subgroup size of five, then $\frac{50}{5}$ (or 10 samples need to be taken during the shift).

On an 8-hour shift there are 480 minutes. Therefore, we need to take a sample every 48 minutes ($\frac{480}{10}$). Thus we take a sample of five pieces every 48 minutes in this case.

Note. Large sample sizes tend to make control limits more sensitive than smaller samples.

VARIOUS PLOTTING ERRORS TO AVOID

There are several ways a control chart can be misplotted. A misplotted chart can and will cause a poor decision to be made on the process. The following are some specific errors to avoid.

1. Instrument errors
   a. Wrong discrimination (use the 10 percent rule).
   b. Wrong instrument for the measurement.

      c. Instrument is not calibrated.

      d. Instrument is not properly used.

      e. Many other possible measurement errors such as heat, dirt, and other problems.

  2. Miscalculations

      a. $\overline{X}$ miscalculated on the $\overline{X}$ chart.

      b. Range miscalculated on the range chart.

      c. p, np, c, or u miscalculated in the case of attribute charts.*

  3. Plotting the point

      a. Plotting the point on the wrong line.

  4. Sampling errors

      a. Using one chart for two processes/machines.

      b. Biased data (as a result of bias in measurement).

      c. Taking samples too often.

      d. Not taking samples often enough.

HOW MEASUREMENT ERRORS AFFECT CHARTS

MEASUREMENT ERROR	HOW IT AFFECTS THE CHART
Wrong selection of the tool	a. Large distance between plotted points.
a. Discrimination.	b. Rounding error in plotting.
b. Application.	c. Appearance of an out-of-control condition when there is none.
c. Technique.	d. Action based on meaningless data.
	e. Misguiding chart results.
	f. Incorrect control limits.
Untrained personnel	a. Plotted points questionable
	b. Calculations questionable
	c. Meaningless data
Poor precision of the measuring instrument	a. Appearance of in control when it is actually out-of-control, and vice versa.
Incorrect subgroup size and inspection frequency	a. No meaningful data with respect to time.
	b. Control conditions are witnessed too late to take action.

---

    * p chart, control chart for fraction defective; np chart, control chart for number defective; c chart, control chart for number of defects; u chart, control chart for defects per unit.

HOW CALCULATION ERRORS AFFECT CHARTS

ERROR	HOW IT AFFECTS THE CHART
$\overline{X}$ calculated too low or too high, and plotted	a. Machine adjustment when not necessary (overcontrol) b. Consistently in- and out-of-control c. Appearance of two universes
R calculated too high or too low, and plotted	a. Action taken to correct dispersion when dispersion is not a problem b. No action taken to correct dispersion when dispersion is a problem

VARIOUS CONTROLS

When Defectives Are Found in a Subgroup

In cases where there are some defectives found in a subgroup on a control chart, the action to be taken is as follows.

Step 1. Identify the defectives found and separate them.
Step 2. Screen all products made since the last time the chart was "in control."
Step 3. Take action on the process.

Assembly Control with SPC

In those cases where an assembly dimension must be controlled, you must look at the building blocks of that dimension.

Example

Parts A, B, and C are mounted together to form an assembly. The overall length of the assembly must be controlled.

Solution Control the individual dimensions of the three parts that cause the overall length to vary.

Example

An important dimension that is created on the assembly line must be controlled.

Solution   Control that dimension at the assembly line.

Raw Material Control

For raw materials, such as casting and forgings, the control still may fall back on the "building blocks" approach. Control of the input and processing requirements is sometimes necessary.

Example: Castings

The control of castings may be done on the parameters that make the casting, such as temperature and alloy content, or attribute charts may be used to control the end result with respect to defects.

In other cases, such as mixtures, a control chart may be used to control the amount of additives to that mixture to produce a favorable end result.

Sequential Control

One must always realize that *input material variation* at any stage of the process can cause considerable problems. At times it becomes necessary to back up and control the variation at a previous operation to produce a favorable end result.

Example

Operation 10 produces a dimension that is located as a datum in several future operations. This dimension, if it varies widely, can cause the future operations on the part to vary also.

If this dimension is identified as such, it can be controlled so that it will not cause problems down the line.

Future operations cannot be brought into control unless the previous operation is in control.

GUIDELINES FOR STARTING AN AVERAGE AND RANGE CONTROL CHART

The following pages discuss how to start a control chart. The examples use an $\overline{X}$–R chart. Once a characteristic has been selected for a control chart application, the following guidelines will be helpful in starting the chart.

Note. Decide firsthand the exact purpose of the chart.

1. *Measurement.* Make sure that the measuring instrument discriminates to at least 10 percent of the total tolerance to be measured.
2. *Characteristic selected.* Make sure that the characteristic is clearly defined with respect to the technique used for measurement. If not, include any required special instructions.
3. *Related information.* Fill in the appropriate information at the top of the control chart.
4. *Sample size.* Select the subgroup size and time interval between subgroups. This, at first, should be stringent. It can be relaxed later.
5. *Data collection.* A minimum of 25 subgroups or 125 pieces are required to start a control chart. With these subgroups, the "trial" control limits and centerline can be calculated.

6. *Summation of subgroups.* Add each subgroup and enter the sum in the "sum" column.
7. *Averages and ranges.* Find the average $\overline{X}$ (X-bar) and the range (R) and enter them in the appropriate space below each subgroup.
8. *Average range.* Find the average range $\overline{R}$ (R-bar), which is the sum of all the subgroup ranges divided by the total number of subgroups.
9. *Grand average (X grand bar) ($\overline{\overline{X}}$).* Calculate the grand average $\overline{\overline{X}}$ (X-double bar), which is the sum of all the averages $\overline{X}$ (X-bars) divided by the total number of X-bars.
10. *Plot trial centerlines.* $\overline{R}$ (R-bar) is the centerline of the range chart. X-grand bar ($\overline{\overline{X}}$) is the centerline of the X-bar chart. These should be centered in the chart and drawn in solid lines.
11. *Range chart "trial" control limits.* Calculate the trial control limits for the range chart first. This will help you to select the scale for the X-bar chart later. The range chart control limits are calculated as follows. The upper control limit (UCL) = R-bar times the D_4 factor (for the selected sample size). The lower control limit (LCL) = R-bar times the D_3 factor.

Notes
1. See tables for these factors on pages 481–482.
2. Use the UCL and centerline to select the scale that fits the chart.
3. For sample size (n) less than 7, the lower control limit for a range chart is 0.
4. Refer to the Appendix table for constants A_2, D_4, and D_3.

12. Draw the trial control limits for the range chart on the chart in dashed lines.
13. *X-bar chart "trial" control limits.* Calculate the trial control limits for the X-bar chart. These are

$$UCL_{\bar{x}} = \overline{\overline{X}} + A_2(\overline{R})$$
$$LCL_{\bar{x}} = \overline{\overline{X}} - A_2(\overline{R})$$

UCL equals X-grand bar + (A2 times R-bar)
LCL equals X-grand bar − (A2 times R-bar)

Note. Multiply the A2 factor times R-bar first, then add or subtract.
14. Draw the "trial" control limits on the X-bar chart. (Select a comfortable scale on the chart.)
15. *Plotting points (X-bar chart).* Plot the X-bar ($\overline{X}$) for each subgroup on the $\overline{X}$ chart just below the subgroup, and connect the points.
16. *Plotting (range chart).* Plot the range for each subgroup on the range chart (directly under each subgroup).
17. *Initial decisions on the chart.*
    a. Look for any out-of-control conditions. (See the discussion of pattern analysis later in this chapter.)

b. *If there are out-of-control conditions,* attempt to find an assignable (special) cause. If assignable cause is found, recalculate the control limits and centerlines for each chart excluding those data points that were out-of-control.

c. *If there are no out-of-control conditions,* use the previously calculated control limits and centerlines to start the chart.

18. *Acting on out-of-control conditions.* Act on all out-of-control conditions. Remember: If you find assignable cause for these points, exclude these data when revising control limits.

19. *Revision of control limits frequently.* Control limits need to be revised frequently. As the process tends to improve or deteriorate, the control limits will subsequently narrow or widen, respectively.

Note. Always mark the chart to indicate

1. Out-of-control conditions and the cause.

2. Reason the chart was stopped, for example, the machine is down.

3. Reason the chart was started again, for example, a new setup.

4. When control limits were recalculated.

5. Any adjustments made to the process.

6. Any other pertinent information about the process.

X-Bar and Range Chart Practice Problems and Solutions

1. The following *X*-bars are given: 3.2, 3.4, 3.8, 2.9, 3.7. Find the grand average.
Solution: The grand average (*X*-double-bar) is simply adding all of the *X*-bars and dividing the sum by the number of *X*-bars.

$$\overline{\overline{X}} \text{ (grand average)} = \frac{3.2 + 3.4 + 3.8 + 2.9 + 3.7}{5 \text{ of them}} = \frac{17}{5} = 3.4$$

2. Three subgroups of five pieces each were measured after the grand average was found to be 6. Compute the control limits if the average range was 1.5 and see if the process is in control or not (*X*-bar chart).

$$\text{Subgroups:} \quad \begin{matrix} 3 & 4 & 4 \\ 4 & 8 & 6 \\ 7 & 7 & 3 \\ 6 & 5 & 4 \\ 4 & 4 & 5 \end{matrix}$$

Solution: Control limits = $\overline{\overline{X}} \pm A_2(\overline{R})$ = $6 \pm 0.577(1.5)$ = $\begin{matrix} \text{UCL } 6.866 \\ \text{LCL } 5.135 \end{matrix}$

UCL 6.866 —
$\overline{\overline{X}}$ 6.000 ———————————————————————————
LCL 5.135 —

Group 1	Group 2	Group 3
4.8	5.6	4.4

No, the process is not in control. It is out of control considerably on the low side.

3. For a range of 1.0, compute the range chart limits and see if the ranges are in control. The sample size (*n*) is 5.
Solution: *R*-chart limits are

$$\text{UCL} = \overline{R}(D_4) = 1.0(2.114) = 2.114$$
$$\text{UCL} = \overline{R}(D_3) = 1.0(0) = 0$$

Range is out of control at group 2 for some reason (see Figure 10.8).

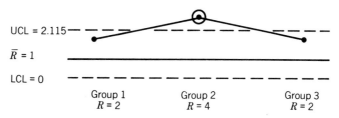

UCL = 2.115

$\overline{R}$ = 1

LCL = 0

Group 1	Group 2	Group 3
R = 2	*R* = 4	*R* = 2

Figure 10.8 One point out at subgroup 2. Note: Values plotted are for example only.

4. On an *X*-bar chart, there is one single point that falls out of the control limits. What do you do?
Solution: A single point falling outside the control limits is not something that should cause wide concern and panic. You should attempt to find the cause, or continue to run the process and monitor the next plotted point closely to see if there is a true problem, or if there is just an outside source of the single point being out.

X-BAR AND SIGMA CHARTS

X-Bar ($\overline{X}$) and sigma (*s*) charts are often used for increased sensitivity to variation (especially when larger sample sizes are used). These charts are more difficult to work with than the $\overline{X}$–*R* charts due to the tedious calculation of the sample standard deviation (*s*).

Sample Standard Deviation(s). The formula is

$$s = \sqrt{\frac{\Sigma(X_i - X)^2}{n - 1}}$$

Note. For population standard deviation, use n, not $n - 1$.

where Σ means the sum

   X_1 is the individual measurements
   X is the average
   n is the sample size

The longhand calculation of (s) is performed in the following manner.

Example   Data:

   5, 6, 7, 6, 8

   Solution

$$n = 5 \qquad \overline{X} = 6.4$$

List the X_i	List $X_i - \overline{X}$	List $(X_i - \overline{X})^2$
5	−1.4	1.96
6	−0.4	0.16
7	0.6	0.36
6	−0.4	0.16
8	1.6	2.56
		Sum (Σ) = 5.20

$$s = \sqrt{\frac{\Sigma(X_1 - \overline{X})^2}{n - 1}}$$

$$= \sqrt{\frac{5.20}{4}}$$

$$= \sqrt{1.30}$$

$$= 1.140 \text{ rounded}$$

In most cases a calculator or computer should be used to avoid the time-consuming mathematics. It must be noted that at times the $\overline{X}$–s chart is not as practical to use as the $\overline{X}$–R chart, for many reasons.

The $\overline{X}$ chart is constructed in the same way as it was described earlier, except that since sigma (s) is used, the control limits are calculated using different factors.

The control limits for an $\overline{X}$ chart using sigma (s) are calculated using the following formula.

$$\text{UCL}_{\overline{x}} = \overline{\overline{X}} + A_1(\overline{s}) \qquad \text{LCL}_{\overline{x}} = \overline{\overline{X}} + A_1(\overline{s})$$

Note. Multiply first, then add or subtract.

where $\overline{\overline{X}}$ is the grand average

   A_1 is a factor depending upon the sample size (see the table in Figure 10.39)
   $\overline{s}$ the average sample standard deviation

The control limits for the sigma (*s*) chart are calculated using the following formula and the table in Figure 10.9.

$$UCL_s = B_4(\bar{s})$$
$$LCL_s = B_3(\bar{s})$$

$\bar{s}$ is the average sample standard deviation. It is the centerline of the ($\bar{s}$) chart.

$$\bar{s} = \frac{\text{sum of the individual sigmas}}{\text{total number of sigmas}}$$

n	2	3	4	5	6	7	8	9	10
B_4	3.27	2.57	2.27	2.09	1.97	1.88	1.82	1.76	1.72
B_3	a	a	a	a	0.03	0.12	0.19	0.24	0.28
A_1	3.76	2.39	1.88	1.60	1.41	1.28	1.18	1.09	1.03

[a] There is no lower control limit for a sigma chart when the sample size (*n*) is less than 6 pieces.

Figure 10.9 Table of factors B_3, B_4, A_1.

Basic Steps to Construction

Step 1. Decide the purpose of the chart.

Step 2. Select the characteristic to be controlled.

Step 3. Collect the data and record the subgroups on the chart. (Select sample size and frequency.)

Step 4. Calculate $\overline{X}$ for each subgroup and record it.

Step 5. Calculate (*s*) for each subgroup and record it.

Step 6. Find $\overline{\overline{X}}$ and record it.

Step 7. Find $\bar{s}$ (the average of the sample standard deviations) and record it.

Step 8. Find the control limits for both charts and draw them on the chart. Select the proper scale.

Step 9. Plot the $\overline{X}$ values on the $\overline{X}$ chart and connect the points.

Step 10. Plot the *s* values on the sigma chart and connect the points.

Step 11. Interpret the chart for "control" the same way as the $\overline{X}$–*R* chart.

Step 12. Revise the control limits if necessary.

Step 13. Continue to use the chart for control.

Capability on an $\overline{X}$–*s* Chart

The approximate standard deviation ($\hat{\sigma}$) called sigma hat can be calculated by

$$\hat{\sigma} = \frac{\bar{s}}{c_4}$$

Refer to Figure 10.10 for the c_4 factors.

If both $\overline{X}$ and s charts are in control, and the individual measurements are normally distributed, process capability can be assessed (see process capability on page 428).

n	2	3	4	5	6	7	8	9	10
c_4	0.798	0.886	0.921	0.940	0.952	0.959	0.965	0.969	0.973

Figure 10.10   Table of c_4 factors.

MEDIAN CONTROL CHARTS

Median control charts are alternative charts to $\overline{X}$–R charts for measured data.

Median. The middle value of grouped data.

The specific advantages of a median chart are that it

1. Is easy to use.
2. Does not require day-to-day calculations.
3. Shows the spread of the process output and gives a picture of the process variation.
4. Shows both the median and the spread.

Both individual values as well as medians are plotted.

Basic Steps to Construction

The following steps can be followed to start a median chart.

Step 1. Gather data (usually 10 or less in the sample).
Odd sizes are more convenient!
Only a single graph is plotted.
Set the scale to indicate the larger of either
a. The specification tolerance (plus some extra space)
OR
b. $1\frac{1}{2}$ to 2 times the range.
The gage used should be at least accurate to 5 percent of the tolerance being measured.

Step 2. Plot *all* of the individual measurements for each group on a vertical line above the group.

Step 3. Circle the median of each group and connect them with lines.

Step 4. Enter the median and range for each subgroup in the data column.

Step 5. Find the average of the sample medians ($\overline{\overline{X}}$) and draw it as the centerline of the chart. Find $\overline{R}$ and record it and calculate the UCL and LCL of range and median (use the table in Figure 10.11).

$$\text{UCL}_R = \overline{R}D_4 \qquad \text{LCL}_R = \overline{R}D_3 \qquad (D_3 \text{ is 0 if } n \text{ is less than 7})$$
$$\text{UCL}_{\tilde{x}} = \overline{\overline{X}} + A_2\overline{R} \qquad \text{LCL}_{\tilde{x}} = \overline{\overline{X}} - A_2\overline{R} \qquad \overline{\overline{X}} \text{ is the average median value}$$

Step 6. Plot the control limits for medians on the chart.

n	2	3	4	5	6	7	8	9	10
D_4	3.27	2.57	2.28	2.11	2.00	1.92	1.86	1.82	1.78
D_3	0	0	0	0	0	0.08	0.14	0.18	0.22
A_2	1.88	1.02	0.73	0.58	0.48	0.42	0.37	0.34	0.36

Figure 10.11   Table of factors for median charts.

Interpretation

Compare the UCL_R and LCL_R with each range. You can mark an index card or piece of plastic with the centerline of range and range control limits, and compare this to the chart values plotted. Put a narrow vertical box around any subgroup that has excessive range.

Mark any median that is beyond the median control limits, and note the spread of medians within the control limits. (Two-thirds of the points should be within the middle third area of the control limits.) Take action on out-of-control conditions.

Capability

1. Estimate sigma ($\hat{\sigma}$) by $\overline{R}/d_2$ for d_2 factors (refer to Figure 10.12).
2. If the process has a *normal distribution*, this estimate of the standard deviation ($\hat{\sigma}$) can be used directly to assess capability as long as the *medians and ranges are in control.*

n	2	3	4	5	6	7	8	9	10
d_2	1.13	1.69	2.06	2.33	2.53	2.70	2.85	2.97	3.08

Figure 10.12   Table of d_2 factors.

CONTROL CHARTS FOR INDIVIDUALS AND MOVING RANGES

Individuals Control Chart

Control charts for individuals use individual readings (X_i) instead of subgroups. Some examples of when an individual's control chart may be used are

1. In short product runs.
2. In destructive testing.

There are some drawbacks to individuals' control charts. Some of these are:

1. You must take care in the interpretation if the distribution of the data is *not* normal.
2. Individuals' charts do not separate the piece-to-piece repeatability of the process.
3. Since you have only one value considered in the subgroup, even if the process is in control, the values of the average $(\overline{\overline{X}})$ and the standard deviation $(\hat{\sigma})$ can have wide variability until you reach about 100 pieces or more.
4. Individuals charts are not as sensitive to changes in the process as the $\overline{X}$–R chart.

It may be better in certain cases to use the $\overline{X}$–R chart with small sample sizes and increased time intervals as opposed to using an individuals control chart.

Steps in Starting the Individual's Chart

Step 1. Collect the data. Individual readings are recorded on the chart from left to right. This can be done on the same format as the X-bar chart except the title of the chart must be changed.

Step 2. Calculate the moving range (Rm) between the individual measurements. The moving range is the difference between the first and second values, the second and third values, the third and fourth values, and so on.
Example: Five Measurements
Data: 7, 8, 10, 9, and 10

The moving range (Rm) between 7 and 8 is 1.
The moving range (Rm) between 8 and 10 is 2.
The moving range (Rm) between 10 and 9 is 1.
The moving range (Rm) between 9 and 10 is 1.

Answer: The four moving ranges are 1, 2, 1, and 1.

Note. There is no moving range for the first number. Since you calculated Rm on each of two values, $n = 2$.

Step 3. Select the scale for the individual's chart (X). This scale could be the largest of two values:

   a. The blueprint tolerance, *or*

   b. About two times the difference between the highest and lowest readings.

    Make sure that you number the scale on the chart.

Step 4. Select the scale for the moving range (Rm) chart. This scale should be the same scale as the individual's chart. Number the selected scale on the chart.

Step 5. Calculate and plot the process average ($\overline{\overline{X}}$), which is the sum of the individual measurements (X_i) divided by the total number of individual measurements (k).

$$\overline{\overline{X}} = \frac{\Sigma X_i}{k}$$

where Σ = sum or add
     X_i = individual measurements
     k = number of measurements

Step 6. Calculate and plot the average moving range ($\overline{Rm}$), which is the sum of the moving ranges (Rm) divided by the total number of moving ranges ($k - 1$).

$$\overline{Rm} = \frac{\Sigma Rm}{k - 1}$$

where Σ = sum or add
       Rm = each moving range
   $k - 1$ = total number of moving ranges

    Remember. There is always one less moving range than there are individual measurements.

Step 7. Calculate and plot the control limits for the individual's chart.

$$\text{UCL}_x = \overline{\overline{X}} + E_2(\overline{Rm})$$
$$\text{LCL}_x = \overline{\overline{X}} - E_2(\overline{Rm})$$

    Note. Always multiply first, then add or subtract. (See the table in Figure 10.13 for E_2 factors.)

n	2	3	4	5
D_4	3.27	2.57	2.28	2.11
D_3	*	*	*	*
E_2	2.66	1.77	1.46	1.29

* For sample sizes less than 7, D_3 is zero.   **Figure 10.13** Table of D_4, D_3, E_2.

Step 8. Calculate and plot the moving range chart control limits

$$UCL_{Rm} = D_4(\overline{Rm})$$
$$LCL_{Rm} = D_3(\overline{Rm})$$

Note. There is no lower control limit for an *Rm* chart if the sample size (*n*) is less than seven pieces.

Step 9. Plot the *X* values measured on the individual's chart and connect the points.

Step 10. Plot the moving ranges (*Rm*) on the moving range chart and connect the points.

General Note. Make sure the chart form:

1. Is filled out at the top with related information.
2. Is identified as an individual's (*X*) and *Rm* chart.
3. Has *X* values recorded in the data blocks.
4. Has *Rm* values recorded in the data blocks.

Interpretation of Individuals' and Moving Range Charts

Control. The moving range chart should be reviewed for points that are beyond control limits. These are signs that special causes exist. Be careful, however, in attempting to act on "trends" on a moving range chart, since this chart uses successive moving ranges.

The individual's chart should be interpreted for

1. Points beyond control limits.
2. The spread between points within the limits.
3. Trends.

Note. The distribution of the data must be normal or some false signals of special causes may occur.

Process capability. The approximate standard deviation ($\hat{\sigma}$), called "sigma hat," can be calculated in the same way as the range chart, except that you are using the average moving range ($\overline{Rm}$). Divide the $\overline{Rm}$ by d_2 to get sigma hat ($\hat{\sigma}$) (refer to Figure 10.14).

n	2	3	4	5	6
d_2	1.13	1.69	2.06	2.33	2.53

Figure 10.14 Table of d_2 factors.

If the process is in control and distributed normally, process capability can be assessed.

THE PRECONTROL TECHNIQUE

There are techniques that can be used to control a process other than those previously discussed in this chapter. It must also be understood that there are certain conditions that must be met for any of them to be effective. Precontrol is a simple method that can be effectively used in certain situations. The following pages basically describe precontrol, how it is used, what it does for the process, and some of its limitations.

A significant difference between precontrol and those process control "tools" previously discussed is the fact that there are no calculations or plotting required. Precontrol uses specification limits and precontrol limits (PCL). It must be understood, however, that the reason why precontrol is easy to use is the fact that it is not as effective as other process controls such as the statistical control charts.

There are certain benefits that are lost when precontrol is used. Some of these are

1. Sensitivity to process variation is reduced.
2. A certain amount of "overcontrol" and "undercontrol" could result, since adjustments are made using individual measurements.
3. Precontrol assists you in meeting a prescribed AQL where other process control techniques do not use AQL's as a goal.
4. Precontrol effectiveness assumes that there is 6 sigma capability and that the process data are normally distributed. Neither of these can be assumed.
5. Precontrol does not assist you in achieving statistical control or measuring process capability.

The preceding discussion is not intended to suggest that precontrol is ineffective. The only intent is to state that precontrol is limited in its ability, and that the choice to use it should be a careful one with the correct goal in mind. Precontrol can be beneficial in many ways.

1. Precontrol is effective in helping keep the process centered, although centering the process is only "half of the battle."
2. Precontrol is very simple to use and to implement.
3. The precontrol technique (if used properly) will decrease the probability of making defective parts to a certain extent.
4. Go–NoGo gaging can be used.
5. Precontrol introduces a relative factor of safety to the process.

Precontrol Limits

Establishing the precontrol limits (PCL) is very simple. First, find one-fourth of the total specification tolerance.

Example

> Stated tolerance = 0.510 − 0.490 diameter
> Total tolerance = 0.020
> One-fourth of total tolerance = one-fourth of 0.20 = 0.005 in.

Next, draw a chart with lines that reflect the upper and lower tolerance limits; the precontrol limits are drawn inside each tolerance limit by one-fourth of the total tolerance.

Example

> Upper specification limit (0.510) _____
> Upper PCL (0.505) ____ ____ ____ ____ ____ ____ ____ ____
> Nominal specification (0.500) _____
> Lower PCL (0.495) ____ ____ ____ ____ ____ ____ ____ ____
> Lower specification limit (0.490) _____

If the process is normally distributed and capable within 6 sigma, there is very little chance that two measurements in a row will fall outside a precontrol limit. If so, adjustment may be necessary.

There is also little chance that two consecutive points will be located such that each one is outside of a different precontrol limit. This is an indication that something must be done to decrease the variation in the process.

Using Precontrol Limits

There are a few general rules to follow when using precontrol limits. To begin, every piece is measured until the rules allow you to switch to frequency gaging. Note that *reset* means to adjust the process toward the center of specification limits.

1. Rules for Resetting the Process.   If the first piece is outside specification limits.
OR
If two pieces in a row are inside the specification limits but outside the same precontrol limit at any time.

2. Rules for Checking the Next Part Prior to a Decision.   If the first piece is inside specification limits but outside precontrol limits.

3. Rules for Continuing the Process.   If the first piece is outside and the second piece is inside a precontrol limit.
OR
If both pieces are inside precontrol limits.

4. Rules Requiring Action to Reduce Variation.   When one piece is outside the upper precontrol limit, and the next piece is outside the lower precontrol limit.

5. Rules for Frequency Gaging. When five pieces in a row fall between precontrol limits. This rule applies at the start and whenever there is a reset. Frequency gaging may continue as long as the average number of checks to a reset is 25. Do not adjust the process until parts exceed a precontrol limit.

a. Frequency gaging may be relaxed if more than 25 checks are made without having to reset the process.
b. Frequency gaging should be increased if there is a need to reset the process prior to 25 checks.
c. An average of 25 checks to a reset indicates that the frequency of gaging is just right.

When used properly, precontrol is a viable tool to reduce the number of defectives from a process, but it should not be confused with the effectiveness of other process control charts such as the *X*-bar and the range chart.

ATTRIBUTE CONTROL CHARTS

These are types of control charts other than the variable charts discussed previously. The following control charts control *attributes*.

Attributes. Attributes are involved in any situation where you have to (or choose to) say that a quality characteristic is good or bad, on or off, Go or NoGo, and the like. An example of this is a part that is gaged with a plug gage (Go–NoGo) or a light that is either off or on.

Various kinds of attribute control charts may be used. The chart used depends on what specific attribute you want to control. The attribute charts covered in this chapter are

- *p* **Chart,** for controlling the *fraction defective.*
- **100***p* **Chart,** for controlling the *percent defective.*
- *c* **Chart,** for controlling the *number of defects.*
- *np* **Chart** for controlling the *number of defectives.*
- *u* **Chart,** for controlling the *number of defects per unit.*

Defect. A single characteristic on a part. Each part can have several defects.
Defective. A part that has one or more defects is still called a defective part.

Example

A washer has many characteristics (we will use only three here). The washer has an inside diameter, outside diameter, and a thickness. If these three characteristics were all out of specification limits, then the washer has three defects. However, it is still only *one defective washer*.

The *p* Chart

The most commonly used attribute control chart is the *p* chart. Remember that *p* means fraction defective, not percent.

Fraction Defective (*p*). The fraction defective (*p*) is the ratio of the number of defectives found divided by the number of parts inspected.

Example

$$\text{Number of parts inspected} = 20$$

$$\text{Number of parts defective} = 3$$

$$p = \frac{3}{20} = 0.15$$

The *p* chart is used in the shop for several reasons, some of which are listed as follows:

1. On the processes that use Go–NoGo gaging, such as plug gages, ring gages, or functional gages.
2. At inspection areas to monitor or control the fraction defective.
3. Any other attribute situation you want to control.

p chart construction

Step 1. Have a clear understanding of the purpose of the chart.

Step 2. Decide what area you want to control.

Step 3. Decide what the sample size will be.

Note. The sample size should be the same if at all possible. The following formulas for control limits only work if it is the same.

Step 4. Record the sample size (*n*) and the number of defectives found (*np*).

Step 5. Calculate *p* for each sample.

$$p = \frac{np}{n} = \frac{\text{number of defectives found}}{\text{number of parts inspected}}$$

Step 6. Find the average fraction defective ($\bar{p}$), which is the centerline of the chart.

$$\bar{p} = \frac{\text{total number of defectives found in all groups}}{\text{total number of pieces inspected}}$$

Note. $\bar{p}$ should not be calculated by adding all of the p values (especially if you have varying sample sizes).

Step 7. Calculate the upper and lower control limits (UCL–LCL).

Formula: If n is constant,

$$\bar{p} \pm 3 \left(\sqrt{\frac{\bar{p}(1 - \bar{p})}{n}} \right)$$

If n is not constant,

$$\bar{p} \pm 3 \left(\sqrt{\frac{\bar{p}(1 - \bar{p})}{\bar{n}}} \right) \quad \text{average sample size}$$

Or, first formula must be used to calculate the control limits for *each point*.

Example

The sample size is constant ($n = 100$) and $\bar{p}$ is calculated to be 0.26.

Control Limits

$$\text{UCL, LCL} = \bar{p} \pm 3 \left(\sqrt{\frac{\bar{p}(1 - \bar{p})}{n}} \right) \quad \text{(since } n \text{ is a constant 100 pieces)}$$

$$= 0.26 \pm 3 \left(\sqrt{\frac{0.26(1 - 0.26)}{100}} \right)$$

$$= 0.26 \pm 3 \left(\sqrt{\frac{0.1924}{100}} \right)$$

$$= 0.26 \pm 3 \,(0.044)$$
$$= 0.26 \pm 0.132$$
$$= \text{UCL} = 0.26 + 0.132 = 0.392$$
$$= \text{LCL} = 0.26 - 0.132 = 0.128$$

Note. If the lower control limit is a minus number, always use 0 instead.

Step 8. Plot the $\bar{p}$ (centerline) and the control limits (UCL = 0.392 and LCL = 0.128) on the control chart.

Note. $\bar{p}$ goes in the center of the chart (solid line) and the scale is then selected to fit the chart.

Step 9. Plot each p on the control chart and connect those points.

Step 10. Look for out-of-control conditions. If you find them and identify the cause, recalculate the $\bar{p}$ and control limits using the remaining data.

Step 11. This shows (since $\bar{p} = 0.26$) that the percent defective on the average is 0.26 $\times$ 100 = 26 percent. Therefore, the yield of good parts that can be expected is 74 percent.

Important. Remember that the formula for control limits using n only works when your sample sizes are all the same size. If your sample sizes are different, you need to

a. Divide by ($\bar{n}$) the average sample size in the formula, *or*

b. Calculate control limits for every plotted point.

Interpretation of the *p* chart.   A point above the upper control limit (UCL) means

1. A mistake *may* have been made in calculating the control limit or plotting the point.
2. The process has worsened either at that point in time or as part of a trend.
3. Something in the measuring system may have changed (the gage, or the inspection method, for instance).

A point below the lower control limit (LCL) means

1. The control limit or the plotted point is in error.
2. The process has improved (this should be studied for improvements that might be made permanent practice).
3. The measuring system (gage, method, inspector) has changed.

p Chart Practice Problems and Solutions

1. Five production runs were completed. Each had some defectives. Compute the control limits for a *p* chart based on the information given in this table.

Run No.	Quantity Made	Quantity Defective	Percent Defective
1	500	4	0.8
2	100	1	1.0
3	300	2	0.7
4	250	2	0.8
5	400	3	0.75
	Sum 1550	Sum 12	

Solution: First calculate $\bar{p}$. *P*-bar is the sum of the quantity defective divided by the sum of the quantity produced (since the quantities produced are not the same).

$$\bar{p} = \frac{\text{sum defective}}{\text{sum produced}} = \frac{12}{1550} = 0.008$$

$$\text{Control limits} = \bar{p} \pm 3 \left(\sqrt{\frac{\bar{p}(1 - \bar{p})}{\bar{n}}} \right)$$

$$= 0.008 \pm 3 \sqrt{\frac{0.008(1 - 0.008)}{310}}$$

$$= 0.008 \pm 3 \sqrt{0.0000256}$$

$$= 0.008 \pm 3 (0.00506)$$

$$= 0.008 \pm 0.01518$$

$$UCL = 0.008 + 0.01518 = 0.023$$

$$LCL = 0.008 - 0.01518 = -0.007 \text{ or } 0$$

2. Using the chart above, would six defective pieces out of a 400-piece run of parts be out of control?
Solution: No, because six defectives out of 400 is 0.015 fraction defective, and the upper control limit is 0.023.

3. Is a *p* chart sensitive to variations in the process?
Solution: No, a *p* chart is simply an attribute chart for controlling the fraction defective produced by a process.

The *np* Chart

The *np* chart is identical to the *p* chart except you are considering the number of defectives (*np*) rather than the fraction defective (*p*). As with the *p* chart, the sample size (*n*) should be constant. The *np* chart is used when:

1. The number of defectives is easier or more meaningful to report.
2. The sample size (*n*) remains constant from period to period.

Steps in using the *np* chart are:

Step 1. Record the constant sample sizes (*n*) on the chart.
Step 2. Record the *number of defectives* found in each sample.
Step 3. Plot the number of defectives found on the chart.
Step 4. Calculate the average (*n$\bar{p}$*).

$$n\bar{p} = \frac{\text{sum of all } np}{\text{total number of subgroups}}$$

Step 5. Calculate the upper and lower control limits.

$$\begin{matrix} UCL \\ LCL \end{matrix} = n\bar{p} \pm 3 \left(\sqrt{n\bar{p} \left(1 - \frac{n\bar{p}}{n}\right)} \right)$$

Interpretation of the *np* chart

Control. The *np* chart has the same interpretation as the *p* chart except that you are talking in terms of the *number of defectives (np)* instead of the fraction defective (*p*).

Capability. The process capability is the average number of defectives (*n$\bar{p}$*).

The 100*p* Chart

The 100*p* chart is the same as the *p* chart except that you convert the fraction defective (*p*) to percent defective (100*p*) by multiplying it times 100.

Example

$$p = \frac{\text{number of defectives found}}{\text{number of units inspected}}$$

$100p = p \times 100$ (answer is in percent, %)

$$\bar{p} = \frac{\text{total number of defectives found}}{\text{total number of parts inspected}}$$

$100\bar{p} = \bar{p} \times 100$ (answer is the average percent defective, %)

Interpretation of the 100p chart.   The interpretation of the $100p$ chart is exactly the same as the p chart except that you are talking in terms of percent defective ($100p$) not fraction defective (p).

Capability.   The capability of the process is $100\bar{p}$, which is the average percent defective.

The c Chart

The c chart is an attribute chart that is used for controlling the number of defects. As mentioned earlier, a defect is a single quality characteristic on the part. One part could have been several defects, but it is only *one defective part*.

The centerline of the c chart is the average number of defects ($\bar{c}$), called c-bar. This is found by adding the total number of defects found and dividing by the number of units inspected.

Example

Five automobiles were inspected and the following number of defects recorded.

AUTO	NUMBER OF DEFECTS
1	12
2	7
3	10
4	3
5	1

Therefore, c for each automobile is 12, 7, 10, 3, and 1. The $\bar{c}$ (or the average number of defects) is found by adding each quantity of defects and dividing by 5.

$$\bar{c} = \frac{12 + 7 + 10 + 3 + 1}{5}$$

$$= \frac{33}{5}$$

$$= 6.6 \quad \text{(this is the centerline of the chart)}$$

The control limits for the c chart are calculated using the following formulas:

$\text{UCL} = \bar{c} + 3\sqrt{\bar{c}}$	$\text{LCL} = \bar{c} - 3\sqrt{\bar{c}}$
$= 6.6 + 3\sqrt{6.6}$	$= 6.6 - 3\sqrt{6.6}$
$= 6.6 + 3\,(2.569)$	$= 6.6 - 3\,(2.569)$
$= 6.6 + 7.707$	$= 6.6 - 7.707$
$= 14.307$	$= -1.107$

Note. If this is a minus number, use 0 instead.

After the centerline and control limits are calculated and plotted, you can plot the individual c values on the chart and connect these points.

When out-of-control conditions are found, you should go back and investigate the particular defects that caused this condition. It is likely that you may find the cause upon investigation.

Interpretation of the c chart

Control.   The c chart is interpreted the same way as the p chart except that you are discussing the number of *defects* (not defectives).

Capability.   The process capability is $\bar{c}$, which is the average number of defects.

c Chart Practice Problems and Solutions

1. Find the centerline of a $\bar{c}$ chart for the following numbers:

ASSEMBLY	NUMBER OF DEFECTS FOUND
1	12
2	8
3	2
4	9
5	4

Solution: Centerline ($\bar{c}$) is equal to the number of defects added, then divided by the number of samples taken.

$$\bar{c} = \frac{12 + 8 + 2 + 9 + 4}{5} = 7$$

2. For the data above, compute the upper and lower control limits of a c chart.
Solution: Control limits (c-chart) $= \bar{c} + 3\sqrt{\bar{c}}$.

Upper limit $= 7 + 3\sqrt{7}$	Lower limit $= 7 - 3\sqrt{7}$
$= 7 + 3\,(2.6)$	$= 7 - 3\,(2.6)$
$= 7 + 7.78$	$= 7 - 7.8$
$= 14.8$ *Answer*	$= 0$ *Answer*

3. After the control chart above was made, five more assemblies were found to have defects. Is the number of defects here in control or not?

ASSEMBLY	NUMBER OF DEFECTS FOUND
6	12
7	10
8	9
9	15
10	16

Solution: No, the defects found at assembly numbers 9 and 10 are out of control.

The *u* Chart

The *u* chart is used to control the number of *defects per unit*. This chart should be used instead of the *c* chart in situations where the sample size (*n*) can vary. The *u* chart has control limits that are "floating" (or recalculated on each point) because the sample size (*n*) is not constant.

Steps in using the *u* chart

Step 1. Collect the data.

Step 2. Record and plot the *u* (defects per unit) for *each subgroup*.

$$u = \frac{c}{n} = \frac{\text{number of defects found}}{\text{number of units inspected}}$$

Note. Units above can mean a part, subassembly, or assembly.

Step 3. Calculate $\bar{u}$, which is the centerline of the chart.

$$\bar{u} = \frac{\text{total number of defects found}}{\text{total number of units inspected}}$$

Step 4. Find and record the square root of each sample size (*n*).

Step 5. Find and record the following for each sample taken.

$$\frac{3(\sqrt{\bar{u}})}{\sqrt{n}}$$

Step 6. Plot the UCL and LCL for each sample.

$$\text{UCL} = \bar{u} + 3\,\frac{\sqrt{\bar{u}}}{\sqrt{n}} \qquad \text{LCL} = \bar{u} - 3\,\frac{\sqrt{\bar{u}}}{\sqrt{n}}$$

Note. If sample sizes vary slightly you may be able to use the square root of $\bar{n}$ (the average sample size) and not calculate the control limits for each point.

Interpretation of the *u* chart. The *u* chart is interpreted the same as the *c* chart except that you are talking in terms of defects found per unit, not just the total defects found.

Example

50 defects were found when five units were inspected.

$$\text{Defects per unit } (u) = \tfrac{50}{5} = 10 \text{ defects per unit}$$

Capability. The capability of the process on a *u* chart is $\bar{u}$, which is the average number of defects per unit.

INTERPRETING CONTROL CHARTS: PATTERN ANALYSIS

There are a wide variety of patterns that may be seen on a control chart, and an even wider variety of special causes that make these patterns appear on a given chart. Each process is different. The amount of *special* causes that exist and the *natural tolerance* of the process depends largely on the condition of that process (wear, tear, methods used, training of operator, etc.).

Example

A new process (new machine, equipment, tooling, etc.) would be expected to have less special and common causes (and a smaller natural tolerance) than a very old process. Therefore, the new process is expected to be better. Most of the time this is true. An exception to this rule is when the wrong process has been selected to produce the end result.

Example

A brand new drill press is selected to produce hole sizes that really require the accuracy of a broaching machine. This process will generally have wider control limits (natural tolerance) than the broach. In addition, the drill press would likely be determined as an incapable process for the required holes.

Control limits. Control limits are established by the inherent variation in a process. These limits are *not* related to blueprint (specification) limits.

In this case, that new drill press could be found to be incapable of producing the required accuracy of the holes; yet it is a brand new machine.

The following examples show some control chart trends for the $\bar{X}$–R charts with some possible causes (not all of them) for these trends.

The Eight Basic Statements of Control

Note. Refer to Figure 10.15. If you can answer "true" to all seven of the statements that follow, the process is in control. A false answer to any statement

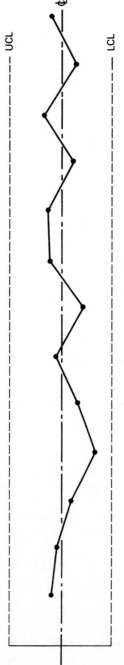

Figure 10.15 State of statistical control.

means that the process is out of control. Use the chart in Figure 10.15 for practice. Several false answers usually mean *trouble*.

1. There are no plotted points that are outside the control limits.	True	False
2. There is a relatively close balance between the total number of points that are above and below the centerline.	True	False
3. There are no consecutive runs of seven or more points on one side of the centerline.	True	False
4. The plotted points, in general, seem to be randomly falling over and under the centerline.	True	False
5. There are no steady trends of points heading directly toward either control limit.	True	False
6. Only a few of all points are near either control limit.	True	False
7. The plotted points do not appear to be "hugging" closely to the centerline with very little distance between them.	True	False
8. There are no straight-line patterns.	True	False

These are the basic features of control. Further control interpretation comes from pattern analysis.

$\overline{X}$-Chart Conditions and Possible Causes

Listed below are some examples of $\overline{X}$ chart conditions and some possible causes.

CONDITION/DESCRIPTION

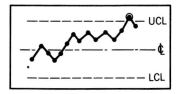

Figure 10.16 Jump shift (run) pattern (high or low).

POSSIBLE CAUSES

1. Change in machine setting.
2. Different operator.
3. Different material, method, process.
4. Minor failure of a machine part.
5. Measuring equipment setting/technique.
6. Adjustment on individuals (overcontrol).
7. New gage.
8. Fixture change.
9. Parts changed.

CONDITION/DESCRIPTION

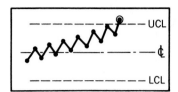

Figure 10.17   Trend pattern (either direction).

POSSIBLE CAUSES

1. Tool wear.
2. Gradual equipment wear.
3. Seasonal effects (temperature/humidity).
4. Dirt/chip buildup on work-holding devices.
5. Operator fatigue.
6. Change in coolant temperature.

CONDITION/DESCRIPTION

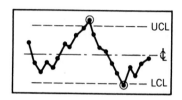

Figure 10.18   Recurring cycles pattern.

POSSIBLE CAUSES

1. Different incoming materials.
2. Cold startup.
3. Seasonal effects (temperature/humidity).
4. Voltage fluctuations.
5. Merging of different processes.
6. Chemical or mechanical properties.
7. Periodic rotation of operators.
8. Measuring equipment not precise.
9. Calculation and plotting mistakes.
10. Machine will not hold setting.

CONDITION/DESCRIPTION

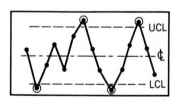

Figure 10.19   Two-universe pattern.

POSSIBLE CAUSES

1. Large differences in material quality.
2. Two or more machines using the same chart.
3. Within the piece variation not considered (such as taper/roundness).
4. Large differences in the method of measurement of the product.

CONDITION/DESCRIPTION

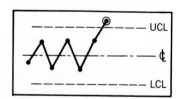

Figure 10.20   One point out (freak).

POSSIBLE CAUSES

1. Power surge.
2. Hardness on a single part.
3. Broken tool.
4. Gage jumped setting.

Range Chart Conditions and Possible Causes

Listed below are some range charge conditions and some of the possible causes for those conditions.

CONDITION/DESCRIPTION

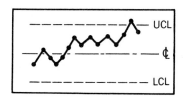

Figure 10.21 Jump shift (run) patterns (above the centerline).

POSSIBLE CAUSES

1. Sudden increase in gear play.
2. Greater variation in incoming material.
3. Inexperienced operator.
4. Miscalculation of ranges.
5. Excessive speeds and feeds.
6. New operator.
7. Change in methods.
8. Long-term increase in process variability.
9. Gage drift.
10. New tools.
11. Fixture change.

CONDITION/DESCRIPTION

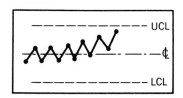

Figure 10.22 Trend pattern (increasing).

POSSIBLE CAUSES

1. Decrease in operator skill due to fatigue.
2. A gradual decline in the homogeneity of incoming material.
3. Some machine part or fixture loosening.
4. Gage drift.
5. Deterioration of maintenance.
6. Tool wear.

CONDITION/DESCRIPTION

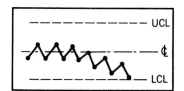

Figure 10.23 Trend pattern (decreasing).

POSSIBLE CAUSES

1. Improved operator skill.
2. A gradual improvement in the homogeneity/uniformity of incoming material.
3. Better maintenance intervals and program.
4. Previous operation is more uniform in its output.

CONDITION/DESCRIPTION

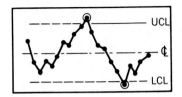

Figure 10.24 Recurring cycles pattern.

POSSIBLE CAUSES

1. Operator fatigue and rejuvenation due to periodic breaks.
2. Lubrication cycles.
3. Rotation of operators, fixtures, and gages.
4. Differences between shifts.
5. Worn tools.
6. Differences between machine needs.

CONDITION/DESCRIPTION

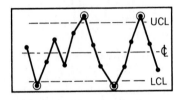

Figure 10.25 Two-universe pattern.

POSSIBLE CAUSES

1. Different machines or operators using the same chart.
2. Materials used from different suppliers.

CONDITION/DESCRIPTION

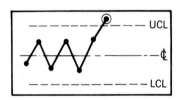

Figure 10.26 One point out (no trend).

POSSIBLE CAUSES

1. Power surge.
2. Hardness of a single part.
3. Tool broke.
4. Gage jumped setting.

PROCESS CAPABILITY

Once a process has been brought into *statistical control* (all *special* causes of variation have been found and eliminated), you can bring out the blueprint (or other specifications) and study the *capability*.

A Capable Process

A capable process

1. Is in statistical control.
2. Is where the process is adequately *centered* between specification limits ($\overline{\overline{X}}$ is near the drawing nominal.)

3. Is where the area between ±4 standard deviations (±4σ) is equal to or less than specification limits.

Note. ±3 standard deviation limits (±3σ) used does not allow $\overline{X}$ to move at all. The extra 1 sigma (1σ) on each end is clearance area.

Standard Deviation (σ)

The standard deviation is a measure of variability. Just as weight is measured in pounds, and length is measured in inches, the variation of a process is measured in terms of standard deviation units. The Greek letter sigma (σ) is used to indicate *one standard deviation.* The area under the normal curve is $\overline{X}$ (the average) plus and minus three standard deviations (see Figure 10.27).

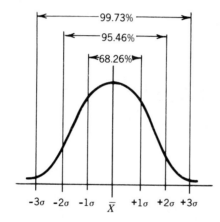

Figure 10.27 Areas under the normal curve.

Example of a process that is capable to +4σ limits (Figure 10.28). The standard deviation can easily be approximated ($\hat{\sigma}'$) on a control chart by dividing

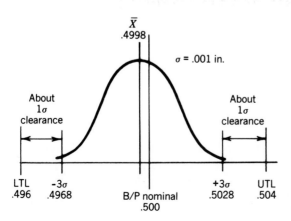

Figure 10.28 Example of ±4 (or 8) sigma capability.

Sample Size (n)	d_2
2	1.128
3	1.693
4	2.059
5	2.326
6	2.534

Figure 10.29 Table of d_2 factors.

the average range ($\overline{R}$) by the factor d_2 (found in the table in Figure 10.29). The d_2 factor depends on the sample size of the control chart (e.g., five pieces per hour).

Example

$$\overline{R} = 0.0023 \qquad n = 5 \text{ per hour}$$

$$\hat{\sigma}' = \frac{\overline{R}}{d_2} = \frac{0.0023}{2.326} = 0.001 \text{ (rounded off)}$$

Definition of a Capable Process

Three primary conditions are needed for a capable process.

1. The process must be in a state of statistical control.
2. The control limits should be well inside the tolerance limits. The goal is usually a process capability ratio (CR) of 0.75 max.

$$CR = \frac{6\sigma}{\text{total tolerance}}$$

where $\hat{\sigma} = \dfrac{\overline{R}}{d_2}$ for control charts, or

σ = one standard deviation for samples
$\overline{R}$ = average range
d_2 = constant for a given sample size

Example

$$\sigma = 0.0005$$

$$\text{Specified tolerance} = \pm 0.0025$$

$$\text{Total tolerance} = 0.005$$

$$CR = \frac{6(0.0005)}{0.005} = 0.6$$

3. Prior to calculating the CR, test the distribution of individuals for normality and centrality (which means that $\bar{X}$ should be close to the nominal dimension).

Distribution Comparisons

Figure 10.30 shows some comparisons between normal distributions and tolerance limits. It is clear that a capable process requires centering and a narrow six sigma spread to the extent that each three sigma limit is well within tolerance limits.

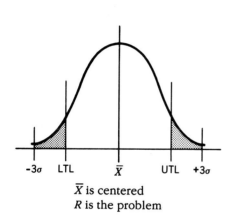

$\bar{X}$ is centered
R is the problem

Figure 10.30a Process spread is too wide.

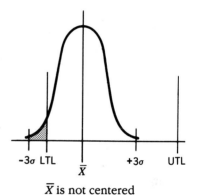

$\bar{X}$ is not centered
R is not the problem

Figure 10.30b Process center needs adjustment.

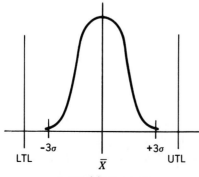

A capable process

Figure 10.30c Process spread is centered and comfortably within tolerance limits.

Some Alternatives for Incapable Processes

When a process is in control* and it is assessed that the process is not capable of producing to specification limits, there are several alternatives that can be considered. Listed below (not in order of selection) are a few of those alternatives.

1. Repair existing equipment.
2. Modify the existing process.
3. Select a better process.
4. 100 percent inspection (this is not advisable and shall only be considered as a last resort).
5. Revise (loosen) drawing tolerances where possible.
6. Purchase new equipment.
7. Selective assembly (this is also not recommended but is applicable in certain special cases).
8. Reconsider the make/buy decision (if the new decision is buy, the source must have a capable process).

There are numerous other alternatives that can be considered. The goal is to achieve a capable process at minimum costs.

Capability Studies: Control Chart Method

There are certain conditions that must be met before you can study the capability of a process and trust the answers you get. These are described in the following steps.

Step 1. Make sure that the process is in stable statistical control before attempting to study the capability.
Step 2. Make sure that the *individual measurements* are normally distributed. This is the point where you do not use the averages on the chart (as they will always be normally distributed).
This is often done by a frequency distribution as shown in Figure 10.31.

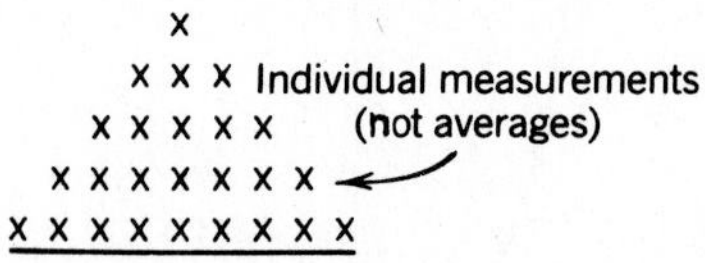

```
            x
          x x x  Individual measurements
        x x x x x    (not averages)
      x x x x x x x ⟵
      x x x x x x x x x
```
Figure 10.31 Frequency distribution of individuals.

Note. The distribution should closely approximate the normal bell-shaped curve. In practice, do not expect them to be perfectly normal.

* If only one point is outside the control limits, it may be that the process is in control but is an isolated cause.

Step 3. Construct a frequency distribution on the individuals and draw lines on the scale to show $\overline{X}$ and each 3 sigma limit (see Figure 10.32).

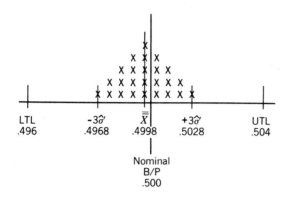

LTL	$-3\hat{\sigma}'$	$\overline{\overline{X}}$	$+3\hat{\sigma}'$	UTL
.496	.4968	.4998	.5028	.504

Nominal
B/P
.500

Figure 10.32 Tolerance lines and 3 sigma lines drawn on the distribution.

Step 4. Draw the nominal dimension and tolerance lines on the same distribution, and calculate $\hat{\sigma}'$, which is $\overline{R}$ divided by d_2.

Step 5a. If the process is centered (as shown in Figure 10.32), a capability ratio (CR) can be used.

$$CR = \frac{6\hat{\sigma}'}{\text{total special tolerance}} = \frac{0.006 \text{ in.}}{0.008} = 0.75$$

Notes

a. For ±4 sigma capability, the ratio should be 0.75 maximum.
b. Capability ratios (CR) are only accurate if the process is normally distributed and adequately centered on the specification.

Step 5b. Another method is the capability index (Cp).

$$Cp = \frac{\text{total special tolerance}}{6\hat{\sigma}'} = \frac{0.008}{0.006} = 1.33$$

Notes

a. For ±4 sigma capability, a Cp index of 1.33 minimum is needed.
b. The Cp index also requires that the distribution be normal and the process be closely centered on the specification limits.

Step 5c. **When the process is not centered on specification limits.** When the process is off-center to specification limits (as shown in Figure 10.33), a Cpk index can be used.

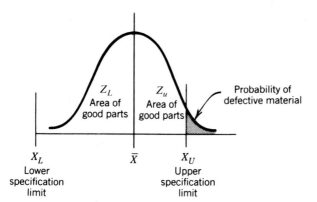

Figure 10.33 At ± 3 (or 6 sigma) capability, there is no room for X to vary.

Using the Cpk Index for Bilateral Tolerances

The Cpk index is a worst-case capability index which applies when processes are off-center of specification nominal, basic, or other ideal values between two specification limits. The Cpk is used in place of the Cp in these cases. Unlike the CR and Cp discussed previously, the Cpk is applicable to specifications with two limits (e.g., limit tolerances, or bilateral tolerances). The Cpk is calculated as follows:

$$\text{Cpk} = \text{the lesser of } \frac{X_U - \overline{X}}{3\sigma} \quad \text{or} \quad \frac{\overline{X} - X_L}{3\sigma} \text{ whichever is smaller}$$

where X_U is the upper specification limit
X_L is the lower specification limit
σ is the standard deviation
$\overline{X}$ is the process average

The resultant Cpk value (like the Cp) should be 1.33 minimum.

Example

The specification limits for the product are 0.500 − 0.508 diameter. The process is normally distributed, the $\overline{X}$ value is 0.505, and the standard deviation (σ) is 0.0008 in. Calculate the Cpk.

Solution

$$\text{Cpk} = \text{the lesser of } \frac{X_U - \overline{X}}{3\sigma} \quad \text{or} \quad \frac{\overline{X} - X_L}{3\sigma} \text{ whichever is smaller}$$

$$\text{Cpk} = \text{the lesser of } \frac{0.508 - 0.505}{3(0.0008)} \quad \text{or} \quad \frac{0.505 - 0.500}{3(0.0008)}$$

$$\frac{0.003}{0.0024} \quad \text{or} \quad \frac{0.005}{0.0024}$$

$$1.25 \quad \textit{Answer} \quad \text{or} \quad 2.08 \quad \textit{Answer}$$

The Cpk is the lesser of 1.25 or 2.08; therefore, it is 1.25 and the process does not meet the goal of 1.33 minimum Cpk.

Using the Cpk Index for Single Limit Specifications

The Cpk must be calculated carefully when dealing with single specification limits such as surface finish (e.g., AA32 max.), runout tolerances (e.g., 0.002 TIR max.), and other similar tolerancing methods which identify only one limit of concern. To calculate the Cpk properly for these tolerances, use the equation

$$\text{Cpk (for single limits)} = \frac{\text{only limit} - \overline{X}}{3\sigma}$$

Example

A runout tolerance of 0.002 in. TIR is specified, the average of the process ($\overline{X}$) is 0.0006 in. runout, and the standard deviation is 0.0001 in. Calculate the Cpk.

Solution

$$\text{Cpk (for single limits)} = \frac{\text{only limit} - \overline{X}}{3\sigma}$$

$$= \frac{0.002 - 0.0006}{0.0003} = \frac{0.0014}{0.0003}$$

$$= 4.67 \quad \textit{Answer}$$

The Cpk of 4.67 is greater than the minimum goal of 1.33. The process is considered capable.

Limitations to Using the CR, Cp, and Cpk

These indexes provide reasonable ratios of process capability when they are properly used. Proper use of the CR and Cp, of course, is to apply them to processes that are centered on specifications. Proper use of the Cpk is to apply it to bilateral limits when processes are off-center and to use the only specified limits when specifications are single sided. All of these indexes assume that the process is normally distributed (individual values, not averages). An understood assumption is that the process average ($\overline{X}$) is within specification boundaries.

Predicting the Percent Yield of a Process

Percent yield is the probability of good product expected from the process based on required specification tolerances and process variability. To compute the yield of a process, one must be able to define specific areas and percentages under the bell-shaped curve (refer to the table of areas under the normal curve on p. 483). This table helps calculate the area (or percentage) of the curve between $\overline{X}$ and a specification limit (based on the standard deviation of the process) by using Z

values. The information necessary to compute a Z value are the average of the process ($\overline{X}$), the upper specification limit (for Z_U, the upper Z score), and the lower specification limit (for Z_L, the lower Z score). Once a Z value is computed (rounded to two decimal places) the area (percent) under the curve can be found in the Appendix (p. 483).

The table is used by finding the whole number and first decimal place of the Z value in the Z column, then finding the column that shows the second decimal place of the Z value. For example, to find the area under the curve for a Z value of 2 (or 2.00), first use 2.0 in the Z column, then use the second decimal place (0) to select the appropriate column (0 column) in this case. Converge these two columns to find the area under the curve (0.47725 or 47.725%).

Note. The values found in the table are decimal fractions. To convert them to percentages, multiply by 100 (or move the decimal point two places to the right). The area found is the *percent yield* of good product expected to be between the process average ($\overline{X}$) and the specification limit.

For specifications with two specified limits, an upper Z value (using the upper tolerance limit) and a lower Z value (using the lower tolerance limit) must be obtained. It is important to note that for single specification limits, only one Z value is needed. The yield for the "other side" will be 49.999%. The following are equations for Z_U and Z_L:

$$Z_L = \frac{\overline{X} - X_L}{\sigma'}$$

$$Z_U = \frac{X_U - \overline{X}}{\sigma'}$$

Example

Percent Yield Calculation   (See Figure 10.34.)

$$Z_L = \frac{\overline{X} - X_L}{\hat{\sigma}'}$$

$$= \frac{0.502 - 0.496}{0.001}$$

$$= \frac{0.006}{0.001}$$

$$= 6$$

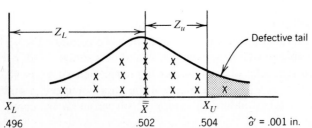

$\hat{\sigma}' = .001$ in.   **Figure 10.34**   Defective tail.

A value of 6 in the table will cause you to read the maximum value of 0.49999 or 49.999 percent.

$$Z_U = \frac{X_U - \overline{X}}{\hat{\sigma}'}$$

$$= \frac{0.504 - 0.502}{0.001}$$

$$= \frac{0.002}{0.001}$$

$$= 2$$

A value of 2 in the table will be 0.47725 or 47.725 percent.

Once you find both Z values in percentage (Z_L = 49.999 percent and Z_U = 47.725 percent) you simply add them to get the probability of good parts. These two values added together represent the percent of area under the normal curve that is expected to be acceptable. Therefore,

$$Z_L + Z_U = 0.97725 \text{ or } 97.725 \text{ percent}$$

This is the probability of good parts (percent yield) if nothing is changed. The probability of defective parts is 100 percent minus 97.725 percent = 2.275 percent in this case. This is the probability of "oversized" parts.

The Reason for ±4 Sigma Capability

All normal distributions have what is called the 6 sigma (6σ) spread. From the center of the distribution ($\overline{X}$) there are three equal standard deviations to the right, and three equal standard deviations to the left (see Figure 10.35).

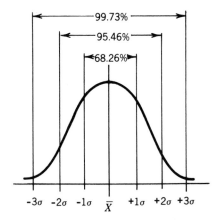

Figure 10.35 Areas under the normal curve.

If you have a normal distribution and you know what one standard deviation is equal to for example, $\sigma = 0.001$ in., you can predict very accurately what the process will do from a given center ($\overline{X}$), which is typically the setup value.

Example

$$\hat{\sigma}' = 0.001 \text{ in.}$$

$$\overline{X}' = 0.500$$

The process is in control. The distribution of the individuals is normal. If the process is set up on 0.500, the worst parts produced within plus or minus 3 sigma (or 99.73 percent of the time) will be as high as 0.503 and as low as 0.497 and most parts will be near 0.500 (see Figure 10.36).

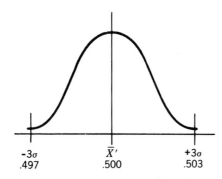

-3σ	$\overline{X}'$	$+3\sigma$
.497	.500	.503

Figure 10.36   The 6 sigma spread of a process.

You can see now that the *natural tolerance* of the process is equal to 6 sigma (6σ), which is 6 times 0.001 in. = 0.006 in.

Natural tolerance. The ability of the process itself. This is what the process can do at best unless you reduce the variation (σ). This natural tolerance has nothing to do with the specification to tolerance (blueprint) other than the fact that you hope it is less than the drawing tolerance.

If the natural tolerance were exactly equal to the specification tolerance, there would be no room for setup error, or external variation (materials variation for example). This is shown in Figure 10.37.

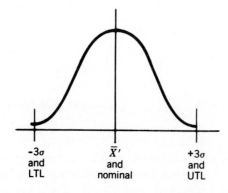

-3σ	$\overline{X}'$	$+3\sigma$
and	and	and
LTL	nominal	UTL

Figure 10.37 Six sigma equals the total tolerance.

If the setup ($\overline{X}'$) moves at all, the curve (or spread) moves with it. Therefore, any move in $\overline{X}'$ here would cause defective parts to be produced.

To solve this problem, we include an extra standard deviation at each end of the curve to allow room for $\overline{X}'$ to move without the probability of making defective parts. This is shown in Figure 10.38.

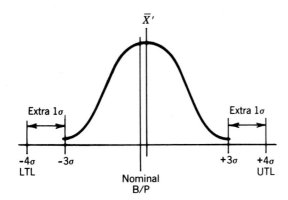

Figure 10.38 ±4 (or 8 sigma) capability.

Remember. The blueprint tolerance is equal to the upper limit minus the lower limit. The natural tolerance is equal to 6 sigma (6σ).

A basic point of view is to find $\hat{\sigma}'$ (which is $\overline{R}/d_2$), then multiply $\hat{\sigma}'$ by 8. If this value is equal to the total tolerance (and the process setup is closely centered on the blueprint nominal size), you have a process that *is capable* to plus or minus 4 sigma ($\pm4\sigma$). Remember, the distribution *must be normal*.

The Potential Capability of a Process

The potential capability of a process can be studied by taking a relatively small sample of *continuous* products from the process, measuring them, and applying basic statistics to get an idea of the *potential* capability. The process potential study is a short-term study that does not give you the long-term accuracy of the control chart method, but does give you an idea of what the process can do.

Steps in a Process Potential Study

Step 1. Identify the process that is to be studied, and decide the purpose of the study.

Step 2. Have the process set up ideally as close to the nominal specification size as possible. This is not a must but does simplify the study.

Step 3. Make sure that the operator of the process understands what you are doing. You are not studying the person, just the process.

Step 4. Have the operator let the process run the consecutive pieces (a minimum of 30 pieces, but 50 to 100 are recommended) without any adjustments.

Step 5. Also have those pieces numbered in the sequence from which they came out of the process (in case further analysis is needed).

Step 6. Inspect the parts and record the sizes on the data sheet.

Step 7. Plot the data on a frequency distribution (check for normality). The process should be a normal distribution for the study to be accurate.

Step 8. Draw on the frequency distribution the nominal size and the specification limits.

Step 9. Calculate the average $(\overline{X})$ and the standard deviation (σ).

Step 10. Draw the average $(\overline{X})$ on the frequency distribution.

Step 11. Calculate the area under the normal curve (refer to the preceding pages on capability and the table of areas under the normal curve on page 483.

Step 12. It is recommended for a short-term study like this that a capability within ±4 sigma be demonstrated.

Process potential studies (on a short-term basis) involve studying a small group of consecutively produced parts (usually 30, 50, and 100 pieces). However, to simplify the example given here, I choose to only use 10 pieces (not recommended for actual use) and a two place decimal for the tolerance.

Example Study

Machine number XX has been chosen to produce a shaft diameter with a tolerance of 0.50 ± 0.01 diameter. It is necessary to study machine XX to see if it is capable of producing this 0.02 total tolerance.

Step 1. Set up the machine to produce the shaft (preferably as close to 0.50 nominal dimension as possible on the setup).

Step 2. Once the machine is set up, run the entire sample (10 pieces here) without any adjustments. *Let the machine run the parts.*

Step 3. Inspect the sample (10 pieces) and record each measurement to at least one more decimal place accuracy.

> Pieces recorded: 0.503 0.509 0.507 0.495 0.498
> 0.506 0.504 0.492 0.496 0.506

Step 4. Calculate the standard deviation (using a statistical calculator or the method shown under standard deviation). Use $n - 1$ in the calculation, since it is a sample. The standard deviation of these numbers is 0.0059. (Also calculate $\overline{X}$ (X-bar), which is 0.5016 in this problem.)

Now we have computed the two main values necessary for finding the potential capability of the machine. It is time now to compare the blueprint tolerance of the shaft to the machine spread. Immediately, we can see that there is a problem because the 6 sigma spread of the machine (which is $6 \times 0.0059 \pm 0.035$) is larger than the total tolerance of 0.02 that we want to produce. For a process (or machine) to be capable, the spread of the machine should be *less than the total tolerance being produced.*

Step 5. The next step involves the use of the table of areas under the normal curve. This table will aid you in predicting the % yield (or the expected amount of good parts that this machine will produce according to the study without altering the machine). The table enables you to compare the central tendency and spread of the data to the tolerance limits on the blueprint (see the Appendix table on page 483).

Step 6. Draw a bell curve (see Figure 10.40) putting the blueprint tolerance limits at each end and, on that same curve, draw a line for the X-bar of the data (in its proper place). Also, note the standard deviation below the curve.

Step 7. Now, you must calculate the area between the tolerance limits (X) and the central tendency ($\overline{X}$), using the standard deviation (σ). Since there are two tolerance limits, there are two areas you must calculate and add them together for the total capability. The formula for each area you need is the same:

$$Z = \frac{X = \overline{X}}{\sigma}$$

(Refer to Figure 10.39.)

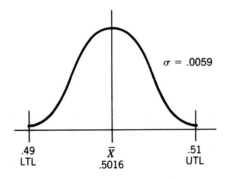

$\sigma = .0059$

.49
LTL

$\overline{X}$
.5016

.51
UTL

Figure 10.39 Capability study.

Z_1 is the area between the upper tolerance limit (0.51) and X-bar (0.5016).
Z_2 is the area between the lower tolerance limit (0.49) and X-bar (0.5016).

So,

$$Z_1 = \frac{X - \overline{X}}{\sigma} \qquad\qquad Z_2 = \frac{X - \overline{X}}{\sigma}$$

$$= \frac{0.51 - 0.5016}{0.0059} \qquad\qquad = \frac{0.49 - 0.5016}{0.0059}$$

$$= \frac{0.0084}{0.0059} \qquad\qquad = \frac{-0.0116}{0.0059} \quad \text{(the minus sign means nothing here)}$$

$$= 1.42 \qquad\qquad\qquad = 1.97$$

Step 8. Now, look at the table in the Appendix to find that the area under the curve for a Z value of 1.42 is 0.4222 (or 42.22 percent) and the area for a Z value of 1.97 is

0.4756 (or 47.56 percent). Therefore, both areas added together are 42.22 + 47.56 percent = 89.78 percent.

For a process to be considered capable within $\pm 3\sigma$ limits, the total area must be 99.730 percent or higher. You can see that in this case, machine XX has a capability of 89.78 percent, and is close to, but not quite capable of producing the shaft tolerance of 0.02 total. Usually, a capability of ± 4 sigma (99.994 percent) is preferred for a potential study. Again, see the table of areas under the normal curve (for the Z values above) in the Appendix.

Process Capability Practice Problems and Solutions

1. You have measured the sample of parts and found the following:

$$\sigma = 0.0009 \qquad X\text{-bar} = 0.5015$$

If the blueprint limits are 0.505–0.495, what is the % yield of the process?

$$\text{Yield} = Z_1 + Z_2$$

Solution: $\quad Z_1 = \dfrac{X_1 - \overline{X}}{\sigma} = \dfrac{0.505 - 0.5015}{0.0009} = 3.88$

$$Z_2 = \dfrac{X_2 - \overline{X}}{\sigma} = \dfrac{0.495 - 0.5015}{0.0009} = 7.22$$

Referring to the table of areas under the normal curve (in the Appendix), we obtain

$$Z_1 = 3.88 = 0.5000 \text{ or } 50 \text{ percent}$$

$$Z_2 = 7.22 = 0.5000 \text{ or } 50 \text{ percent}$$

$Z_1 + Z_2 = 100$ percent (but you never say 100 percent; say 99.999 percent).

Note. You never say 100 percent in any case, because you are never 100 percent sure of anything.

2. Change the blueprint limits above to 0.503–0.500 and predict the yield again.

Solution: $\quad Z_1 = 0.503 - 0.5015 \div 0.0009 = 1.67 = 0.4525$ (in table)

$$Z_2 = 0.500 - 0.5015 \div 0.0009 = 1.67 = 0.4525 \text{ (in table)}$$

$$Z_1 + Z_2 = 0.4525 + 0.4525 = 0.905 = 90.5 \text{ percent}$$

The yield is 90.5 percent here; therefore, the process is not capable, because for a process to be capable within ± 3 sigma limits, the yield must be 99.730 percent or higher.

3. Now, change sigma above to 0.0001 and compute the yield.

$$Z_1 = 0.503 - 0.5015 \div 0.0001 = 15$$

$$Z_2 = 0.500 - 0.5015 \div 0.0001 = 15$$

Both of these Z values in the table are equal to 0.5000 or 50 percent, so when they are added, your yield will be 99.999 percent. The process is now capable.

REVIEW QUESTIONS

1. (True or False) A key word associated with the process control is repeatability.
2. Variables tend to follow the _____ distribution.
3. Calculate the process average ($\overline{X}$) of the following data: 2, 0, 4, 5, −1.
4. Calculate the range (R) of the data in Question 3.
5. Calculate the sample standard deviation (σ_{n-1}) for the data in Question 3.
6. Find the mode of the following data: 5, 6, 7, 6, 5, 7, 8, 4, 5, 5, 7
7. Find the median of the data in Question 6.
8. If the standard deviation is 9, what is the variance?
9. Calculate the UCL of an $\overline{X}$ chart where $n = 3$, $\overline{\overline{X}} = 0.500$, and $\overline{R} = 0.002$.
10. Calculate the UCL of a range chart where $\overline{R} = 0.004$ and $n = 5$.
11. Calculate the UCL of a p chart where $\overline{p} = 0.05$ and $n = 30$.
12. Calculate the UCL of a c chart where $\overline{c} = 20$.
13. Calculate the LCL of a u chart where $n = 30$ and $\overline{u} = 12$.
14. Calculate the capability ratio (CR) where the specification tolerance is 0.750–0.758 and the standard deviation is 0.001 in.
15. In Question 14, find the Cpk if $\overline{X}$ is 0.755.
16. What is the Cpk for a single specification limit of AA32 surface finish where the process average ($\overline{X}$) is 12 surface finish and the standard deviation is 3? (*Note:* Try to calculate the Cpk of this problem and see the difference in the results.)

11

Problem-Solving Techniques

To identify and solve problems effectively, there are three primary considerations: the problem-solving team, steps, and the techniques that are used. Every
problem has one or more true causes and usually, several symptoms. The elimination of the true causes of problems will correct the problems themselves, and
often will prevent the problem from recurring. A fact of life is that the symptoms
of a problem are generally easier to find than causes, and subsequently, symptoms
are often solved, but the problem remains.

PROBLEM-SOLVING TEAMS

The best time to solve a problem is when it occurs, and the best team to solve the
problem is a *natural work team*. A natural work team is a group of people with
different responsibilities who are closest to the process or the problem. Companies should not just gather a team of individuals to solve a variety of company
problems. The best results come when the problem-solving team is selected by
the problem statement.

The *direct members* of a problem-solving team are those who work the
process where the problem originated on a daily basis. In the shop, the direct
members of a team are the operator, the shop supervisor, process setup personnel, and the inspector (if applicable). This membership applies whether it is a
process or a product problem. The *indirect members* of a problem-solving team
are those personnel who have technical responsibility for given subjects related to
the support of the product or process. Indirect problem-solving team members

may be on more than one team at a time. For the shop, they include the manufacturing engineer, quality engineer, and often the industrial engineer. For problems in the machine shop, these direct and indirect members are "naturally" the ones closest to the problem.

There are times when *"consultants"* from other departments are also requested to participate in the solution of a problem. These consultants may be, for example, a tooling engineer, metallurgist, chemist, or a statistician. The outside consultant requested depends on the expertise required by the team to obtain the answers necessary in the problem-solving process.

PROBLEM-SOLVING STEPS

There are five steps to effective problem solving, which are used by problem-solving teams:

1. *Identify the problem* and prepare a clear problem statement.
2. *Quantify the problem* (measure the impact and extent).
3. *Identify the cause(s)* (not symptoms).
4. *Take action to correct the cause(s).*
5. *Follow up* to ensure that the action taken was effective.

PROBLEM-SOLVING TECHNIQUES

There are a variety of tools that can be used at any step of the problem-solving process. Some of these tools are statistical methods, covered in this chapter and Chapter 10. Other tools, though essentially nonstatistical, are used to structure the problem-solving process. Some of these tools are

1. Pareto analysis (refer to Chapter 8)
2. Brainstorming
3. Cause-and-effect analysis
4. Scatter diagrams
5. Statistical control charts (refer to Chapter 10)
6. Statistical experimental designs (not covered in this book)

The Brainstorming Technique

Brainstorming is used considerably by problem solving teams. It is used to define problem areas, probable causes, probable solutions, and for many other things that require ideas. The technique is very simple, but the uses for brainstorming are many.

It is the brainstorming technique that brings the collection of minds together and directs it toward one particular idea, or solution. Everyone participates in the brainstorming, and everyone has his or her own thoughts on the subject.

Brainstorming is also a lot of fun. The team members are allowed (in one phase of the brainstorming) to come up with any idea, no matter how wild, and enter it into the analysis. Sometimes, the wildest of ideas actually has merit. If you do not believe it, consider the success of the Pet Rocks that sold so successfully.

Collective thinking is a very powerful tool in problem solving. When one person might ordinarily be assigned to solve a problem (with only their single outlook on that problem), the task may be overwhelming; but if several people are studying the problem, there is a good chance that they will more readily come up with a solution. This is especially true when the problem exists in their own area of experience.

Some rules for brainstorming

1. Every member shall have a turn to voice his/her idea (only one idea per turn).
2. If no thought comes to mind when it is their turn, they simply say "pass," and it goes to the next member.
3. Ideas can be any possible idea (no matter how wild) during the idea gathering part of brainstorming.
4. No one is allowed to ridicule a person's idea. This can cause ideas to be kept inside and damage the brainstorming process.
5. During the initial brainstorming, there should be no critique made about a member's idea. They are simply listed.
6. The brainstorming is finished when every member in the room has said "pass" in sequence.
7. The ideas are listed by the team leader or delegate as they are given. (Large poster paper is good for this.)
8. By team voting, all of the ideas are narrowed down to a few, then they are voted on again to determine the priorities of the remaining ideas.
9. The end result is that one, or a few ideas (or probable causes), are selected. Brainstorming is used to define problem areas, probably causes, and solutions.

Cause-and-Effect Analysis

Major areas of probable causes.   In the cause-and-effect analysis, there are six major areas that are considered to be the areas of probable causes for problems:

1. Manpower.
2. Methods.
3. Machines.
4. Materials.
5. Measurements.
6. Environment.

Manpower. This category states that the probable cause of the problem is related to the people involved. Some of the reasons why people create problems are

1. Inadequate training.
2. Not following work instructions (methods).
3. Health (eyesight, coordination, and similar problems).
4. Personal problems (inability to concentrate on job needs).
5. Personal tools (poor).
6. Boredom (may be caused by lack of motivation).
7. Fatigue.
8. Lack of communication.

Methods. This category states that the probable cause for the problem is related to the methods by which something is done. The cause for poorly assembled units could be the method (or technique) used to assemble them, or that work instructions are inadequate, or that personnel skills are not suitably matched to those required for the job.

Materials. This category states that the materials used could be the cause of the problem. The raw materials used to make the product could be

1. Too hard.
2. Too soft.
3. Dimensionally poor.
4. Internally defective.
5. Externally defective.

Machines. This category states that the machines could be the cause of the problem. Some machine-related causes include

1. Worn out and/or incapable of the tolerances needed.
2. Poor tooling.
3. Hydraulic problems.
4. Coolant problems.
5. Wrong machine for the job.

Measurements. This category states that measurements could be the cause of the problem. Some measurement-related causes when using variables data include

1. Gage problems, such as wrong tool, wrong gaging method, poor discrimination, and other errors identified in Chapter 4 which can cause defective products to appear to be acceptable and acceptable products to appear to be defective.
2. Observer measuring problems, such as parallax, manipulation, flinching, rounding, feel, training, and several others also covered in Chapter 4.

Some measurement-related causes when using attributes (gaging and visual inspection) data include:

1. Gage problems, such as poor design, too much wear allowance, wrong datums contacted, and wrong-size gaging members.
2. Observer problems, such as no standards, no gage or inspection instructions, fatigue, and boredom.

Environment. This category states that the environment could be the cause of the problem. Some environmental-related causes include temperatures, dust, dirt, noise, distractions, locations, and other causes that have to do with the environment in which work is performed.

One of the best ways for the team to assess the possible cause(s) of a problem is the use of cause-and-effect analysis.

Cause-and-effect analysis (the fishbone technique). The cause-and-effect analysis is often referred to as the *fishbone technique*. It is called this because of the way it is structured. The six main cause areas—manpower, methods, machines, materials, measurements, and environment—are put on "lines" that flow into the actual problem area. The structure appears to be the bone structure of a fish, as shown in Figure 11.1.

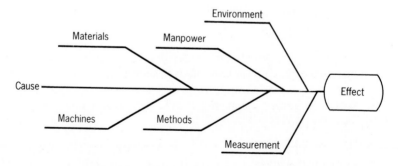

Figure 11.1 Fishbone diagram: cause and effect.

Each of the main causes has other branches from it where the detailed causes are listed. The fishbone technique displays a picture of the probable causes and a definition of the problem. It is very useful in problem solving, and should be kept as a permanent record. Once the selection has been made, the primary cause (according to the voting of the team) is circled on the diagram.

Scatter Diagrams

Scatter diagrams are a simple tool that can be used to see if one variable correlates with another.

Correlation. How an independent variable correlates with a dependent variable. In other words, if the independent variable changes, is there a proportional change (positive or negative) in the dependent variable?

Example

Let's say that two measuring tools are being considered—a micrometer and a vernier caliper. These tools are both graduated in 0.001 in. increments. Measuring the same parts with both tools should show a similar result.

TABLE 11.1[a]

	Micrometer	Caliper
P/N 1	0.500	0.500
P/N 2	0.502	0.502
P/N 3	0.501	0.501
P/N 4	0.500	0.500
P/N 5	0.499	0.499
P/N 6	0.500	0.500
P/N 7	0.502	0.502
P/N 8	0.501	0.501
P/N 9	0.500	0.500
P/N 10	0.503	0.503

[a] Notice that the values in the columns correlate perfectly.

The readings given in Table 11.1 could be called the independent variable Y for the micrometer and dependent variable X for the caliper. If these readings were plotted on a scale, the graphical representation in Figure 11.2 would result. A straight line going upward to the right means that there is perfect positive correlation between the Y and X variable. So a given value of Y does cause a given value of X to occur.

A straight line going downward to the right in this case means negative correlation. This means that a positive value of Y will cause a given *negative* value of X to occur.

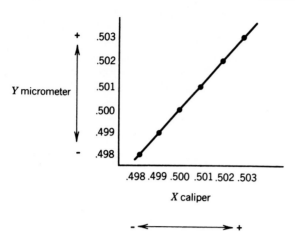

Figure 11.2 Scatter diagram: perfect correlation.

Sometimes a scatter diagram will show no correlation between the dependent and independent variables. This will show on the diagram as random points that are not in a straight line (not even close to a straight line) (see Figure 11.3).

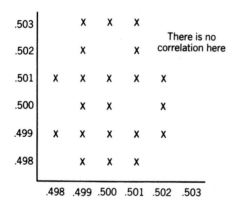

Figure 11.3 Scatter diagram: no correlation.

There are formulas that can be used to analyze the correlation between variables further than the simple scatter diagram, but in many cases the scatter diagram is effective and practical. It gives you some ideas of how a specific change in one variable can affect another variable.

REVIEW QUESTIONS

1. A child is coughing due to a common cold. The child is given cough medicine for the cough. Is this medicine treating the cause of the cold or a symptom?
2. One of the most important problem-solving steps is _____
_____ .

3. A technical problem exists in a chemical cleaning process. A problem-solving team is established consisting of the following members. Is this a natural work team? Chemical engineer, process operator, accountant, process supervisor

4. (True or False) During the initial brainstorming of problem causes, each idea is discussed and voted upon by the team members as it is suggested.

5. Name two of the six categories of problem causes.

6. A helpful diagram that shows the six cause categories is often nicknamed the _____ diagram.

7. Pareto analysis identifies the vital _____ versus the trivial many problems or causes.

8. An effective problem-solving tool that identifies if there is a relationship between two variables is a _____ _____ .

9. Give an example of a problem cause from the environmental category.

10. Give an example of a problem cause from the measurement category.

11. Under which cause category would poor raw stock be listed?

12. What is the cause category that has to do with the way processes are performed?

Glossary

Acceptable quality level. The maximum percentage or proportion of variant units in a lot or batch that, for the purposes of acceptance sampling, can be considered satisfactory as a process average. The term *variant unit* should be replaced by more specific terms, such as *nonconforming unit* or *defective unit,* where appropriate. (2)*

Acceptance number. The maximum number of defects or defective units in the sample that will permit acceptance of the inspection lot or batch. (1)*

Accuracy (of measurement). An unbiased true value. The difference between the average of several measurements and the true value.

Addendum. The distance from the pitch diameter to the crest on an external thread (for example).

Allowance. The intended dimensional different between mating parts (whether it be clearance or interference).

Alloy. Two or more metals, or metals and nonmetals, combined for various reasons to form a substance called an alloy.

Aluminum. A lightweight, shiny, nonmagnetic metal that resists oxidation.

Anneal. Heating a metal and letting it gradually cool for softening and making it less brittle.

* Ref (1): Department of Defense, Mil-Std-109B, U.S. Government Printing Office. Ref. (2): ASQC, *Glossary and Tables for Statistical Quality Control;* reprinted by permission.

Attribute. A characteristic or property that is appraised in terms of whether it does or does not exist, (e.g., Go or Not Go) with respect to a given requirement. (1)

Attribute gages. Gages that measure on a Go–NoGo basis. An example is a plug gage for a hole.

Average outgoing quality (AOQ). The expected quality of outgoing product following the use of an acceptance sampling plan for a given value of incoming product quality. (2)

Average outgoing quality limit (AOQL). For a given acceptance sampling plan, the maximum AOQ over all possible levels of incoming quality. (2)

Batch. A definite quantity of some product or material produced under conditions that are considered uniform. *Note:* A batch is usually smaller than a lot or population. (2)

Bearing. May support a shaft in rotation; the mating part is generally called a journal.

Bevel. An angled edge other than a right angle.

Bias in measurement. This bias occurs when one uses a micrometer to measure a part, and tends to squeeze the thimble tighter to get the desired reading, not the *actual* reading.

Bolt circle (pattern). A collection of holes on a circular centerline related to a common center.

Bore. Making a hole using a lathe, boring bar, drill press, and so on, or the size of the hole (i.e., the bore).

Boss. A projection, usually on a casting or forging.

Brass. An alloy, essentially made of copper and zinc.

Brinell. A hardness testing method for metals.

Broach. A cutting tool for removing material to shape an outside surface or a hole.

Bronze. An alloy made mostly of copper and some tin.

Buff. Polishing, sometimes with a wheel made of fabric that has abrasives added.

Burnish. To rub a material with a tool for compacting or smoothing or for turning an edge.

Burr. A rough edge on a part produced by machining.

Calibration. A comparison of two instruments or measuring devices—one of which is a standard of known accuracy traceable to national standards—to detect, correlate, report, or eliminate by adjustment any discrepancy in accuracy of the instrument or measuring device being compared with the standard. (1)

Calibration interval. A specified amount of time between calibrations where the accuracy of gages (or test equipment) is considered valid.

Cam. A rotating or sliding member that imparts motion to a roller or pin moving against its edge.

Case harden. Hardening the outer surface of steel by heating then quenching. *See also* Quench.

Casting. A part produced by pouring molten material into a mold.

Center drill. To drill holes in the ends of a part that is to be mounted on centers.

Characteristic. A property that helps to differentiate between items of a given sample or population. *Note:* The differentiation may be either quantitative (by variables) or qualitative (by attributes). (2)

Chill. For example, hardening the outer surface of a casting by quick cooling.

Clearance number. As associated with a continuous sampling plan, the number of successively inspected units of product that must be found acceptable during the 100 percent inspection sequence before action to change the amount of inspection can be taken. (2)

Coin. Forming a part, or a surface of a part, by stamping using a mold or die.

Cold rolling (steel). Steel that is rolled while it is cold, producing smooth, accurate stock.

Collar. Used on shafts to prevent axial movement or to hold something in place.

Consumer's risk (beta risk), B. The probability of accepting a bad lot.

Cope. The top part of a casting mold. (The drag is the bottom part.)

Core. Using sand to form a shape that will produce an opening inside a casting when it is poured.

Counterbore. A larger diameter drilled on the same centerline of an existing hole.

Countersink. A bit or drill for making a countersink or a funnel-shaped enlargement at the outer end of a drilled hole. It is generally used to seat the head of a conical head screw, or make it easier to start a bolt.

Critical defect. This classification of defect is one that experience or judgment indicates is likely to cause unsafe conditions for those who use, maintain, or depend on the product; or a defect likely to present performance of the function of a major end-item.

Dedendum. The distance from the pitch diameter of an external thread to the root of the thread (for example).

Defect. A departure of a quality characteristic from its intended level or state that occurs with a severity sufficient to cause an associated product or service not to satisfy intended normal, or reasonably foreseeable, usage requirements. (2)

Defective (defective unit). A unit of product or service containing at least one defect, or having several imperfections that in combination cause the unit not to satisfy intended normal, or reasonably foreseeable, usage requirements. (2)

Defects per hundred units. The number of defects per hundred units of any given quantity of product is the number of defects contained therein divided by the total number of units of product, the quotient multiplied by 100 (one or more defects being possible in any unit of product). Expressed as an equation,

$$\text{defects per 100 units} = \frac{\text{number of defects} \times 100}{\text{number of units}} \tag{1}$$

Deviation. Written authorization, granted prior to the manufacture of an item, to depart from a particular performance or design requirement of a contract, specification, or referenced document, for a specific number of units or specific period of time. (1)

Deviation (measurement sense). The difference between a measurement or quasi-measurement and its stated value or intended level. (2)

Die. A tool made of hard metal, used to cut or form a required shape in sheet metal, forgings, and so on.

Die casting. Metal poured under pressure into a metal die to make a very accurate casting.

Differential measurement. The use of a device that transforms actual movement into a known value, a dial indicating gage, for example.

Direct measurement. Where the standard is directly applied to the part, and a reading can be taken. A steel rule to measure length is an example.

Discrimination. The direct distance between two lines on a scale.

Discrimination rule. Never attempt to "read between the lines"; use an instrument with the proper discrimination for the measurement.

Draft. Tapered shapes in parts to allow them to be easily taken out of a mold or die.

Drag. The lower part of a flask used in casting. *See also* Cope.

Draw. To deform or stretch a metal.

Drill. To cut a hole using a drill bit.

Drill press. A machine used to drill holes.

Face. A surface at right angles to the centerline of rotation, or any of the plane surfaces that bound a geometric solid.

FAO. Term meaning "to finish all over."

Fillet. An internal radius.

Fin. Metal squeezed out between dies or molds.

Fit. The degree of tightness (or looseness) between mating parts.

Fixture. A device for holding the workpiece.

Flange. A rim or rib of strength, for guiding, or for attachment to another object.

Flask. A box made of two or more parts for holding sand in a mold.

Flute. The groove on twist drills and taps, and so on.

Forge. Heating, then hammering or pressing metals into a desired shape.

Gasket. A thin piece of material put between mating parts generally to form a seal.

Gate. An opening in a sand mold through which the molten metal enters.

Graduations. Accurate divisions on a scale or dial face.

Grind. Accurate removal of metal by means of an abrasive wheel (stone, etc.).

Harden. Heating metal to a specified temperature and then quenching (cooling) it in oil or water, or other cooling material.

Heat treating. Changing the properties of metals by heating, then cooling them.

Inspection. The process of measuring, examining, testing, gaging, or otherwise comparing the unit with the applicable requirements. *Note:* The term *requirements* sometimes is used broadly to include standards of good workmanship. (2)

Inspection by attributes. Inspection whereby either the unit of product or characteristics thereof are classified simply as defective or nondefective, or the number of defects in the unit of product is counted, with respect to a given requirement. (1)

Inspection by variables. Inspection wherein certain quality characteristics of sample are evaluated with respect to a continuous numerical scale and expressed as precise points along this scale. Variable inspection records the degree of conformance or nonconformance of the unit with specified requirements for the quality characteristics involved. (1)

Inspection level. A feature of a sampling scheme relating the size of the sample to that of the lot. *Note:* Selection of an *inspection level* may be based on simplicity and cost of a unit of product, inspection cost, destructiveness of inspection, or quality consistency between lots. In some of the more widely used sampling systems, lot or batch size are grouped into convenient sets, and the sample size appropriate to each set is furnished for a designated *inspection level*. (2)

Inspection lot. A collection of similar units, or a specific quantity of similar material, offered for inspection and subject to a decision with respect to acceptance.

Inspection record. Recorded data concerning the results of inspection action.

Inspection reduced. A feature of a sampling scheme permitting smaller sample sizes than are used in normal inspection. *Reduced inspection* is used in some sampling schemes when experience with the level of submitted quality is sufficiently good and other stated conditions apply. (2)

Inspection, tightened. A feature of a sampling scheme using stricter acceptance criteria than those used in normal inspection. *Tightened inspection* is used in some sampling schemes as a protective measure to increase the probability of rejecting lots when experience shows the level of submitted quality has deteriorated significantly. (2)

Interchangeable. Parts made to dimensions so that they will fit and function when interchanged among the same upper-level part number assemblies.

Jig. Maintains correct positional relationship between workpiece and tool.

Journal. That portion of a rotating shaft that turns in a bearing.

Kerf. The wasted material that is removed by a saw.

Key. A small piece of metal seated in a slot on a shaft and hub to prevent rotation.

Keyseat. The slot on a shaft that holds the key.

Keyway. The slot inside a shaft or hub that holds the key.

Knurl. A particular pattern of small ridges or beads on a metal surface to aid in gripping (e.g., a handle).

Lap. To produce a fine surface finish by sliding the lapping material (with or without abrasive powder) over the surface.

Lathe. A machine that is used to shape a diameter on the workpiece by rotating the piece against a cutting tool.

Linear distance. A straight-line distance between two planes.

Lot. A definite quantity of a product or material accumulated under conditions that are considered uniform for sampling purposes. (2)

Lot size (N). The number of units in the lot. (2)

Lug. A projection of metal, not round as a boss, but some other shape, usually to contain a tapped hole for a bolt.

Major defect. This is a defect other than critical that may cause the product to fail, cause poor performance, shortened life, or prevent interchangeability.

Malleable casting. A casting that is made less brittle and tougher by annealing. *See also* Annealing.

Measured surface. That surface of a measuring tool that is movable and from which the measurement is taken: for example, the spindle of a micrometer.

Measurement error. Something that can be determined and is simply the difference between the measured value and the actual value.

Measurement pressure. Should be positive but not excessive; the most important factor (in most cases) is that the pressure used on the workpiece be the same pressure as that used when the tool was calibrated.

Measurement standard. A standard of measurement that is a true value and is recognized by all as a basis for comparison.

Measuring and test equipment. All devices used to measure, gage, test, inspect, diagnose, or otherwise examine materials, supplies, and equipment to determine compliance with technical requirements. (1)

Metrology. The science of measurement.

Mill. Removal of material by means of passing the part under rotating cutters.

Minor defect. A defect that is not likely to reduce materially the usability of the unit. **Example:** A scratch on the side of a dishwasher.

Mold. A cavity formed by sand or other material that forms a cavity where molten metal is poured.

Nonconforming unit. A unit of product or service containing at least one nonconformity. (2)

Nonconformity. A departure of a quality characteristic from its intended level or state that occurs with a severity sufficient to cause an associated product or service not to meet a specification requirement. (2)

Normal inspection. Inspection, under a sampling plan, which is used when there is no evidence that the quality of the product being submitted is better or poorer than the specified quality level. (1)

Normalize. Heating material to a specified temperature and cooling it in air.

One hundred percent inspection. Inspection in which specified characteristics of each unit of product are examined or tested to determine conformance with requirements. (1)

Operating characteristic curve (OC curve). 1. For isolated or unique lots or a lot from an isolated sequence: A curve showing, for a given sampling plan, the probability of accepting a lot as a function of the lot quality (type A) 2. For a continuous stream of lots: A curve showing, for a given sampling plan, the probability of accepting a lot as a function of the process average (type B) 3. For continuous sampling plans: A curve showing the proportion of submitted product over the long run accepted during the sampling phases of the plan as a function of the product quality. 4. For special plans: A curve showing, for a given sampling plan, the probability of continuing to permit the process to continue without adjustment as a function of the process quality. (2)

Original inspection. The first inspection of a lot as distinguished from the inspection of a lot that has been resubmitted after previous nonacceptance. (2)

Pad. A raised area used generally to provide a support surface.

Parallax error. The apparent shifting of an object caused by shifting of the observer. An example is the act of viewing an indicator dial face from an angle, when it should be viewed directly.

Pattern. A model (usually made of wood) used to form a mold for casting.

Peen. To draw, bend, or flatten by hammering with a peen, or as if a peen were used.

Percent defective. The percent defective of any given quantity of units of product is 100 times the number of defective units of product contained therein divided by the total number of units of product, that is,

$$\text{percent defective} = \frac{\text{number of defectives} \times 100}{\text{number of units inspected}} \tag{1}$$

Plane. Removal of material to make a flat surface by means of planing.

Plate. To put a thin layer of metal on a surface (such as a chrome plate).

Polish. To produce a fine surface finish (e.g., by means of a very fine abrasive).

Precision. The closeness of agreement between randomly selected individual measurements or test results. (2)

Probability of acceptance (Pa). The probability that a lot will be accepted under a given sampling plan. (2)

Process average. The average percent of defective or average number of defects per hundred units of product submitted by the supplier for original inspection. (1)

Producer's risk (alpha risk), α. The probability of rejecting a good lot.

Profile. Usually, an irregular shape cut on a part by means of a template (or guide).

Punch. To cut, emboss, or perforate a part by means of forcing a rigid tool into the workpiece.

Quality. The totality of features and characteristics of a product or service that bear on its ability to satisfy given needs. (2)

Quality assurance. All those planned or systematic actions necessary to provide adequate confidence that a product or service will satisfy given needs. (2)

Quality control. The operational techniques and the activities that sustain a quality of product or service that will satisfy given needs; also, the use of such techniques and activities. (2)

Quality management. The totality of functions involved in the determination and achievement of quality. (2)

Quench. To cool a heated piece of metal suddenly by immersion in water, oil, or other coolants.

Rack. A bar with gear teeth to engage the teeth in a gear.

Random sampling. The process of selecting units for a sample of size n in such a manner that all combinations of n units under consideration have an equal or ascertainable chance of being selected as the sample. (2)

Ream. To slightly enlarge or shape a hole with a reamer.

Reduced inspection. Inspection under a sampling plan using the same quality level as for normal inspection but requiring a smaller sample. (1)

Reference surface. The surface of a measuring tool that is fixed: for example, a micrometer anvil.

Rejection number. The minimum number of variants or variant units in the sample that will cause the lot or batch to be designated as not acceptable. (2)

Reliability. The probability that an item will perform its intended function for a specified interval under stated conditions. (1)

Relief. An offset surface on a part to provide clearance for machining.

Resubmitted lot. A lot that previously has been designated as not acceptable and that is submitted again for acceptance inspection after having been further tested, sorted, reprocessed, and so on. (2)

Rib. Additional members on a part used to brace or support.

Sample. A group of units, portion of material, or observations taken from a larger collection of units, quantity of material, or observations that serves to provide information that may be used as a basis for making a decision concerning the larger quantity. (2)

Sample percent defective. This percent is found by dividing the number of defectives found in the sample by the sample size and then multiplying by 100, or in a formula,

$$\frac{D}{n} \times 100$$

Sample size (*n*). One or more units selected at *random* from a lot without regard for their quality.

Sampling, double. Sampling inspection in which the inspection of the first sample of size n_1, leads to a decision to accept a lot; not to accept it; or to take a second sample of size n_2. (2)

Sampling, multiple. Sampling inspection in which, after each sample is inspected, the decision is made to accept a lot; not to accept it; or to take another sample to reach the decision. There may be a prescribed maximum number of samples, after which a decision to accept or not to accept the lot must be reached. (2)

Sampling plan. A statement of the sample size or sizes to be used and the associated acceptance and rejection criteria. (1)

Sampling, sequential. Sampling inspection in which, after each unit is inspected, the decision is made to accept the lot, not to accept it, or to inspect another unit. (2)

Sampling, single. Sampling inspection in which the decision to accept or not to accept a lot is based on the inspection of a single sample of size *n*. (2)

Sandblast. Blowing sand at high speed with air or steam pressure to clean cut, or engrave castings, forgings, and other materials.

Screening inspection. Inspection in which each item of product is inspected for designated characteristics and all defective items are removed. (1)

Secondary reference standards. Standards used to perform standards/equipment calibration. They are a lower (second)-level standard, usually compared calibrated to primary standards.

Shape. Removal of metal by means of a shaper.

Shear. To cut metal by forcing it to separate across its cross-sectional area.

Shim. Thin pieces of material used to take up space between mating parts, or to level or adjust.

Specification limits. Limits that define the conformance boundaries for an individual unit of a manufacturing or service operation. (2)

Spline. A keyway or keyways cut into a hole, especially to maintain alignment.

Spotface. A shallow counterbore usually used to provide a good seat for a bolt head.

Sprue. The hole in the sand that leads to the gate where the molten metal enters the casting.

Tap. Cutting internal threads with a tap.

Taper. Gradual decrease of diameter, thickness, or width in an object.

Taper reamer. A reamer that produces accurate tapered holes when required.

Temper. Reheating steel to bring it to a particular degree of hardness or softness.

Template. A guide used to cut shapes into the workpiece or assist in inspecting the piece.

Testing. A means of determining the capability of an item to meet specified requirements by subjecting the item to a set of physical, chemical, environmental, or operating actions and conditions. (2)

Tightened inspection. Switching to tightened inspection is usually done when quality levels are observed to be getting worse.

Tolerance. The amount of permissible variation in a dimension.

Trepan. To cut a hole into a workpiece without having to waste the material in the middle.

Turn. To produce a diameter on a lathe.

Unit. A quantity of product, material, or service forming a cohesive entity on which a measurement or observation may be made. (2)

Universe. A group of populations, often reflecting different characteristics of the items or material under consideration. (2)

Upset. Forming a particular shape on the end of a bar (such as a head) by hammering it between two dies.

Variable gage. Gages that are capable of measuring the actual size of a part. A dial-indicating gage is an example.

Variables, method of. Measurement of quality by the method of variables consists of measuring and recording the numerical magnitude of a quality characteristic for each of the units in the group under consideration. This involves reference to a continuous scale of some kind. (2)

Variant. An item or event that is classified differently from others of its kind or type. (2)

Web. *Same as* Rib.

Working standards. Standards used to perform equipment calibration. They are a lower (third)-level standard, usually compared calibrated to secondary standards.

Appendix Tables and Charts

TABLE A-1   SAMPLE AVERAGE AND RANGE CONTROL CHART (ASQC Quality Press. Reprinted with permission.)

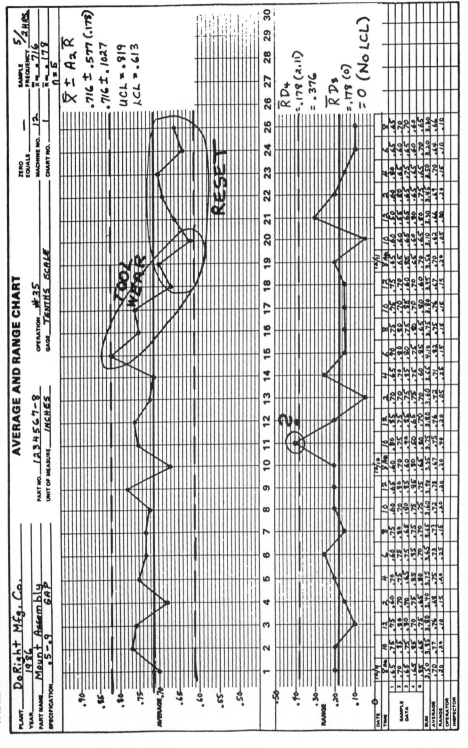

TABLE A-2 SAMPLE AVERAGE AND SIGMA CONTROL CHART (ASQC Quality Press. Reprinted with permission.)

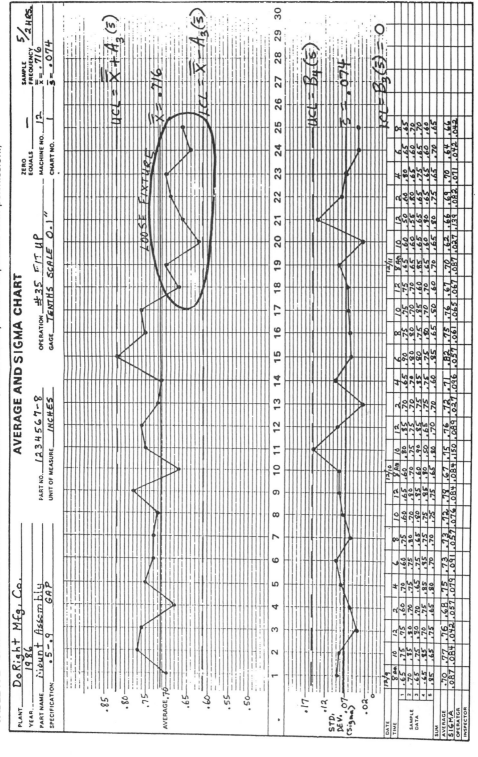

465

TABLE A-3 SAMPLE INDIVIDUALS AND MOVING RANGE CONTROL CHART (ASQC Quality Press. Reprinted with permission.)

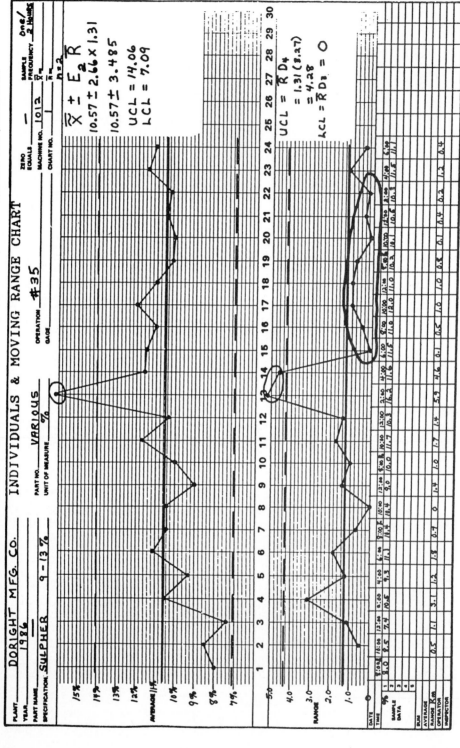

TABLE A-4 SAMPLE MEDIANS CONTROL CHART (ASQC Quality Press. Reprinted with permission.)

TABLE A-5   SAMPLE PRE-CONTROL CHART (ASQC Quality Press. Reprinted with permission.)

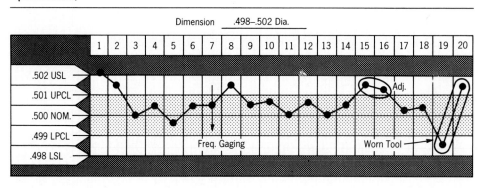

TABLE A-6 SAMPLE POTENTIAL PROCESS CAPABILITY STUDY

Process Capability Study

Frequency	3.30	3.31	3.32	3.33	3.34	3.35	3.36	3.37	3.38	3.39	3.40
9					X						
8				X	X	X					
7				X	X	X	X				
6			X	X	X	X	X				
5			X	X	X	X	X	X			
4			X	X	X	X	X	X			
3		X	X	X	X	X	X	X	X		
2		X	X	X	X	X	X	X	X		
1	X	X	X	X	X	X	X	X	X		X

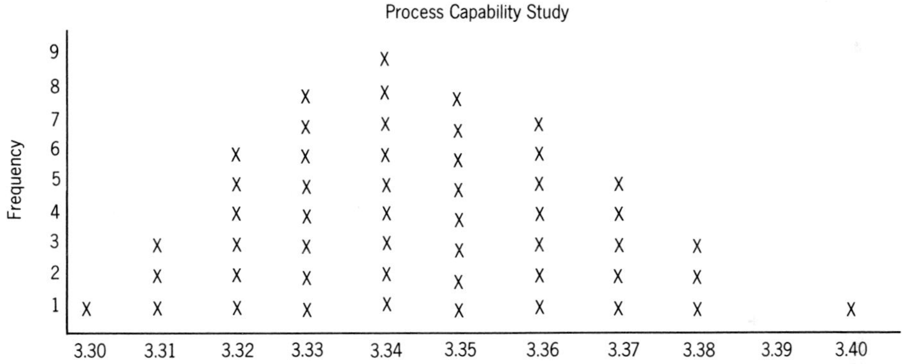

Specifications

Nominal = 3.35
Upper tolerance = 3.40
Lower tolerance = 3.30

Percent yield

Zupper = 2.38
Zlower = 2.10
Yield (%) = 97.34%
% out of tolerance = 2.66%

Statistics

$\bar{X}$ = 3.344
Sigma = .021
Six Sigma = .126
CR = 1.26
Cpk = 0.70

TABLE A-7   SAMPLE "p" CONTROL CHART FOR PROPORTION DEFECTIVE (ASQC Quality Press. Reprinted with permission.)

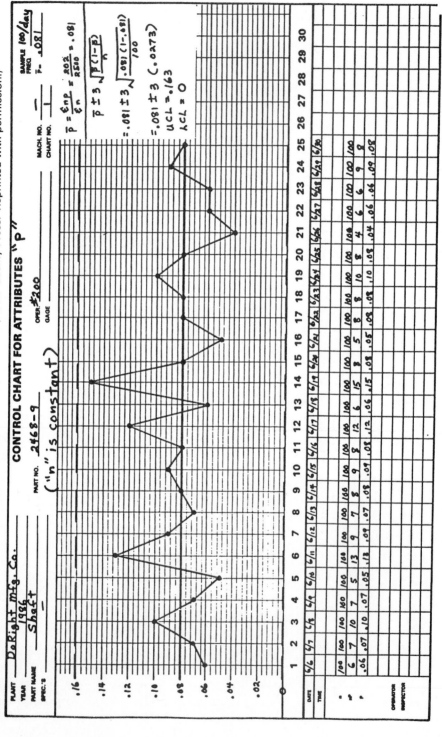

TABLE A-8 SAMPLE "np" CONTROL CHART FOR NUMBER OF DEFECTIVES (ASQC Quality Press. Reprinted with permission.)

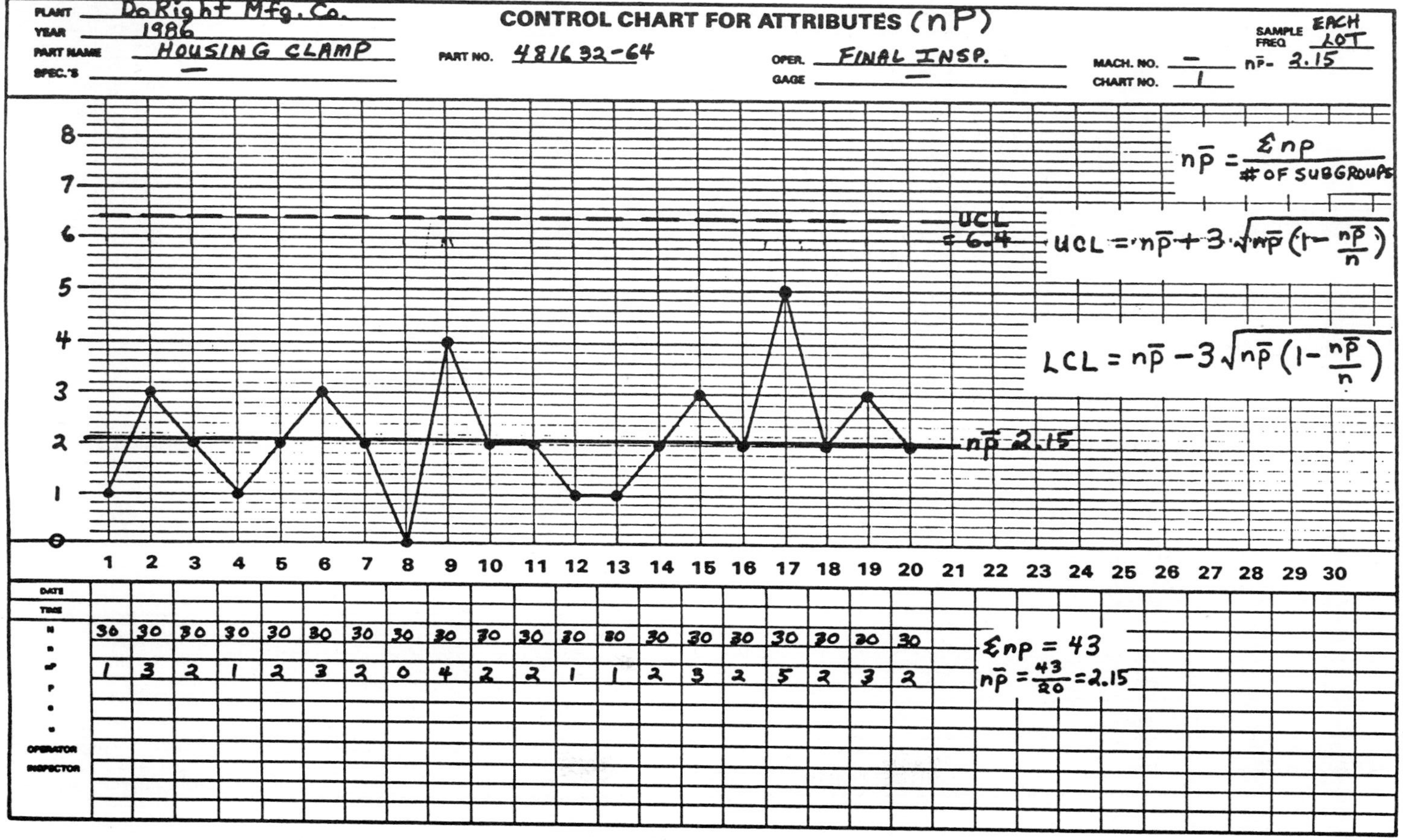

CONTROL CHART FOR ATTRIBUTES (nP)

PLANT _DoRight Mfg. Co._
YEAR _1986_
PART NAME _HOUSING CLAMP_
SPEC'S _—_

PART NO. _48/632-64_

OPER. _FINAL INSP._
GAGE _—_

MACH. NO. _—_
CHART NO. _/_

SAMPLE FREQ _EACH LOT_
$n\bar{p} = 2.15$

$n\bar{p} = \dfrac{\Sigma np}{\text{\# of SUBGROUPS}}$

$UCL = n\bar{p} + 3\sqrt{n\bar{p}\left(1 - \dfrac{n\bar{p}}{n}\right)}$

$LCL = n\bar{p} - 3\sqrt{n\bar{p}\left(1 - \dfrac{n\bar{p}}{n}\right)}$

UCL = 6.4

$n\bar{p} = 2.15$

$\Sigma np = 43$

$n\bar{p} = \dfrac{43}{20} = 2.15$

DATE																				
TIME																				
	36	30	30	30	30	30	30	30	30	30	30	30	30	30	30	30	30	30	30	30
	1	3	2	1	3	3	2	0	4	2	2	1	1	3	2	5	2	3	3	2
OPERATOR																				
INSPECTOR																				

471

TABLE A-9 SAMPLE "c" CONTROL CHART FOR DEFECTS (ASQC Quality Press. Reprinted with permission.)

472

TABLE A-10 SAMPLE "u" CONTROL CHART FOR DEFECTS PER UNIT (ASQC Quality Press. Reprinted with permission.)

TABLE A-11   BASIC TRIGONOMETRIC FUNCTIONS

ANGLE	SIN	COS	TAN	COT	ANGLE
0	.0000	1.0000	.0000	–	90
1	.0175	.9998	.0175	57.2900	89
2	.0349	.9994	.0349	28.6363	88
3	.0523	.9986	.0524	19.0811	87
4	.0698	.9976	.0699	14.3007	86
5	.0872	.9962	.0875	11.4301	85
6	.1045	.9945	.1051	9.5144	84
7	.1219	.9925	.1228	8.1443	83
8	.1392	.9903	.1405	7.1154	82
9	.1564	.9877	.1584	6.3138	81
10	.1736	.9848	.1763	5.6713	80
11	.1908	.9816	.1944	5.1446	79
12	.2079	.9781	.2126	4.7046	78
13	.2250	.9744	.2309	4.3315	77
14	.2419	.9703	.2493	4.0108	76
15	.2588	.9659	.2679	3.7321	75
16	.2756	.9613	..2867	3.4874	74
17	.2924	.9563	.3057	3.2709	73
18	.3090	.9511	.3249	3.0777	72
19	.3256	.9455	.3443	2.9042	71
20	.3420	.9397	.3640	2.7475	70
21	.3584	.9336	.3839	2.6051	69
22	.3746	.9272	.4040	2.4751	68
23	.3907	.9205	.4245	2.3559	67
24	.4067	.9135	.4452	2.2460	66
25	.4226	.9063	.4663	2.1445	65
26	.4384	.8988	.4877	2.0503	64
27	.4540	.8910	.5095	1.9626	63
28	.4695	.8829	.5317	1.8807	62
29	.4848	.8746	.5543	1.8040	61
30	.5000	.8660	.5774	1.7321	60
31	.5150	.8572	.6009	1.6643	59
32	.5299	.8480	.6249	1.6003	58
33	.5446	.8387	.6494	1.5399	57
34	.5592	.8290	.6745	1.4826	56
35	.5736	.8192	.7002	1.4281	55
36	.5878	.8090	.7265	1.3764	54
37	.6018	.7986	.7536	1.3270	53
38	.6157	.7880	.7813	1.2799	52
39	.6293	.7771	.8098	1.2349	51
40	.6428	.7660	.8391	1.1918	50
41	.6561	.7547	.8693	1.1504	49
42	.6691	.7431	.9004	1.1106	48
43	.6820	.7314	.9325	1.0724	47
44	.6947	.7193	.9657	1.0355	46
45	.7071	.7071	1.0000	1.0000	45
ANGLE	COS	SIN	COT	TAN	ANGLE

THE THREE BASIC FORMULAS:

$$SIN = \frac{OPP}{HYP} \qquad COS = \frac{ADJ}{HYP}$$

$$TAN = \frac{OPP}{ADJ}$$

READING THE TABLE:

USE WORDS AT TOP WITH ANGLES AT LEFT COLUMN.

USE WORDS AT BOTTOM WITH ANGLES IN RIGHT COLUMN.

BASIC STEPS TO SOLVE A RIGHT TRIANGLE:

(1) DRAW THE PROBLEM
(2) NAME THE SIDES YOU ARE WORKING WITH.
(3) CHOOSE THE FORMULA TO USE THAT HAS ALL OF THESE THINGS IN IT. (THERE'S ONLY ONE ALWAYS.)
(4) ENTER WHAT YOU KNOW INTO THE FORMULA.
(5) DO ALL THE SIMPLE MATH YOU CAN AT THIS POINT.
(6) CHANGE THE FORMULA TO GET WHAT YOU ARE LOOKING FOR ON ONE SIDE OF THE EQUAL SIGN. (SEE SHOP MATH CHAPTER).
(7) FIND THE ANGLE IN THE TABLE THAT CORRESPONDS TO THE FUNCTION YOU FOUND.
(8) SOLVE THE FORMULA BY DOING THE FINAL MATH NECESSARY.

EXAMPLES OF READING THE TABLE:
(1) Sin 17 degrees = .2924
(2) Sin 52 degrees = .7880
(3) Tan 41 degrees = .8693
(4) .7771 is the function for Cos 39 degrees and Sin 51 degrees.
   All of the above functions (numbers) serve two angles that are complimentary; that is added together equals 90° (39° + 51° = 90°)

TABLE A-12   DECIMAL EQUIVALENTS FOR
COMMON FRACTIONS

HALVES	THIRTY-SECONDS	SIXTY-FOURTHS
1/2 = 0.500	17/32 = 0.53125	37/64 = 0.578125
FOURTHS	19/32 = 0.59375	39/64 = 0.609375
1/4 = 0.250	21/32 = 0.65625	41/64 = 0.640625
3/4 = 0.750	23/32 = 0.71875	43/64 = 0.671875
EIGHTHS	25/32 = 0.78125	45/64 = 0.703125
1/8 = 0.125	27/32 = 0.84375	47/64 = 0.734375
3/8 = 0.375	29/32 = 0.90625	49/64 = 0.765625
5/8 = 0.625	31/32 = 0.96875	51/64 = 0.796875
7/8 = 0.875		53/64 = 0.828125
	SIXTY-FOURTHS	55/64 = 0.859375
SIXTEENTHS	1/64 = 0.015625	57/64 = 0.890625
1/16 = 0.0625	3/64 = 0.046875	59/64 = 0.921875
3/16 = 0.1875	5/64 = 0.078125	61/64 = 0.953125
5/16 = 0.3125	7/64 = 0.109375	63/64 = 0.984375
7/16 = 0.4375	9/64 = 0.140625	1 = 1.000000
9/16 = 0.5625	11/64 = 0.171875	
11/16 = 0.6875	13/64 = 0.203125	
13/16 = 0.8125	15/64 = 0.234375	
15/16 = 0.9375	17/64 = 0.265625	
THIRTY-SECONDS	19/64 = 0.296875	
1/32 = 0.03125	21/64 = 0.328125	
3/32 = 0.09375	23/64 = 0.359375	
5/32 = 0.15625	25/64 = 0.390625	
7/32 = 0.21875	27/64 = 0.421875	
9/32 = 0.28125	29/64 = 0.453125	
11/32 = 0.34375	31/64 = 0.484375	
13/32 = 0.40625	33/64 = 0.515625	
15/32 = 0.46875	35/64 = 0.546875	

TABLE A-13 CONVERSION CHART FOR MILLIMETERS TO INCHES

MM	INCHES	MM	INCHES	MM	INCHES	MM	INCHES
.01	.00039	.34	.01339	.67	.02638	1.0	.03937
.02	.00079	.35	.01378	.68	.02677	2.0	.07874
.03	.00118	.36	.01417	.69	.02717	3.0	.11811
.04	.00157	.37	.01457	.70	.02756	4.0	.15748
.05	.00197	.38	.01496	.71	.02795	5.0	.19685
.06	.00236	.39	.01535	.72	.02835	6.0	.23622
.07	.00276	.40	.01575	.73	.02874	7.0	.27599
.08	.00315	.41	.01614	.74	.02913	8.0	.31496
.09	.00354	.42	.01654	.75	.02953	9.0	.35433
.10	.00394	.43	.01693	.76	.02992	10.0	.39370
.11	.00433	.44	.01732	.77	.03032	11.0	.43307
.12	.00472	.45	.01772	.78	.03071	12.0	.47244
.13	.00512	.46	.01811	.79	.03110	13.0	.51181
.14	.00551	.47	.01850	.80	.03150	14.0	.55118
.15	.00591	.48	.01890	.81	.03189	15.0	.59055
.16	.00630	.49	.01929	.82	.03228	16.0	.62992
.17	.00669	.50	.01969	.83	.03268	17.0	.66929
.18	.00709	.51	.02008	.84	.03307	18.0	.70866
.19	.00748	.52	.02047	.85	.03346	19.0	.74803
.20	.00787	.53	.02087	.86	.03386	20.0	.78740
.21	.00827	.54	.02126	.87	.03425	21.0	.82677
.22	.00866	.55	.02165	.88	.03465	22.0	.86614
.23	.00906	.56	.02205	.89	.03504	23.0	.90551
.24	.00945	.57	.02244	.90	.03543	24.0	.94488
.25	.00984	.58	.02283	.91	.03583	25.0	.98425
.26	.01024	.59	.02323	.92	.03622	26.0	1.02362
.27	.01063	.60	.02362	.93	.03661	27.0	1.06299
.28	.01102	.61	.02402	.94	.03701	28.0	1.10236
.29	.01142	.62	.02441	.95	.03740	29.0	1.14173
.30	.01181	.63	.02480	.96	.03780	30.0	1.18110
.31	.01220	.64	.02520	.97	.03819	31.0	1.22047
.32	.01260	.65	.02559	.98	.03858	32.0	1.25984
.33	.01299	.66	.02598	.99	.03898	33.0	1.29921

EXAMPLE: 0.61 MILLIMETERS IS EQUAL TO .02402"

PRACTICE PROBLEM: A DRAWING CALLS FOR A CERTAIN LENGTH TO BE 28 ± 0.5 MM. WHAT ARE THE EXACT HIGH AND LOW LIMITS IN INCHES ?

SOLUTION: "28MM EQUALS 1.10236"   UPPER LIMIT = 1.10236

"0.5MM EQUALS .01969"   + .01969

1.12205

LOWER LIMIT = 1.10236

- .01969

1.08267

TABLE A-14 ENGLISH–METRIC
CONVERSIONS

Measures of Length

1 millimeter (mm) = 0.03937 inch
1 centimeter (cm) = 0.39370 inch
1 meter (m) = 39.37008 inches
        = 3.2808 feet
        = 1.0936 yards
1 kilometer (km) = 0.6214 mile
1 inch = 25.4 millimeters (mm)
      = 2.54 centimeters (cm)
1 foot = 304.8 millimeters (mm)
      = 0.3048 meter (m)
1 yard = 0.9144 meter (m)
1 mile = 1.609 kilometers (km)

Measures of Area

1 square millimeter = 0.00155 square inch
1 square centimeter = 0.155 square inch
1 square meter = 10.764 square feet
        = 1.196 square yards
1 square kilometer = 0.3861 square mile
1 square inch = 645.2 square millimeters
        = 6.452 square centimeters
1 square foot = 929 square centimeters
        = 0.0929 square meter
1 square yard = 0.836 square meter
1 square mile = 2.5899 square kilometers

Measures of Capacity (Dry)

1 cubic centimeter (cm^3) = 0.061 cubic inch
1 liter = 0.0353 cubic foot
      = 61.023 cubic inches
1 cubic meter (m^3) = 35.315 cubic feet
        = 1.308 cubic yards
1 cubic inch = 16.38706 cubic centimeters (cm^3)
1 cubic foot = 0.02832 cubic meter (m^3)
        = 28.317 liters
1 cubic yard = 0.7646 cubic meter (m^3)

Measures of Capacity (Liquid)

1 liter = 1.0567 U.S. quarts
      = 0.2642 U.S. gallon
      = 0.2200 Imperial gallon
1 cubic meter (m^3) = 264.2 U.S. gallons
        = 219.969 Imperial gallons
1 U.S. quart = 0.946 liter
1 Imperial quart = 1.136 liters
1 U.S. gallon = 3.785 liters
1 Imperial gallon = 4.546 liters

Measures of Weight

1 gram (g) = 15.432 grains
        = 0.03215 ounce troy
        = 0.03527 ounce avoirdupois
1 kilogram (kg) = 35.274 ounces avoirdupois
        = 2.2046 pounds
1000 kilograms (kg) = 1 metric ton (t)
        = 1.1023 tons of 2000 pounds
        = 0.9842 ton of 2240 pounds
1 ounce avoirdupois = 28.35 grams (g)
1 ounce troy = 31.103 grains
1 pound = 453.6 grams
        = 0.4536 kilogram (kg)
1 ton of 2240 pounds = 1016 kilograms (kg)
        = 1.016 metric tons
1 grain = 0.0648 gram (g)
1 metric ton = 0.9842 ton of 2240 pounds
        = 2204.6 pounds

TABLE A-15   TEMPERATURE CONVERSIONS

Use this table to convert Fahrenheit degrees (F°) directly to Centigrade degrees (C°) and vice versa. It covers the range of temperatures used in most hardening, tempering and annealing operations.

Lower, higher and intermediate conversions can be made by substituting a known Fahrenheit (F°) or Centigrade (C°) temperature figure in either of the following formulas.

$$F° = \frac{C° \times 9}{5} + 32 \qquad\qquad C° = \frac{F° - 32}{9} \times 5$$

F°	C°	F°	C°	F°	C°	F°	C°	F°	C°
− 160	− 107	340	171	840	449	1340	727	1840	1004
− 140	− 96	360	182	860	460	1360	738	1860	1016
− 120	− 84	380	193	880	471	1380	749	1880	1027
− 100	− 73	400	204	900	482	1400	760	1900	1038
− 80	− 62	420	216	920	493	1420	771	1920	1049
− 60	− 51	440	227	940	504	1440	782	1940	1060
− 40	− 40	460	238	960	516	1460	793	1960	1071
− 20	− 29	480	249	980	527	1480	804	1980	1082
0	− 18	500	260	1000	538	1500	816	2000	1093
20	− 7	520	271	1020	549	1520	827	2020	1104
40	4	540	282	1040	560	1540	838	2040	1116
60	16	560	293	1060	571	1560	849	2060	1127
80	27	580	304	1080	582	1580	860	2080	1138
100	38	600	316	1100	593	1600	871	2100	1149
120	49	620	327	1120	604	1620	882	2120	1160
140	60	640	338	1140	616	1640	893	2140	1171
160	71	660	349	1160	627	1660	904	2160	1182
18ʋ	82	680	360	1180	638	1680	916	2180	1193
200	93	700	371	1200	649	1700	927	2200	1204
220	104	720	382	1220	660	1720	938	2220	1216
240	116	740	393	1240	671	1740	949	2240	1227
260	127	760	404	1260	682	1760	960	2260	1238
280	138	780	416	1280	693	1780	971	2280	1249
300	149	800	427	1300	704	1800	982	2300	1260
320	160	820	438	1320	716	1820	993	2320	1271

TABLE A-16 SQUARES AND SQUARE ROOTS

X	x^2	$\sqrt{x}$	X	x^2	$\sqrt{x}$	X	x^2	$\sqrt{x}$
1	1	1.000	28	784	5.292	55	3025	7.416
2	4	1.414	29	841	5.385	56	3136	7.483
3	9	1.732	30	900	5.477	57	3249	7.550
4	16	2.000	31	961	5.568	58	3364	7.616
5	25	2.236	32	1024	5.657	59	3481	7.681
6	36	2.449	33	1089	5.745	60	3600	7.746
7	49	2.646	34	1156	5.831	61	3721	7.810
8	64	2.828	35	1225	5.916	62	3844	7.874
9	81	3.000	36	1296	6.000	63	3969	7.937
10	100	3.162	37	1369	6.083	64	4096	8.000
11	121	3.317	38	1444	6.164	65	4225	8.062
12	144	3.464	39	1521	6.245	66	4356	8.124
13	169	3.606	40	1600	6.325	67	4489	8.185
14	196	3.742	41	1681	6.403	68	4624	8.246
15	225	3.873	42	1764	6.481	69	4761	8.307
16	256	4.000	43	1849	6.557	70	4900	8.367
17	289	4.123	44	1936	6.633	71	5041	8.426
18	324	4.243	45	2025	6.708	72	5184	8.485
19	361	4.359	46	2116	6.782	73	5329	8.544
20	400	4.472	47	2209	6.856	74	5476	8.602
21	441	4.583	48	2304	6.928	75	5625	8.660
22	484	4.690	49	2401	7.000	80	6400	8.944
23	529	4.796	50	2500	7.071	85	7225	9.220
24	576	4.899	51	2601	7.141	90	8100	9.487
25	625	5.000	52	2704	7.211	95	9025	9.747
26	676	5.099	53	2809	7.280	100	10000	10.000
27	729	5.196	54	2916	7.348			

TABLE A-17 BOLT CIRCLE CHART

Used for converting equally spaced
holes on a bolt circle to
coordinates for inspection purposes.

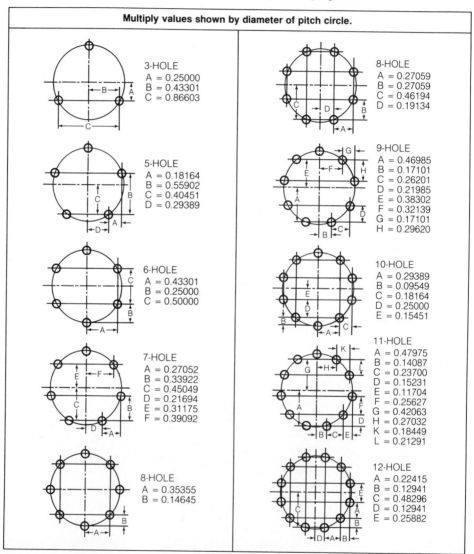

Multiply values shown by diameter of pitch circle.

3-HOLE
A = 0.25000
B = 0.43301
C = 0.86603

5-HOLE
A = 0.18164
B = 0.55902
C = 0.40451
D = 0.29389

6-HOLE
A = 0.43301
B = 0.25000
C = 0.50000

7-HOLE
A = 0.27052
B = 0.33922
C = 0.45049
D = 0.21694
E = 0.31175
F = 0.39092

8-HOLE
A = 0.35355
B = 0.14645

8-HOLE
A = 0.27059
B = 0.27059
C = 0.46194
D = 0.19134

9-HOLE
A = 0.46985
B = 0.17101
C = 0.26201
D = 0.21985
E = 0.38302
F = 0.32139
G = 0.17101
H = 0.29620

10-HOLE
A = 0.29389
B = 0.09549
C = 0.18164
D = 0.25000
E = 0.15451

11-HOLE
A = 0.47975
B = 0.14087
C = 0.23700
D = 0.15231
E = 0.11704
F = 0.25627
G = 0.42063
H = 0.27032
K = 0.18449
L = 0.21291

12-HOLE
A = 0.22415
B = 0.12941
C = 0.48296
D = 0.12941
E = 0.25882

(Courtesy of L.S. Starrett Co.)

NOTES:
1. Pitch circle, hole circle, and bolt circle mean the same thing.
2. This chart is very useful for conveniently converting bolt circle diameters to coordinates for inspection of a bolt circle on the surface plate.
3. Keep this chart handy for use as needed. It is recognized throughout industry as an acceptable tool for use in the shop.

TABLE A-18   FACTORS FOR COMPUTATION OF 3 SIGMA
CONTROL LIMITS

| SAMPLE SIZE | AVERAGE CHARTS FACTORS FOR THE CONTROL LIMITS | | | RANGE CHARTS | | | | | |
| | | | | FACTORS FOR CENTRAL LINE | | FACTORS FOR CONTROL LIMITS | | | |
	A	A_1	A_2	d_2	d_3	D_1	D_2	D_3	D_4
2	2.121	3.760	1.880	1.128	0.853	0	3.686	0	3.267
3	1.732	2.394	1.023	1.693	0.888	0	4.358	0	2.575
4	1.500	1.880	0.729	2.059	0.880	0	4.698	0	2.282
5	1.342	1.596	0.577	2.326	0.864	0	4.918	0	2.114
6	1.225	1.410	0.483	2.534	0.848	0	5.078	0	2.004
7	1.134	1.277	0.419	2.704	0.833	0.205	5.203	0.076	1.924
8	1.061	1.175	0.373	2.847	0.820	0.387	5.307	0.136	1.864
9	1.000	1.094	0.337	2.970	0.808	0.546	5.394	0.184	1.816
10	0.949	1.028	0.308	3.078	0.797	0.687	5.469	0.223	1.777
11	0.905	0.973	0.285	3.173	0.787	0.812	5.534	0.256	1.744
12	0.866	0.925	0.266	3.258	0.778	0.924	5.592	0.284	1.716
13	0.832	0.884	0.249	3.336	0.770	1.026	5.646	0.308	1.692
14	0.802	0.848	0.235	3.407	0.762	1.121	5.693	0.329	1.671
15	0.775	0.816	0.223	3.472	0.755	1.207	5.737	0.348	1.652
16	0.750	0.788	0.212	3.532	0.749	1.285	5.779	0.364	1.636
17	0.728	0.762	0.203	3.588	0.743	1.359	5.817	-.379	1.621
18	0.707	0.738	0.194	3.640	0.738	1.426	5.854	0.392	1.608
19	0.688	0.717	0.187	3.689	0.733	1.490	5.888	0.404	1.596
20	0.671	0.697	0.180	3.735	0.729	1.548	5.922	0.414	1.586
21	0.655	0.679	0.173	3.778	0.724	1.606	5.950	0.425	1.575
22	0.640	0.662	0.167	3.819	0.720	1.659	5.979	0.434	1.5˝
23	0.626	0.647	0.162	3.858	0.716	1.710	6.006	0.443	1.557
24	0.612	0.632	0.157	3.895	0.712	1.759	6.031	0.452	1.548
25	0.600	0.619	0.153	3.931	0.709	1.804	6.058	0.459	1.541

TABLE A-19 FACTORS FOR CONTROL LIMITS FOR STANDARD DEVIATION CHARTS

Sample Size	FACTORS FOR CENTRAL LINE c_2	FACTORS FOR CONTROL LIMITS B_1	B_2	B_3	B_4
2	0.5642	0	1.843	0	3.267
3	0.7236	0	1.858	0	2.568
4	0.7979	0	1.808	0	2.266
5	0.8407	0	1.756	0	2.089
6	0.8686	0.026	1.711	0.030	1.970
7	0.8882	0.105	1.672	0.118	1.882
8	0.9027	0.167	1.638	0.185	1.815
9	0.9139	0.219	1.609	0.239	1.761
10	0.9227	0.262	1.584	0.284	1.716
11	0.9300	0.299	1.561	0.321	1.679
12	0.9359	0.331	1.541	0.354	1.646
13	0.9410	0.359	1.523	0.382	1.618
14	0.9453	0.384	1.507	0.406	1.594
15	0.9490	0.406	1.492	0.428	1.572
16	0.9523	0.427	1.478	0.448	1.552
17	0.9551	0.445	1.465	0.466	1.534
18	0.9576	0.461	1.454	0.482	1.518
19	0.9599	0.477	1.443	0.497	1.503
20	0.9619	0.491	1.433	0.510	1.490
21	0.9638	0.504	1.424	0.523	1.477
22	0.9655	0.516	1.415	0.534	1.466
23	0.9670	0.527	1.407	0.545	1.455
24	0.9684	0.538	1.399	0.555	1.445
25	0.9696	0.548	1.392	0.565	1.435

TABLE A-20 AREAS UNDER THE NORMAL CURVE

Example: If *Z* = 1.52 this is .4357 or 43.57% of the area.

Z	0	1	2	3	4	5	6	7	8	9
0.0	.0000	.0040	.0080	.0120	.0160	.0199	.0239	.0279	.0319	.0359
0.1	.0398	.0438	.0478	.0517	.0557	.0596	.0636	.0675	.0714	.0753
0.2	.0793	.0832	.0871	.0910	.0948	.0987	.1026	.1064	.1103	.1141
0.3	.1179	.1217	.1255	.1293	.1331	.1368	.1406	.1443	.1480	.1517
0.4	.1554	.1591	.1628	.1664	.1700	.1736	.1772	.1808	.1884	.1879
0.5	.1915	.1950	.1985	.2019	.2054	.2088	.2123	.2157	.2190	.2224
0.6	.2257	.2291	.2324	.2357	.2389	.2422	.2454	.2486	.2518	.2549
0.7	.2580	.2612	.2642	.2673	.2704	.2734	.2764	.2794	.2823	.2852
0.8	.2881	.2910	.2939	.2967	.2995	.3023	.3051	.3078	.3106	.3133
0.9	.3159	.3186	.3212	.3238	.3264	.3289	.3315	.3340	.3365	.3389
1.0	.3413	.3438	.3461	.3485	.3508	.3531	.3554	.3577	.3599	.3621
1.1	.3643	.3665	.3686	.3708	.3729	.3749	.3770	.3790	.3810	.3830
1.2	.3849	.3869	.3888	.3907	.3925	.3944	.3962	.3980	.3997	.4015
1.3	.4032	.4049	.4066	.4082	.4099	.4115	.4131	.4147	.4162	.4177
1.4	.4192	.4207	.4222	.4236	.4251	.4265	.4279	.4292	.4306	.4319
1.5	.4332	.4345	.4357	.4370	.4382	.4394	.4406	.4418	.4429	.4441
1.6	.4452	.4463	.4474	.4484	.4495	.4505	.4515	.4525	.4535	.4545
1.7	.4554	.4564	.4573	.4582	.4591	.4599	.4608	.4616	.4625	.4633
1.8	.4641	.4649	.4656	.4664	.4671	.4678	.4686	.4693	.4699	.4706
1.9	.4713	.4719	.4726	.4732	.4738	.4744	.4750	.4756	.4761	.4767
2.0	.4772	.4778	.4783	.4788	.4793	.4798	.4803	.4808	.4812	.4817
2.1	.4821	.4826	.4830	.4834	.4838	.4842	.4846	.4850	.4854	.4857
2.2	.4861	.4864	.4868	.4871	.4875	.4878	.4881	.4884	.4887	.4890
2.3	.4893	.4896	.4898	.4901	.4904	.4906	.4909	.4911	.4913	.4916
2.4	.4918	.4920	.4922	.4925	.4927	.4929	.4931	.4932	.4934	.4936
2.5	.4938	.4940	.4941	.4943	.4945	.4946	.4948	.4949	.4951	.4952
2.6	.4953	.4955	.4956	.4957	.4959	.4960	.4961	.4962	.4963	.4964
2.7	.4965	.4966	.4967	.4968	.4969	.4970	.4971	.4972	.4973	.4974
2.8	.4974	.4975	.4976	.4977	.4977	.4978	.4979	.4979	.4980	.4981
2.9	.4981	.4982	.4982	.4983	.4984	.4984	.4985	.4985	.4986	.4986
3.0	.49865	.4987	.4987	.4988	.4988	.4989	.4989	.4989	.4990	.4990
3.1	.49903	.4991	.4991	.4991	.4992	.4992	.4992	.4992	.4993	.4993
3.2	.49931	.4993	.4994	.4994	.4994	.4994	.4994	.4995	.4995	.4995
3.3	.49952	.4995	.4995	.4996	.4996	.4996	.4996	.4996	.4996	.4997
3.4	.49966	.4997	.4997	.4997	.4997	.4997	.4997	.4997	.4998	.4998
3.5	.49977	.4998	.4998	.4998	.4998	.4998	.4998	.4998	.4998	.4998
3.6	.49984	.4998	.4999	.4999	.4999	.4999	.4999	.4999	.4999	.4999
3.7	.49989	.4999	.4999	.4999	.4999	.4999	.4999	.4999	.4999	.4999
3.8	.49993	.4999	.4999	.4999	.4999	.4999	.4999	.4999	.4999	.4999
3.9	.49995	.4999	.4999	.4999	.4999	.4999	.4999	.4999	.4999	.4999
4.0	.4999683									
4.5	.4999966									
5.0	.4999997133									

$$Z = \frac{x - \bar{x}}{\sigma}$$

$\bar{X}$ X

TABLE A-21　DEPTH OF THREADS (SINGLE THREADS)

*NOT APPLICABLE

(USING THE TURN METHOD)

EXAMPLE: A ¼-16 THREAD PLUG GOES 8 FULL TURNS INTO A HOLE. THE THREADS ARE .500" DEEP.

THREADS PER INCH	NUMBER OF TURNS								
	5	6	7	⑧	9	10	11	12	13
9	.555	.656	.777	.888	1"	*	*	*	*
10	.500	.600	.700	.800	.900	1"	*	*	*
11	.454	.545	.636	.727	.818	.909	1"	*	*
12	.416	.500	.583	.666	.750	.833	.916	1"	*
13	.384	.461	.538	.615	.692	.769	.846	.923	1"
14	.357	.428	.500	.571	.642	.714	.785	.856	.928
⑯	.312	.375	.437	⑤⓪⓪	.562	.625	.687	.750	.812
18	.277	.333	.388	.444	.500	.555	.610	.666	.721
20	.250	.300	.350	.400	.450	.500	.550	.600	.650
24	.208	.250	.291	.332	.374	.416	.458	.499	.540
28	.178	.214	.250	.285	.321	.357	.393	.428	.464
32	.156	.187	.218	.250	.281	.312	.343	.374	.406
36	.138	.166	.194	.222	.249	.277	.305	.332	.360
40	.125	.150	.175	.200	.225	.250	.275	.300	.325

THREADS PER INCH	NUMBER OF TURNS						
	14	15	16	17	18	19	20
16	.875	.937	1"	*	*	*	*
18	.777	.832	.888	.943	1"	*	*
20	.700	.750	.800	.850	.900	.950	1"
24	.582	.624	.666	.707	.749	.790	.832
28	.500	.535	.571	.607	.643	.678	.714
32	.437	.468	.499	.530	.562	.593	.624
36	.388	.415	.443	.471	.499	.526	.554
40	.350	.375	.400	.425	.450	.475	.500

TABLE A-22 FINDING THE ACTUAL DIAMETRAL ZONE OF A PRODUCED HOLE USING THE ACTUAL COORDINATES

ΔY	.001	.002	.003	.004	.005	.006	.007	.008	.009	.010	.011
.020	.0400	.0402	.0404	.0408	.0412	.0418	.0424	.0431	.0439	.0447	.0456
.019	.0380	.0382	.0385	.0388	.0393	.0398	.0405	.0412	.0420	.0429	.0439
.018	.0360	.0362	.0365	.0369	.0374	.0379	.0386	.0394	.0402	.0412	.0422
.017	.0340	.0342	.0345	.0349	.0354	.0360	.0368	.0376	.0385	.0394	.0405
.016	.0321	.0322	.0325	.0330	.0335	.0342	.0349	.0358	.0367	.0377	.0388
.015	.0301	.0303	.0306	.0310	.0316	.0323	.0331	.0340	.0350	.0360	.0372
.014	.0281	.0283	.0286	.0291	.0297	.0305	.0313	.0322	.0333	.0344	.0356
.013	.0261	.0263	.0267	.0272	.0278	.0286	.0295	.0305	.0316	.0328	.0340
.012	.0241	.0243	.0247	.0253	.0260	.0268	.0278	.0288	.0300	.0312	.0325
.011	.0221	.0224	.0228	.0234	.0242	.0250	.0261	.0272	.0284	.0297	.0311
.010	.0201	.0204	.0209	.0215	.0224	.0233	.0244	.0256	.0269	.0283	.0297
.009	.0181	.0184	.0190	.0197	.0206	.0216	.0228	.0241	.0254	.0269	.0284
.008	.0161	.0165	.0171	.0179	.0189	.0200	.0213	.0226	.0241	.0256	.0272
.007	.0141	.0146	.0152	.0161	.0172	.0184	.0198	.0213	.0228	.0244	.0261
.006	.0122	.0126	.0134	.0144	.0156	.0170	.0184	.0200	.0216	.0233	.0250
.005	.0102	.0108	.0117	.0128	.0141	.0156	.0172	.0189	.0206	.0224	.0242
.004	.0082	.0089	.0100	.0113	.0128	.0144	.0161	.0179	.0197	.0215	.0234
.003	.0063	.0072	.0085	.0100	.0117	.0134	.0152	.0171	.0190	.0209	.0228
.002	.0045	.0056	.0072	.0089	.0108	.0126	.0146	.0165	.0184	.0204	.0224
.001	.0028	.0045	.0063	.0082	.0102	.0122	.0141	.0161	.0181	.0201	.0221

ΔX ← → .001 .002 .003 .004 .005 .006 .007 .008 .009 .010 .011

FORMULA:

$$\text{ACTUAL DIA. ZONE} = 2\sqrt{X^2 + Y^2}$$

WHERE X = THE DIFFERENCE BETWEEN THE BASIC AND THE ACTUAL COORDINATE DIMENSION IN "X" DIRECTION
Y = THE DIFFERENCE BETWEEN THE BASIC AND THE ACTUAL DIMENSION IN "Y" DIRECTION

EXAMPLE: The hole is .003 past its basic dimension in "x" direction, and .005 past its basic dimension in "y" direction.
Actual zone = .0117 Dia.

TABLE A-23 Mil-Std-414 TABLES FOR AQL CONVERSION AND SAMPLE SIZE CODE LETTERS

Sample Size Code Letters[1]

Lot Size	Inspection Levels				
	I	II	III	IV	V
3 to 8	B	B	B	B	C
9 to 15	B	B	B	B	D
16 to 25	B	B	B	C	E
26 to 40	B	B	B	D	F
41 to 65	B	B	C	E	G
66 to 110	B	B	D	F	H
111 to 180	B	C	E	G	I
181 to 300	B	D	F	H	J
301 to 500	C	E	G	I	K
501 to 800	D	F	H	J	L
801 to 1,300	E	G	I	K	L
1,301 to 3,200	F	H	J	L	M
3,201 to 8,000	G	I	L	M	N
8,001 to 22,000	H	J	M	N	O
22,001 to 110,000	I	K	N	O	P
110,001 to 550,000	I	K	O	P	Q
550,001 and over	I	K	P	Q	Q

[1]Sample size code letters given in body of table are applicable when the indicated inspection levels are to be used.

AQL Conversion Table

For specified AQL values falling within these ranges	Use this AQL value
— to 0.049	0.04
0.050 to 0.069	0.065
0.070 to 0.109	0.10
0.110 to 0.164	0.15
0.165 to 0.279	0.25
0.280 to 0.439	0.40
0.440 to 0.699	0.65
0.700 to 1.09	1.0
1.10 to 1.64	1.5
1.65 to 2.79	2.5
2.80 to 4.39	4.0
4.40 to 6.99	6.5
7.00 to 10.9	10.0
11.00 to 16.4	15.0

TABLE A-24 Mil-Std-414 TABLE FOR STANDARD DEVIATION METHOD SINGLE SPECIFICATION LIMIT—FORM 1 (MASTER TABLE FOR *NORMAL AND TIGHTENED* INSPECTION FOR PLANS BASED ON VARIABILITY UNKNOWN)

Sample size code letter	Sample size	.04	.065	.10	.15	.25	.40	.65	1.00	1.50	2.50	4.00	6.50	10.00	15.00
		k	k	k	k	k	k	k	k	k	k	k	k	k	k
B	3	→	→	→	→	→	→	→	▼	▼	1.12	.958	.765	.566	.341
C	4	→	→	→	→	→	→	→	1.45	1.34	1.17	1.01	.814	.617	.393
D	5	→	→	→	→	→	→	1.65	1.53	1.40	1.24	1.07	.874	.675	.455
E	7	→	→	→	→	2.00	1.88	1.75	1.62	1.50	1.33	1.15	.955	.755	.536
F	10	→	→	→	2.24	2.11	1.98	1.84	1.72	1.58	1.41	1.23	1.03	.828	.611
G	15	2.64	2.53	2.42	2.32	2.20	2.06	1.91	1.79	1.65	1.47	1.30	1.09	.886	.664
H	20	2.69	2.58	2.47	2.36	2.24	2.11	1.96	1.82	1.69	1.51	1.33	1.12	.917	.695
I	25	2.72	2.61	2.50	2.40	2.26	2.14	1.98	1.85	1.72	1.53	1.35	1.14	.936	.712
J	30	2.73	2.61	2.51	2.41	2.28	2.15	2.00	1.86	1.73	1.55	1.36	1.15	.946	.723
K	35	2.77	2.65	2.54	2.45	2.31	2.18	2.03	1.89	1.76	1.57	1.39	1.18	.969	.745
L	40	2.77	2.66	2.55	2.44	2.31	2.18	2.03	1.89	1.76	1.58	1.39	1.18	.971	.746
M	50	2.83	2.71	2.60	2.50	2.35	2.22	2.08	1.93	1.80	1.61	1.42	1.21	1.00	.774
N	75	2.90	2.77	2.66	2.55	2.41	2.27	2.12	1.98	1.84	1.65	1.46	1.24	1.03	.804
O	100	2.92	2.80	2.69	2.58	2.43	2.29	2.14	2.00	1.86	1.67	1.48	1.26	1.05	.819
P	150	2.96	2.84	2.73	2.61	2.47	2.33	2.18	2.03	1.89	1.70	1.51	1.29	1.07	.841
Q	200	2.97	2.85	2.73	2.62	2.47	2.33	2.18	2.04	1.89	1.70	1.51	1.29	1.07	.845
		.065	.10	.15	.25	.40	.65	1.00	1.50	2.50	4.00	6.50	10.00	15.00	

Acceptable Quality Levels (normal inspection)

Acceptable Quality Levels (tightened inspection)

All AQL values are in percent defective.
↓ Use first sampling plan below arrow, that is, both sample size as well as k value. When sample size equals or exceeds lot size, every item in the lot must be inspected.

TABLE A-25 Mil-Std-414 TABLE FOR STANDARD DEVIATION METHOD DOUBLE SPECIFICATION LIMIT AND FORM 2—SINGLE SPECIFICATION LIMIT (MASTER TABLE FOR *NORMAL AND TIGHTENED* INSPECTION FOR PLANS BASED ON VARIABILITY UNKNOWN)

Sample size code letter	Sample size	\.04 M	\.065 M	\.10 M	\.15 M	\.25 M	\.40 M	\.65 M	1.00 M	1.50 M	2.50 M	4.00 M	6.50 M	10.00 M	15.00 M
								Acceptable Quality Levels (normal inspection)							
B	3	↓	↓	↓	↓	↓	↓	↓	▼	▼	7.59	18.86	26.94	33.69	40.47
C	4	↓	↓	↓	↓	↓	↓	↓	1.53	5.50	10.92	16.45	22.86	29.45	36.90
D	5	↓	↓	↓	↓	↓	1.06	1.33	3.32	5.83	9.80	14.39	20.19	26.56	33.99
E	7	↓	↓	↓	↓	0.422	1.06	2.14	3.55	5.35	8.40	12.20	17.35	23.29	30.50
F	10	↓	↓	↓	0.349	0.716	1.30	2.17	3.26	4.77	7.29	10.54	15.17	20.74	27.57
G	15	0.099	0.186	0.312	0.503	0.818	1.31	2.11	3.05	4.31	6.56	9.46	13.71	18.94	25.61
H	20	0.135	0.228	0.365	0.544	0.846	1.29	2.05	2.95	4.09	6.17	8.92	12.99	18.03	24.53
I	25	0.155	0.250	0.380	0.551	0.877	1.29	2.00	2.86	3.97	5.97	8.63	12.57	17.51	23.97
J	30	0.179	0.280	0.413	0.581	0.879	1.29	1.98	2.83	3.91	5.86	8.47	12.36	17.24	23.58
K	35	0.170	0.264	0.388	0.535	0.847	1.23	1.87	2.68	3.70	5.57	8.10	11.87	16.65	22.91
L	40	0.179	0.275	0.401	0.566	0.873	1.26	1.88	2.71	3.72	5.58	8.09	11.85	16.61	22.86
M	50	0.163	0.250	0.363	0.503	0.789	1.17	1.71	2.49	3.45	5.20	7.61	11.23	15.87	22.00
N	75	0.147	0.228	0.330	0.467	0.720	1.07	1.60	2.29	3.20	4.87	7.15	10.63	15.13	21.11
O	100	0.145	0.220	0.317	0.447	0.689	1.02	1.53	2.20	3.07	4.69	6.91	10.32	14.75	20.66
P	150	0.134	0.203	0.293	0.413	0.638	0.949	1.43	2.05	2.89	4.43	6.57	9.88	14.20	20.02
Q	200	0.135	0.204	0.294	0.414	0.637	0.945	1.42	2.04	2.87	4.40	6.53	9.81	14.12	19.92
		.065	.10	.15	.25	.40	.65	1.00	1.50	2.50	4.00	6.50	10.00	15.00	
						Acceptability Quality Levels (tightened inspection)									

All AQL and table values are in percent defective.

↓ Use first sampling plan below arrow, that is, both sample size as well as M value. When sample size equals or exceeds lot size, every item in the lot must be inspected.

488

TABLE A-26 Mil-Std-414 TABLE FOR ESTIMATING THE LOT PERCENT DEFECTIVE USING THE STANDARD DEVIATION METHOD

Q_U or Q_L	Sample Size															
	3	4	5	7	10	15	20	25	30	35	40	50	75	100	150	200
1.90	0.00	0.00	0.00	0.93	1.75	2.21	2.40	2.51	2.57	2.62	2.65	2.70	2.76	2.79	2.82	2.83
1.91	0.00	0.00	0.00	0.87	1.68	2.14	2.34	2.44	2.51	2.56	2.59	2.63	2.69	2.72	2.75	2.77
1.92	0.00	0.00	0.00	0.81	1.62	2.08	2.27	2.38	2.45	2.49	2.52	2.57	2.63	2.66	2.69	2.70
1.93	0.00	0.00	0.00	0.76	1.56	2.02	2.21	2.32	2.38	2.43	2.46	2.51	2.57	2.60	2.62	2.64
1.94	0.00	0.00	0.00	0.70	1.50	1.96	2.15	2.25	2.32	2.37	2.40	2.45	2.51	2.54	2.56	2.58
1.95	0.00	0.00	0.00	0.65	1.44	1.90	2.09	2.19	2.26	2.31	2.34	2.39	2.45	2.48	2.50	2.52
1.96	0.00	0.00	0.00	0.60	1.38	1.84	2.03	2.14	2.20	2.25	2.28	2.33	2.39	2.42	2.44	2.46
1.97	0.00	0.00	0.00	0.56	1.33	1.78	1.97	2.08	2.14	2.19	2.22	2.27	2.33	2.36	2.39	2.40
1.98	0.00	0.00	0.00	0.51	1.27	1.73	1.92	2.02	2.09	2.13	2.17	2.21	2.27	2.30	2.33	2.34
1.99	0.00	0.00	0.00	0.47	1.22	1.67	1.86	1.97	2.03	2.08	2.11	2.16	2.22	2.25	2.27	2.29
2.00	0.00	0.00	0.00	0.43	1.17	1.62	1.81	1.91	1.98	2.03	2.06	2.10	2.16	2.19	2.22	2.23
2.01	0.00	0.00	0.00	0.39	1.12	1.57	1.76	1.86	1.93	1.97	2.01	2.05	2.11	2.14	2.17	2.18
2.02	0.00	0.00	0.00	0.36	1.07	1.52	1.71	1.81	1.87	1.92	1.95	2.00	2.06	2.09	2.11	2.13
2.03	0.00	0.00	0.00	0.32	1.03	1.47	1.66	1.76	1.82	1.87	1.90	1.95	2.01	2.04	2.06	2.08
2.04	0.00	0.00	0.00	0.29	0.98	1.42	1.61	1.71	1.77	1.82	1.85	1.90	1.96	1.99	2.01	2.03
2.05	0.00	0.00	0.00	0.26	0.94	1.37	1.56	1.66	1.73	1.77	1.80	1.85	1.91	1.94	1.96	1.98
2.06	0.00	0.00	0.00	0.23	0.90	1.33	1.51	1.61	1.68	1.72	1.76	1.80	1.86	1.89	1.92	1.93
2.07	0.00	0.00	0.00	0.21	0.86	1.28	1.47	1.57	1.63	1.68	1.71	1.76	1.81	1.84	1.87	1.88
2.08	0.00	0.00	0.00	0.18	0.82	1.24	1.42	1.52	1.59	1.63	1.66	1.71	1.77	1.79	1.82	1.84
2.09	0.00	0.00	0.00	0.16	0.78	1.20	1.38	1.48	1.54	1.59	1.62	1.66	1.72	1.75	1.78	1.79
2.10	0.00	0.00	0.00	0.14	0.74	1.16	1.34	1.44	1.50	1.54	1.58	1.62	1.68	1.71	1.73	1.75
2.11	0.00	0.00	0.00	0.12	0.71	1.12	1.30	1.39	1.46	1.50	1.53	1.58	1.63	1.66	1.69	1.70
2.12	0.00	0.00	0.00	0.10	0.67	1.08	1.26	1.35	1.42	1.46	1.49	1.54	1.59	1.62	1.65	1.66
2.13	0.00	0.00	0.00	0.08	0.64	1.04	1.22	1.31	1.38	1.42	1.45	1.50	1.55	1.58	1.61	1.62
2.14	0.00	0.00	0.00	0.07	0.61	1.00	1.18	1.28	1.34	1.38	1.41	1.46	1.51	1.54	1.57	1.58
2.15	0.00	0.00	0.00	0.06	0.58	0.97	1.14	1.24	1.30	1.34	1.37	1.42	1.47	1.50	1.53	1.54
2.16	0.00	0.00	0.00	0.05	0.55	0.93	1.10	1.20	1.26	1.30	1.34	1.38	1.43	1.46	1.49	1.50
2.17	0.00	0.00	0.00	0.04	0.52	0.90	1.07	1.16	1.22	1.27	1.30	1.34	1.40	1.42	1.45	1.46
2.18	0.00	0.00	0.00	0.03	0.49	0.87	1.03	1.13	1.19	1.23	1.26	1.30	1.36	1.39	1.41	1.42
2.19	0.00	0.00	0.00	0.02	0.46	0.83	1.00	1.09	1.15	1.20	1.23	1.27	1.32	1.35	1.38	1.39
2.20	0.000	0.000	0.000	0.015	0.437	0.803	0.968	1.061	1.120	1.161	1.192	1.233	1.287	1.314	1.340	1.352
2.21	0.000	0.000	0.000	0.010	0.413	0.772	0.936	1.028	1.087	1.128	1.158	1.199	1.253	1.279	1.305	1.318
2.22	0.000	0.000	0.000	0.006	0.389	0.743	0.905	0.996	1.054	1.095	1.125	1.166	1.219	1.245	1.271	1.283
2.23	0.000	0.000	0.000	0.003	0.366	0.715	0.875	0.965	1.023	1.063	1.093	1.134	1.186	1.212	1.238	1.250
2.24	0.000	0.000	0.000	0.002	0.345	0.687	0.845	0.935	0.992	1.032	1.061	1.102	1.154	1.180	1.205	1.218
2.25	0.000	0.000	0.000	0.001	0.324	0.660	0.816	0.905	0.962	1.002	1.031	1.071	1.123	1.148	1.173	1.186
2.26	0.000	0.000	0.000	0.000	0.304	0.634	0.789	0.876	0.933	0.972	1.001	1.041	1.092	1.117	1.142	1.155
2.27	0.000	0.000	0.000	0.000	0.285	0.609	0.762	0.848	0.904	0.943	0.972	1.011	1.062	1.087	1.112	1.124
2.28	0.000	0.000	0.000	0.000	0.267	0.585	0.735	0.821	0.876	0.915	0.943	0.982	1.033	1.058	1.082	1.094
2.29	0.000	0.000	0.000	0.000	0.250	0.561	0.710	0.794	0.849	0.887	0.915	0.954	1.004	1.029	1.053	1.065

Q_U or Q_L	3	4	5	7	10	15	20	25	30	35	40	50	75	100	150	200
2.30	0.000	0.000	0.000	0.000	0.233	0.538	0.685	0.769	0.823	0.861	0.888	0.927	0.977	1.001	1.025	1.037
2.31	0.000	0.000	0.000	0.000	0.218	0.516	0.661	0.743	0.797	0.834	0.862	0.900	0.949	0.974	0.997	1.009
2.32	0.000	0.000	0.000	0.000	0.203	0.495	0.637	0.719	0.772	0.809	0.836	0.874	0.923	0.947	0.971	0.982
2.33	0.000	0.000	0.000	0.000	0.189	0.474	0.614	0.695	0.748	0.784	0.811	0.848	0.897	0.921	0.944	0.956
2.34	0.000	0.000	0.000	0.000	0.175	0.454	0.592	0.672	0.724	0.760	0.787	0.824	0.872	0.895	0.915	0.930
2.35	0.000	0.000	0.000	0.000	0.163	0.435	0.571	0.650	0.701	0.736	0.763	0.799	0.847	0.870	0.893	0.905
2.36	0.000	0.000	0.000	0.000	0.151	0.416	0.550	0.628	0.678	0.714	0.740	0.776	0.823	0.846	0.869	0.880
2.37	0.000	0.000	0.000	0.000	0.139	0.398	0.530	0.606	0.656	0.691	0.717	0.753	0.799	0.822	0.845	0.856
2.38	0.000	0.000	0.000	0.000	0.128	0.381	0.510	0.586	0.635	0.670	0.695	0.730	0.777	0.799	0.822	0.833
2.39	0.000	0.000	0.000	0.000	0.118	0.364	0.491	0.566	0.614	0.648	0.674	0.709	0.754	0.777	0.799	0.810
2.40	0.000	0.000	0.000	0.000	0.109	0.348	0.473	0.546	0.594	0.628	0.653	0.687	0.732	0.755	0.777	0.787
2.41	0.000	0.000	0.000	0.000	0.100	0.332	0.455	0.527	0.575	0.608	0.633	0.667	0.711	0.733	0.755	0.766
2.42	0.000	0.000	0.000	0.000	0.091	0.317	0.437	0.509	0.555	0.588	0.613	0.646	0.691	0.712	0.734	0.744
2.43	0.000	0.000	0.000	0.000	0.083	0.302	0.421	0.491	0.537	0.569	0.593	0.627	0.670	0.692	0.713	0.724.
2.44	0.000	0.000	0.000	0.000	0.076	0.288	0.404	0.474	0.519	0.551	0.575	0.608	0.651	0.672	0.693	0.703
2.45	0.000	0.000	0.000	0.000	0.069	0.275	0.389	0.457	0.501	0.533	0.556	0.589	0.632	0.653	0.673	0.684
2.46	0.000	0.000	0.000	0.000	0.063	0.262	0.373	0.440	0.484	0.516	0.539	0.571	0.613	0.634	0.654	0.664
2.47	0.000	0.000	0.000	0.000	0.057	0.249	0.359	0.425	0.468	0.499	0.521	0.553	0.595	0.615	0.635	0.646
2.48	0.000	0.000	0.000	0.000	0.051	0.237	0.344	0.409	0.452	0.482	0.505	0.536	0.577	0.597	0.617	0.627
2.49	0.000	0.000	0.000	0.000	0.046	0.226	0.331	0.394	0.436	0.466	0.488	0.519	0.560	0.580	0.600	0.609
2.50	0.000	0.000	0.000	0.000	0.041	0.214	0.317	0.380	0.421	0.451	0.473	0.503	0.543	0.563	0.582	0.592
2.51	0.000	0.000	0.000	0.000	0.037	0.204	0.304	0.366	0.407	0.436	0.457	0.487	0.527	0.546	0.565	0.575
2.52	0.000	0.000	0.000	0.000	0.033	0.193	0.292	0.352	0.392	0.421	0.442	0.472	0.511	0.530	0.549	0.558
2.53	0.000	0.000	0.000	0.000	0.029	0.184	0.280	0.339	0.379	0.407	0.428	0.457	0.495	0.514	0.533	0.542
2.54	0.000	0.000	0.000	0.000	0.026	0.174	0.268	0.326	0.365	0.393	0.413	0.442	0.480	0.499	0.517	0.527
2.55	0.000	0.000	0.000	0.000	0.023	0.165	0.257	0.314	0.352	0.379	0.400	0.428	0.465	0.484	0.502	0.511
2.56	0.000	0.000	0.000	0.000	0.020	0.156	0.246	0.302	0.340	0.366	0.386	0.414	0.451	0.469	0.487	0.496
2.57	0.000	0.000	0.000	0.000	0.017	0.148	0.236	0.291	0.327	0.354	0.373	0.401	0.437	0.455	0.473	0.482
2.58	0.000	0.000	0.000	0.000	0.015	0.140	0.226	0.279	0.316	0.341	0.361	0.388	0.424	0.441	0.459	0.468
2.59	0.000	0.000	0.000	0.000	0.013	0.133	0.216	0.269	0.304	0.330	0.349	0.375	0.410	0.428	0.445	0.454
2.60	0.000	0.000	0.000	0.000	0.011	0.125	0.207	0.258	0.293	0.318	0.337	0.363	0.398	0.415	0.432	0.441
2.61	0.000	0.000	0.000	0.000	0.009	0.118	0.198	0.248	0.282	0.307	0.325	0.351	0.385	0.402	0.419	0.428
2.62	0.000	0.000	0.000	0.000	0.008	0.112	0.189	0.238	0.272	0.296	0.314	0.339	0.373	0.390	0.406	0.415
2.63	0.000	0.000	0.000	0.000	0.007	0.105	0.181	0.229	0.262	0.285	0.303	0.328	0.361	0.378	0.394	0.402
2.64	0.000	0.000	0.000	0.000	0.005	0.099	0.172	0.220	0.252	0.275	0.293	0.317	0.350	0.366	0.382	0.390
2.65	0.000	0.008	0.000	0.000	0.005	0.094	0.165	0.211	0.243	0.265	0.282	0.307	0.339	0.355	0.371	0.379
2.66	0.000	0.000	0.000	0.000	0.004	0.088	0.157	0.202	0.233	0.256	0.273	0.296	0.328	0.344	0.359	0.367
2.67	0.000	0.000	0.000	0.000	0.003	0.083	0.150	0.194	0.224	0.246	0.263	0.286	0.317	0.333	0.348	0.356
2.68	0.000	0.000	0.000	0.000	0.002	0.078	0.143	0.186	0.216	0.237	0.254	0.277	0.307	0.322	0.338	0.345
2.69	0.000	0.000	0.000	0.000	0.002	0.073	0.136	0.179	0.208	0.229	0.245	0.267	0.297	0.312	0.327	0.335

TABLE A-26 Mil-Std-414 TABLE FOR ESTIMATING THE LOT PERCENT DEFECTIVE USING THE STANDARD DEVIATION METHOD (*continued*)

Q_U or Q_L	Sample Size															
	3	4	5	7	10	15	20	25	30	35	40	50	75	100	150	200
3.50	0.000	0.000	0.000	0.000	0.000	0.000	0.000	0.002	0.003	0.005	0.007	0.009	0.013	0.015	0.018	0.019
3.51	0.000	0.000	0.000	0.000	0.000	0.000	0.000	0.002	0.003	0.005	0.006	0.009	0.013	0.015	0.017	0.018
3.52	0.000	0.000	0.000	0.000	0.000	0.000	0.000	0.002	0.003	0.005	0.006	0.008	0.012	0.014	0.017	0.018
3.53	0.000	0.000	0.000	0.000	0.000	0.000	0.000	0.001	0.003	0.004	0.006	0.008	0.012	0.014	0.016	0.017
3.54	0.000	0.000	0.000	0.000	0.000	0.000	0.000	0.001	0.003	0.004	0.005	0.008	0.011	0.013	0.015	0.016
3.55	0.000	0.000	0.000	0.000	0.000	0.000	0.000	0.001	0.003	0.004	0.005	0.007	0.011	0.012	0.015	0.016
3.56	0.000	0.000	0.000	0.000	0.000	0.000	0.000	0.001	0.002	0.004	0.005	0.007	0.010	0.012	0.014	0.015
3.57	0.000	0.000	0.000	0.000	0.000	0.000	0.000	0.001	0.002	0.003	0.005	0.006	0.010	0.011	0.013	0.014
3.58	0.000	0.000	0.000	0.000	0.000	0.000	0.000	0.001	0.002	0.003	0.004	0.006	0.009	0.011	0.013	0.014
3.59	0.000	0.000	0.000	0.000	0.000	0.000	0.000	0.001	0.002	0.003	0.004	0.006	0.009	0.010	0.012	0.013
3.60	0.000	0.000	0.000	0.000	0.000	0.000	0.000	0.001	0.002	0.003	0.004	0.006	0.008	0.010	0.012	0.013
3.61	0.000	0.000	0.000	0.000	0.000	0.000	0.000	0.001	0.002	0.003	0.004	0.005	0.008	0.010	0.011	0.012
3.62	0.000	0.000	0.000	0.000	0.000	0.000	0.000	0.001	0.002	0.003	0.003	0.005	0.008	0.009	0.011	0.012
3.63	0.000	0.000	0.000	0.000	0.000	0.000	0.000	0.001	0.001	0.002	0.003	0.005	0.007	0.009	0.010	0.011
3.64	0.000	0.000	0.000	0.000	0.000	0.000	0.000	0.001	0.001	0.002	0.003	0.004	0.007	0.008	0.010	0.011
3.65	0.000	0.000	0.000	0.000	0.000	0.000	0.000	0.001	0.001	0.002	0.003	0.004	0.007	0.008	0.010	0.010
3.66	0.000	0.000	0.000	0.000	0.000	0.000	0.000	0.000	0.001	0.002	0.003	0.004	0.006	0.008	0.009	0.010
3.67	0.000	0.000	0.000	0.000	0.000	0.000	0.000	0.000	0.001	0.002	0.003	0.004	0.006	0.007	0.009	0.010
3.68	0.000	0.000	0.000	0.000	0.000	0.000	0.000	0.000	0.001	0.002	0.002	0.004	0.006	0.007	0.008	0.009
3.69	0.000	0.000	0.000	0.000	0.000	0.000	0.000	0.000	0.001	0.002	0.002	0.003	0.005	0.007	0.008	0.009

TABLE A-26   Mil-Std-414 TABLE FOR ESTIMATING THE LOT PERCENT DEFECTIVE USING THE STANDARD DEVIATION METHOD (*continued*)

Q_U or Q_L	Sample Size															
	3	4	5	7	10	15	20	25	30	35	40	50	75	100	150	200
3.70	0.000	0.000	0.000	0.000	0.000	0.000	0.000	0.000	0.001	0.002	0.002	0.003	0.005	0.006	0.008	0.008
3.71	0.000	0.000	0.000	0.000	0.000	0.000	0.000	0.000	0.001	0.001	0.002	0.003	0.005	0.006	0.007	0.008
3.72	0.000	0.000	0.000	0.000	0.000	0.000	0.000	0.000	0.001	0.001	0.002	0.003	0.005	0.006	0.007	0.008
3.73	0.000	0.000	0.000	0.000	0.000	0.000	0.000	0.000	0.001	0.001	0.002	0.003	0.005	0.006	0.007	0.007
3.74	0.000	0.000	0.000	0.000	0.000	0.000	0.000	0.000	0.001	0.001	0.002	0.003	0.004	0.005	0.007	0.007
3.75	0.000	0.000	0.000	0.000	0.000	0.000	0.000	0.000	0.001	0.001	0.002	0.002	0.004	0.005	0.006	0.007
3.76	0.000	0.000	0.000	0.000	0.000	0.000	0.000	0.000	0.001	0.001	0.001	0.002	0.004	0.005	0.006	0.007
3.77	0.000	0.000	0.000	0.000	0.000	0.000	0.000	0.000	0.001	0.001	0.001	0.002	0.004	0.005	0.006	0.006
3.78	0.000	0.000	0.000	0.000	0.000	0.000	0.000	0.000	0.000	0.001	0.001	0.002	0.004	0.004	0.005	0.006
3.79	0.000	0.000	0.000	0.000	0.000	0.000	0.000	0.000	0.000	0.001	0.001	0.002	0.003	0.004	0.005	0.006
3.80	0.000	0.000	0.000	0.000	0.000	0.000	0.000	0.000	0.000	0.001	0.001	0.002	0.003	0.004	0.005	0.006
3.81	0.000	0.000	0.000	0.000	0.000	0.000	0.000	0.000	0.000	0.001	0.001	0.002	0.003	0.004	0.005	0.005
3.82	0.000	0.000	0.000	0.000	0.000	0.000	0.000	0.000	0.000	0.001	0.001	0.002	0.003	0.004	0.005	0.005
3.83	0.000	0.000	0.000	0.000	0.000	0.000	0.000	0.000	0.000	0.001	0.001	0.002	0.003	0.004	0.004	0.005
3.84	0.000	0.000	0.000	0.000	0.000	0.000	0.000	0.000	0.000	0.001	0.001	0.001	0.003	0.003	0.004	0.005
3.85	0.000	0.000	0.000	0.000	0.000	0.000	0.000	0.000	0.000	0.001	0.001	0.001	0.002	0.003	0.004	0.004
3.86	0.000	0.000	0.000	0.000	0.000	0.000	0.000	0.000	0.000	0.001	0.001	0.002	0.003	0.004	0.004	0.004
3.87	0.000	0.000	0.000	0.000	0.000	0.000	0.000	0.000	0.000	0.001	0.001	0.002	0.003	0.004	0.004	0.004
3.88	0.000	0.000	0.000	0.000	0.000	0.000	0.000	0.000	0.000	0.001	0.001	0.002	0.003	0.004	0.004	0.004
3.89	0.000	0.000	0.000	0.000	0.000	0.000	0.000	0.000	0.000	0.001	0.001	0.002	0.003	0.003	0.003	0.004
3.90	0.000	0.000	0.000	0.000	0.000	0.000	0.000	0.000	0.000	0.001	0.001	0.002	0.003	0.003	0.003	0.004

TABLE A-27 Mil-Std-414 TABLE FOR RANGE METHOD SINGLE SPECIFICATION LIMIT—FORM 1 (MASTER TABLE FOR *NORMAL AND TIGHTENED* INSPECTION FOR PLANS BASED ON VARIABILITY UNKNOWN)

Sample size code letter	Sample size	Acceptable Quality Levels (normal inspection)													
		.04	.065	.10	.15	.25	.40	.65	1.00	1.50	2.50	4.00	6.50	10.00	15.00
		k	k	k	k	k	k	k	k	k	k	k	k	k	k
B	3	→	→	→	→	→	→	→	▼	▼	.587	.502	.401	.296	.178
C	4	→	→	→	→	→	→	→	.651	.598	.525	.450	.364	.276	.176
D	5	→	→	→	→	→	→	.663	.614	.565	.498	.431	.352	.272	.184
E	7	→	→	→	→	.702	.659	.613	.569	.525	.465	.405	.336	.266	.189
F	10	→	→	→	.916	.863	.811	.755	.703	.650	.579	.507	.424	.341	.252
G	15	1.09	1.04	.999	.958	.903	.850	.792	.738	.684	.610	.536	.452	.368	.276
H	25	1.14	1.10	1.05	1.01	.951	.896	.835	.779	.723	.647	.571	.484	.398	.305
I	30	1.15	1.10	1.06	1.02	.959	.904	.843	.787	.730	.654	.577	.490	.403	.310
J	35	1.16	1.11	1.07	1.02	.964	.908	.848	.791	.734	.658	.581	.494	.406	.313
K	40	1.18	1.13	1.08	1.04	.978	.921	.860	.803	.746	.668	.591	.503	.415	.321
L	50	1.19	1.14	1.09	1.05	.988	.931	.893	.812	.754	.676	.598	.510	.421	.327
M	60	1.21	1.16	1.11	1.06	1.00	.948	.885	.826	.768	.689	.610	.521	.432	.336
N	85	1.23	1.17	1.13	1.08	1.02	.962	.899	.839	.780	.701	.621	.530	.441	.345
O	115	1.24	1.19	1.14	1.09	1.03	.975	.911	.851	.791	.711	.631	.539	.449	.353
P	175	1.26	1.21	1.16	1.11	1.05	.994	.929	.868	.807	.726	.644	.552	.460	.363
Q	230	1.27	1.21	1.16	1.12	1.06	.996	.931	.870	.809	.728	.645	.553	.462	.364
		.065	.10	.15	.25	.40	.65	1.00	1.50	2.50	4.00	6.50	10.00	15.00	
		Acceptable Quality Levels (tightened inspection)													

All AQL values are in percent defective.

↓ Use first sampling plan below arrow, that is, both sample size as well as k value. When sample size equals or exceeds lot size, every item in the lot must be inspected.

TABLE A-28 Mil-Std-414 TABLE FOR RANGE METHOD DOUBLE SPECIFICATION LIMIT AND FORM 2—SINGLE SPECIFICATION LIMIT (MASTER TABLE FOR *NORMAL AND TIGHTENED* INSPECTION FOR PLANS BASED ON VARIABILITY UNKNOWN)

Acceptable Quality Levels (normal inspection) — all values are M values

Sample size code letter	Sample size	c factor	.04	.065	.10	.15	.25	.40	.65	1.00	1.50	2.50	4.00	6.50	10.00	15.00
B	3	1.910	↓	↓	↓	↓	↓	↓	↓	↓	↓	7.59	18.86	26.94	33.69	40.47
C	4	2.234	↓	↓	↓	↓	↓	↓	↓	1.53	5.50	10.92	16.45	22.86	29.45	36.90
D	5	2.474	↓	↓	↓	↓	↓	↓	1.42	3.44	5.93	9.90	14.47	20.27	26.59	33.95
E	7	2.830	↓	↓	↓	↓	.28	.89	1.99	3.46	5.32	8.47	12.35	17.54	23.50	30.66
F	10	2.405	↓	↓	↓	.23	.58	1.14	2.05	3.23	4.77	7.42	10.79	15.49	21.06	27.90
G	15	2.379	.061	.136	.253	.430	.786	1.30	2.10	3.11	4.44	6.76	9.76	14.09	19.30	25.92
H	25	2.358	.125	.214	.336	.506	.827	1.27	1.95	2.82	3.96	5.98	8.65	12.59	17.48	23.79
I	30	2.353	.147	.240	.366	.537	.856	1.29	1.96	2.81	3.92	5.88	8.50	12.36	17.19	23.42
J	35	2.349	.165	.261	.391	.564	.883	1.33	1.98	2.82	3.90	5.85	8.42	12.24	17.03	23.21
K	40	2.346	.160	.252	.375	.539	.842	1.25	1.88	2.69	3.73	5.61	8.11	11.84	16.55	22.38
L	50	2.342	.169	.261	.381	.542	.838	1.25	1.60	2.63	3.64	5.47	7.91	11.57	16.20	22.26
M	60	2.339	.158	.244	.356	.504	.781	1.16	1.74	2.47	3.44	5.17	7.54	11.10	15.64	21.63
N	85	2.335	.156	.242	.350	.493	.755	1.12	1.67	2.37	3.30	4.97	7.27	10.73	15.17	21.05
O	115	2.333	.153	.230	.333	.468	.718	1.06	1.58	2.25	3.14	4.76	6.99	10.37	14.74	20.57
P	175	2.331	.139	.210	.303	.427	.655	.972	1.46	2.08	2.93	4.47	6.60	9.89	14.15	19.88
Q	230	2.330	.142	.215	.308	.432	.661	.976	1.47	2.08	2.92	4.46	6.57	9.84	14.10	19.82
Acceptable Quality Levels (tightened inspection)			.065	.10	.15	.25	.40	.65	1.00	1.50	2.50	4.00	6.50	10.00	15.00	

All AQL and table values are in percent defective.

↓ Use first sampling plan below arrow, that is, both sample size as well as M value. When sample size equals or exceeds lot size, every item in the lot must be inspected.

494

TABLE A-29 Mil-Std-414 TABLE FOR ESTIMATING THE LOT PERCENT DEFECTIVE USING THE RANGE METHOD

Q_U or Q_L	Sample Size															
	3	4	5	7	10	15	25	30	35	40	50	60	85	115	175	230
0	50.00	50.00	50.00	50.00	50.00	50.00	50.00	50.00	50.00	50.00	50.00	50.00	50.00	50.00	50.00	50.00
.1	47.24	46.67	46.44	46.29	46.20	46.13	46.08	46.07	46.06	46.05	46.05	46.04	46.03	46.03	46.02	46.02
.2	44.46	43.33	42.90	42.60	42.42	42.29	42.19	42.17	42.16	42.15	42.13	42.12	42.10	42.10	42.08	42.08
.3	41.63	40.00	39.37	38.95	38.70	38.51	38.38	38.34	38.32	38.31	38.28	38.27	38.26	38.24	38.23	38.22
.31	41.35	39.67	39.02	38.59	38.33	38.14	38.00	37.96	37.94	37.93	37.90	37.89	37.88	37.86	37.85	37.84
.32	41.06	39.33	38.67	38.23	37.96	37.77	37.63	37.59	37.57	37.55	37.53	37.51	37.50	37.48	37.47	37.46
.33	40.77	39.00	38.32	37.87	37.60	37.39	37.25	37.21	37.19	37.18	37.15	37.14	37.12	37.11	37.09	37.09
.34	40.49	38.67	37.97	37.51	37.23	37.02	36.88	36.84	36.82	36.80	36.77	36.76	36.74	36.73	36.71	36.71
.35	40.20	38.33	37.62	37.15	36.87	36.65	36.50	36.46	36.44	36.43	36.40	36.39	36.37	36.36	36.34	36.33
.36	39.91	38.00	37.28	36.79	36.50	36.29	36.13	36.09	36.07	36.05	36.03	36.01	35.99	35.97	35.96	35.96
.37	39.62	37.67	36.93	36.43	36.14	35.92	35.76	35.72	35.70	35.68	35.65	35.64	35.62	35.61	35.59	35.59
.38	39.33	37.33	36.58	36.07	35.78	35.55	35.39	35.35	35.33	35.31	35.28	35.27	35.25	35.24	35.22	35.22
.39	39.03	37.00	36.23	35.72	35.41	35.19	35.02	34.98	34.96	34.94	34.92	34.90	34.88	34.87	34.85	34.85
.40	38.74	36.67	35.88	35.36	35.05	34.82	34.66	34.62	34.59	34.58	34.55	34.53	34.51	34.49	34.48	34.48
.41	38.45	36.33	35.54	35.01	34.69	34.46	34.29	34.25	34.23	34.21	34.18	34.17	34.14	34.12	34.11	34.11
.42	38.15	36.00	35.19	34.65	34.33	34.10	33.93	33.89	33.86	33.85	33.82	33.80	33.78	33.77	33.75	33.74
.43	37.85	35.67	34.85	34.30	33.98	33.74	33.57	33.53	33.50	33.48	33.45	33.44	33.41	33.39	33.38	33.38
.44	37.56	35.33	34.50	33.95	33.62	33.38	33.21	33.17	33.14	33.12	33.09	33.08	33.05	33.03	33.02	33.02
.45	37.26	35.00	34.16	33.60	33.27	33.02	32.85	32.81	32.78	32.76	32.73	32.72	32.69	32.67	32.66	32.66
.46	36.96	34.67	33.81	33.24	32.91	32.66	32.49	32.45	32.42	32.40	32.37	32.36	32.33	32.31	32.30	32.30
.47	36.66	34.33	33.47	32.89	32.56	32.31	32.13	32.09	32.06	32.04	32.01	32.00	31.97	31.95	31.94	31.94
.48	36.35	34.00	33.12	32.55	32.21	31.96	31.78	31.74	31.71	31.69	31.66	31.64	31.62	31.61	31.59	31.58
.49	36.05	33.67	32.78	32.20	31.86	31.60	31.42	31.38	31.35	31.33	31.30	31.29	31.26	31.24	31.23	31.23
.50	35.75	33.33	32.44	31.85	31.51	31.25	31.07	31.03	31.00	30.98	30.95	30.94	30.91	30.89	30.88	30.87
.51	35.44	33.00	32.10	31.51	31.16	30.90	30.72	30.68	30.65	30.63	30.60	30.59	30.55	30.55	30.53	30.52
.52	35.13	32.67	31.76	31.16	30.81	30.55	30.37	30.33	30.30	30.28	30.25	30.24	30.21	30.19	30.18	30.17
.53	34.82	32.33	31.42	30.82	30.46	30.21	30.02	29.98	29.95	29.93	29.90	29.89	29.86	29.84	29.83	29.83
.54	34.51	32.00	31.08	30.47	30.12	29.86	29.68	29.64	29.61	29.59	29.56	29.54	29.52	29.50	29.48	29.48
.55	34.20	31.67	30.74	30.13	29.78	29.52	29.33	29.29	29.26	29.24	29.21	29.20	29.17	29.15	29.14	29.14
.56	33.88	31.33	30.40	29.79	29.44	29.18	28.99	28.95	28.92	28.90	28.87	28.86	28.83	28.81	28.80	28.79
.57	33.57	31.00	30.06	29.45	29.09	28.83	28.65	28.61	28.58	28.56	28.53	28.52	28.49	28.47	28.46	28.45
.58	33.25	30.67	29.73	29.11	28.76	28.50	28.31	28.27	28.24	28.22	28.19	28.18	28.15	28.13	28.12	28.12
.59	32.93	30.33	29.39	28.77	28.42	28.16	27.97	27.93	27.91	27.89	27.86	27.84	27.82	27.80	27.78	27.78
.60	32.61	30.00	29.05	28.44	28.08	27.82	27.64	27.60	87.57	27.55	27.52	27.51	27.48	27.46	27.45	27.45
.61	32.28	29.67	28.72	28.10	27.75	27.49	27.31	27.27	27.24	27.22	27.19	27.17	27.15	27.14	27.12	27.11
.62	31.96	29.33	28.39	27.77	27.41	27.16	26.97	26.93	26.91	26.89	26.86	26.84	26.82	26.81	26.79	26.78
.63	31.63	29.00	28.05	27.44	27.08	26.82	26.64	26.60	26.58	26.56	26.53	26.51	26.49	26.48	26.46	26.45
.64	31.30	28.67	27.72	27.11	26.75	26.50	26.32	26.28	26.25	26.23	26.20	26.19	26.16	26.14	26.13	26.13
.65	30.97	28.33	27.39	26.78	26.42	26.17	25.99	25.95	25.92	25.90	25.87	25.86	25.84	25.83	25.81	25.80
.66	30.63	28.00	27.06	26.45	26.10	25.84	25.67	25.63	25.60	25.58	25.55	25.54	25.52	25.50	25.48	25.48
.67	30.30	27.67	26.73	26.12	25.77	25.52	25.34	25.30	25.28	25.26	25.23	25.22	25.20	25.18	25.16	25.16
.68	29.96	27.33	26.40	25.79	25.45	25.20	25.02	24.98	24.96	24.94	24.91	24.90	24.88	24.87	24.85	24.84
.69	29.61	27.00	26.07	25.47	25.12	24.88	24.71	24.67	24.64	24.62	24.59	24.58	24.56	24.55	24.53	24.53

[1] Values tabulated are read in percent.

TABLE A-29 Mil-Std-414 TABLE FOR ESTIMATING THE LOT PERCENT DEFECTIVE USING THE RANGE METHOD (continued)

Q_U or Q_L	Sample Size															
	3	4	5	7	10	15	25	30	35	40	50	60	85	115	175	230
.70	29.27	26.67	25.74	25.14	24.80	24.56	24.39	24.35	24.32	24.31	24.28	24.27	24.25	24.24	24.22	24.21
.71	28.92	26.33	25.41	24.82	24.48	24.24	24.07	24.03	24.01	23.99	23.97	23.95	23.93	23.91	23.90	23.90
.72	28.57	26.00	25.09	24.50	24.17	23.93	23.76	23.72	23.70	23.68	23.66	23.64	23.62	23.60	23.59	23.59
.73	28.22	25.67	24.76	24.18	23.85	23.61	23.45	23.41	23.39	23.37	23.35	23.33	23.32	23.30	23.29	23.29
.74	27.86	25.33	24.44	23.86	23.54	23.30	23.14	23.10	23.08	23.07	23.04	23.03	23.01	23.00	22.98	22.98
.75	27.50	25.00	24.11	23.55	23.22	22.99	22.84	22.80	22.78	22.76	22.74	22.72	22.71	22.69	22.68	22.68
.76	27.13	24.67	23.79	23.23	22.91	22.69	22.53	22.49	22.47	22.46	22.43	22.42	22.41	22.39	22.38	22.38
.77	26.77	24.33	23.47	22.92	22.60	22.38	22.23	22.19	22.17	22.16	22.13	22.12	22.11	22.09	22.08	22.08
.78	26.39	24.00	23.15	22.60	22.30	22.08	21.93	21.90	21.88	21.86	21.85	21.83	21.81	21.80	21.78	21.78
.79	26.02	23.67	22.83	22.29	21.99	21.78	21.64	21.60	21.58	21.57	21.54	21.53	21.52	21.50	21.49	21.49
.80	25.64	23.33	22.51	21.98	21.69	21.48	21.34	21.30	21.28	21.27	21.26	21.24	21.22	21.22	21.20	21.20
.81	25.25	23.00	22.19	21.68	21.39	21.18	21.04	21.01	20.99	20.98	20.97	20.95	20.93	20.93	20.91	20.91
.82	24.86	22.67	21.87	21.37	21.09	20.89	20.75	20.72	20.70	20.69	20.68	20.66	20.64	20.64	20.62	20.62
.83	24.47	22.33	21.56	21.06	20.79	20.59	20.46	20.43	20.41	20.40	20.38	20.37	20.36	20.35	20.34	20.34
.84	24.07	22.00	21.24	20.76	20.49	20.30	20.17	20.15	20.13	20.12	20.10	20.09	20.08	20.06	20.06	20.06
.85	23.67	21.67	20.93	20.46	20.20	20.01	19.89	19.87	19.85	19.84	19.82	19.81	19.79	19.79	19.78	19.78
.86	23.26	21.33	20.62	20.16	19.90	19.73	19.60	19.58	19.57	19.56	19.54	19.54	19.52	19.51	19.50	19.50
.87	22.84	21.00	20.31	19.86	19.61	19.44	19.32	19.31	19.29	19.28	19.26	19.25	19.24	19.24	19.22	19.22
.88	22.42	20.67	20.00	19.57	19.33	19.16	19.04	19.03	19.01	19.00	18.98	18.98	18.97	18.96	18.95	18.95
.89	21.99	20.33	19.69	19.27	19.04	18.88	18.77	18.75	18.74	18.73	18.71	18.70	18.69	18.69	18.68	18.68
.90	21.55	20.00	19.38	18.98	18.75	18.60	18.50	18.48	18.47	18.46	18.44	18.43	18.42	18.42	18.41	18.41
.91	21.11	19.67	19.07	18.69	18.47	18.32	18.22	18.21	18.20	18.19	18.17	18.17	18.17	18.16	18.15	18.15
.92	20.66	19.33	18.77	18.40	18.19	18.05	17.96	17.95	17.93	17.92	17.92	17.90	17.89	17.89	17.88	17.88
.93	20.20	19.00	18.46	18.11	17.91	17.78	17.69	17.68	17.67	17.66	17.65	17.65	17.63	17.63	17.62	17.62
.94	19.74	18.67	18.16	17.82	17.64	17.51	17.43	17.42	17.41	17.40	17.39	17.39	17.37	17.37	17.36	17.36
.95	19.25	18.33	17.86	17.54	17.36	17.24	17.17	17.16	17.15	17.14	17.13	17.13	17.12	17.12	17.11	17.11
.96	18.76	18.00	17.56	17.26	17.09	16.98	16.91	16.90	16.89	16.88	16.88	16.87	16.86	16.86	16.86	16.86
.97	18.25	17.67	17.25	16.97	16.82	16.71	16.65	16.64	16.63	16.63	16.62	16.62	16.61	16.61	16.60	16.60
.98	17.74	17.33	16.96	16.70	16.55	16.45	16.39	16.38	16.38	16.37	16.37	16.37	16.36	16.36	16.36	16.36
.99	17.21	17.00	16.66	16.42	16.28	16.19	16.14	16.13	16.13	16.12	16.12	16.12	16.11	16.11	16.11	16.11
1.00	16.67	16.67	16.36	16.14	16.02	15.94	15.89	15.88	15.88	15.88	15.87	15.87	15.87	15.87	15.87	15.87
1.01	16.11	16.33	16.07	15.87	15.76	15.68	15.64	15.63	15.63	15.63	15.63	15.63	15.62	15.62	15.62	15.62
1.02	15.53	16.00	15.78	15.60	15.50	15.43	15.40	15.39	15.39	15.39	15.39	15.39	15.38	15.38	15.38	15.38
1.03	14.93	15.67	15.48	15.33	15.24	15.18	15.15	15.15	15.15	15.15	15.15	15.15	15.15	15.15	15.15	15.15
1.04	14.31	15.33	15.19	15.06	14.98	14.94	14.91	14.91	14.91	14.91	14.91	14.91	14.91	14.91	14.91	14.91
1.05	13.66	15.00	14.91	14.79	14.73	14.69	14.67	14.67	14.67	14.67	14.67	14.68	14.68	14.68	14.68	14.68
1.06	12.98	14.67	14.62	14.53	14.48	14.45	14.44	14.44	14.44	14.44	14.44	14.44	14.45	14.45	14.45	14.45
1.07	12.27	14.33	14.33	14.27	14.23	14.21	14.20	14.21	14.21	14.21	14.21	14.21	14.22	14.22	14.22	14.22
1.08	11.51	14.00	14.05	14.01	13.98	13.97	13.97	13.98	13.98	13.98	13.98	13.99	13.99	13.99	14.00	14.00
1.09	10.71	13.67	13.76	13.75	13.74	13.73	13.74	13.75	13.75	13.75	13.76	13.76	13.77	13.77	13.78	13.78

Q	Sample Size															
	3	4	5	7	10	15	25	30	35	40	50	60	85	115	175	230
1.90	0.00	0.00	0.00	0.67	1.45	1.99	2.38	2.47	2.53	2.57	2.64	2.68	2.74	2.77	2.81	2.83
1.91	0.00	0.00	0.00	0.62	1.38	1.93	2.32	2.41	2.47	2.51	2.58	2.61	2.67	2.70	2.74	2.76
1.92	0.00	0.00	0.00	0.56	1.32	1.86	2.25	2.34	2.41	2.45	2.51	2.55	2.61	2.64	2.68	2.70
1.93	0.00	0.00	0.00	0.51	1.26	1.80	2.19	2.28	2.34	2.38	2.45	2.49	2.55	2.58	2.61	2.63
1.94	0.00	0.00	0.00	0.46	1.20	1.74	2.13	2.22	2.28	2.32	2.39	2.43	2.49	2.52	2.55	2.57
1.95	0.00	0.00	0.00	0.42	1.15	1.68	2.07	2.16	2.22	2.26	2.33	2.37	2.43	2.46	2.49	2.51
1.96	0.00	0.00	0.00	0.37	1.09	1.62	2.01	2.10	2.16	2.20	2.27	2.31	2.37	2.40	2.43	2.45
1.97	0.00	0.00	0.00	0.33	1.04	1.57	1.95	2.04	2.10	2.14	2.21	2.25	2.31	2.34	2.38	2.40
1.98	0.00	0.00	0.00	0.30	0.99	1.51	1.90	1.99	2.05	2.09	2.15	2.19	2.25	2.28	2.32	2.34
1.99	0.00	0.00	0.00	0.26	0.94	1.46	1.84	1.93	1.99	2.03	2.10	2.14	2.20	2.23	2.26	2.28
2.00	0.00	0.00	0.00	0.23	0.89	1.41	1.79	1.88	1.94	1.98	2.05	2.08	2.14	2.17	2.21	2.23
2.01	0.00	0.00	0.00	0.20	0.84	1.36	1.74	1.83	1.89	1.93	1.99	2.03	2.09	2.12	2.16	2.18
2.02	0.00	0.00	0.00	0.17	0.80	1.31	1.69	1.78	1.83	1.87	1.94	1.98	2.04	2.07	2.10	2.12
2.03	0.00	0.00	0.00	0.14	0.75	1.26	1.64	1.73	1.78	1.82	1.89	1.93	1.99	2.02	2.05	2.07
2.04	0.00	0.00	0.00	0.12	0.71	1.21	1.59	1.68	1.73	1.77	1.84	1.88	1.94	1.97	2.00	2.02
2.05	0.00	0.00	0.00	0.10	0.67	1.17	1.54	1.63	1.69	1.73	1.79	1.83	1.89	1.92	1.95	1.97
2.06	0.00	0.00	0.00	0.08	0.63	1.12	1.49	1.58	1.64	1.68	1.74		1.84	1.87	1.91	1.93
2.07	0.00	0.00	0.00	0.06	0.60	1.08	1.45	1.54	1.59	1.63	1.70	1.74	1.79	1.82	1.86	1.88
2.08	0.00	0.00	0.00	0.05	0.56	1.04	1.40	1.49	1.55	1.59	1.65	1.69	1.75	1.78	1.81	1.83
2.09	0.00	0.00	0.00	0.03	0.53	1.00	1.36	1.45	1.50	1.54	1.61	1.64	1.70	1.73	1.77	1.79
2.10	0.00	0.00	0.00	0.02	0.49	0.96	1.32	1.41	1.46	1.50	1.56	1.60	1.66	1.69	1.72	1.74
2.11	0.00	0.00	0.00	0.01	0.46	0.92	1.28	1.36	1.42	1.46	1.52	1.56	1.61	1.64	1.68	1.70
2.12	0.00	0.00	0.00	0.00	0.43	0.88	1.24	1.32	1.38	1.42	1.48	1.52	1.57	1.60	1.64	1.66
2.13	0.00	0.00	0.00	0.00	0.40	0.85	1.20	1.28	1.34	1.38	1.44	1.48	1.53	1.56	1.60	1.62
2.14	0.00	0.00	0.00	0.00	0.38	0.81	1.16	1.25	1.30	1.34	1.40	1.44	1.49	1.52	1.56	1.58
2.15	0.00	0.00	0.00	0.00	0.35	0.78	1.13	1.21	1.26	1.30	1.36	1.40	1.45	1.48	1.52	1.54
2.16	0.00	0.00	0.00	0.00	0.32	0.75	1.09	1.17	1.22	1.26	1.32	1.36	1.41	1.44	1.48	1.50
2.17	0.00	0.00	0.00	0.00	0.30	0.71	1.06	1.13	1.18	1.22	1.29	1.32	1.38	1.41	1.44	1.46
2.18	0.00	0.00	0.00	0.00	0.28	0.68	1.02	1.10	1.15	1.19	1.25	1.28	1.34	1.37	1.40	1.41
2.19	0.00	0.00	0.00	0.00	0.26	0.65	0.99	1.06	1.11	1.15	1.22	1.25	1.30	1.33	1.37	1.39
2.20	0.000	0.000	0.000	0.000	0.236	0.625	0.954	1.030	1.083	1.122	1.178	1.214	1.267	1.299	1.330	1.346
2.21	0.000	0.000	0.000	0.000	0.217	0.597	0.922	0.997	1.058	1.089	1.144	1.180	1.233	1.265	1.295	1.311
2.22	0.000	0.000	0.000	0.000	0.199	0.570	0.891	0.966	1.018	1.056	1.111	1.147	1.199	1.231	1.261	1.277
2.23	0.000	0.000	0.000	0.000	0.182	0.544	0.861	0.935	0.986	1.025	1.079	1.115	1.167	1.197	1.228	1.244
2.24	0.000	0.000	0.000	0.000	0.166	0.519	0.831	0.905	0.956	0.994	1.048	1.083	1.135	1.165	1.195	1.211
2.25	0.000	0.000	0.000	0.000	0.150	0.495	0.802	0.875	0.926	0.964	1.018	1.052	1.104	1.134	1.163	1.179
2.26	0.000	0.000	0.000	0.000	0.136	0.471	0.775	0.847	0.897	0.935	0.987	1.022	1.073	1.103	1.132	1.148
2.27	0.000	0.000	0.000	0.000	0.123	0.449	0.748	0.819	0.869	0.906	0.958	0.993	1.043	1.073	1.103	1.118
2.28	0.000	0.000	0.000	0.000	0.111	0.427	0.722	0.792	0.841	0.878	0.930	0.964	1.014	1.044	1.073	1.088
2.29	0.000	0.000	0.000	0.000	0.099	0.406	0.697	0.766	0.814	0.851	0.902	0.936	0.986	1.015	1.044	1.059

Q_U or Q_L	Sample Size															
	3	4	5	7	10	15	25	30	35	40	50	60	85	115	175	230
2.30	0.000	0.000	0.000	0.000	0.089	0.386	0.672	0.741	0.789	0.825	0.875	0.909	0.959	0.988	1.016	1.031
2.31	0.000	0.000	0.000	0.000	0.079	0.367	0.648	0.716	0.763	0.799	0.849	0.882	0.931	0.960	0.988	1.003
2.32	0.000	0.000	0.000	0.000	0.070	0.348	0.624	0.691	0.739	0.774	0.823	0.856	0.905	0.934	0.962	0.976
2.33	0.000	0.000	0.000	0.000	0.061	0.330	0.601	0.668	0.715	0.750	0.798	0.831	0.879	0.908	0.935	0.950
2.34	0.000	0.000	0.000	0.000	0.054	0.313	0.579	0.645	0.691	0.720	0.774	0.807	0.854	0.882	0.909	0.924
2.35	0.000	0.000	0.000	0.000	0.047	0.296	0.558	0.623	0.669	0.703	0.750	0.782	0.829	0.857	0.884	0.899
2.36	0.000	0.000	0.000	0.000	0.040	0.280	0.538	0.602	0.646	0.680	0.728	0.759	0.806	0.833	0.860	0.874
2.37	0.000	0.000	0.000	0.000	0.035	0.265	0.518	0.580	0.624	0.658	0.705	0.736	0.782	0.809	0.836	0.850
2.38	0.000	0.000	0.000	0.000	0.029	0.250	0.498	0.560	0.604	0.637	0.683	0.714	0.759	0.787	0.813	0.827
2.39	0.000	0.000	0.000	0.000	0.025	0.236	0.479	0.541	0.584	0.616	0.662	0.693	0.737	0.764	0.791	0.804
2.40	0.000	0.000	0.000	0.000	0.021	0.223	0.461	0.521	0.564	0.596	0.641	0.671	0.715	0.742	0.769	0.782
2.41	0.000	0.000	0.000	0.000	0.017	0.210	0.443	0.503	0.545	0.577	0.621	0.651	0.695	0.721	0.747	0.760
2.42	0.000	0.000	0.000	0.000	0.014	0.198	0.426	0.485	0.526	0.557	0.601	0.631	0.674	0.701	0.726	0.739
2.43	0.000	0.000	0.000	0.000	0.011	0.186	0.410	0.467	0.508	0.539	0.582	0.611	0.654	0.679	0.705	0.718
2.44	0.000	0.000	0.000	0.000	0.009	0.175	0.393	0.450	0.491	0.521	0.564	0.593	0.635	0.660	0.685	0.698
2.45	0.000	0.000	0.000	0.000	0.007	0.165	0.378	0.434	0.473	0.503	0.545	0.573	0.616	0.641	0.665	0.678
2.46	0.000	0.000	0.000	0.000	0.005	0.154	0.362	0.417	0.456	0.486	0.528	0.556	0.597	0.622	0.646	0.659
2.47	0.000	0.000	0.000	0.000	0.004	0.145	0.348	0.403	0.441	0.470	0.511	0.538	0.579	0.604	0.627	0.640
2.48	0.000	0.000	0.000	0.000	0.003	0.136	0.333	0.387	0.425	0.454	0.494	0.522	0.562	0.586	0.609	0.622
2.49	0.000	0.000	0.000	0.000	0.002	0.127	0.321	0.372	0.409	0.438	0.478	0.504	0.545	0.569	0.593	0.605
2.50	0.000	0.000	0.000	0.000	0.001	0.118	0.307	0.358	0.395	0.423	0.463	0.489	0.528	0.552	0.575	0.587
2.51	0.000	0.000	0.000	0.000	0.001	0.111	0.294	0.345	0.381	0.409	0.447	0.473	0.512	0.536	0.558	0.570
2.52	0.000	0.000	0.000	0.000	0.000	0.103	0.282	0.331	0.367	0.394	0.432	0.458	0.497	0.519	0.542	0.553
2.53	0.000	0.000	0.000	0.000	0.000	0.096	0.270	0.319	0.354	0.381	0.418	0.444	0.481	0.503	0.526	0.537
2.54	0.000	0.000	0.000	0.000	0.000	0.089	0.258	0.306	0.340	0.367	0.404	0.428	0.466	0.488	0.510	0.522
2.55	0.000	0.000	0.000	0.000	0.000	0.083	0.247	0.294	0.328	0.354	0.390	0.415	0.451	0.473	0.495	0.506
2.56	0.000	0.000	0.000	0.000	0.000	0.077	0.237	0.283	0.316	0.341	0.377	0.401	0.437	0.459	0.480	0.491
2.57	0.000	0.000	0.000	0.000	0.000	0.071	0.227	0.272	0.304	0.328	0.364	0.388	0.424	0.445	0.466	0.477
2.58	0.000	0.000	0.000	0.000	0.000	0.066	0.217	0.261	0.292	0.317	0.352	0.376	0.411	0.432	0.452	0.463
2.59	0.000	0.000	0.000	0.000	0.000	0.061	0.207	0.251	0.282	0.305	0.340	0.363	0.397	0.418	0.439	0.449
2.60	0.000	0.000	0.000	0.000	0.000	0.056	0.198	0.240	0.271	0.294	0.328	0.351	0.385	0.406	0.426	0.436
2.61	0.000	0.000	0.000	0.000	0.000	0.052	0.189	0.231	0.260	0.283	0.317	0.339	0.372	0.393	0.413	0.423
2.62	0.000	0.000	0.000	0.000	0.000	0.048	0.181	0.221	0.250	0.273	0.306	0.327	0.360	0.381	0.400	0.410
2.63	0.000	0.000	0.000	0.000	0.000	0.044	0.173	0.212	0.241	0.263	0.295	0.316	0.349	0.368	0.388	0.398
2.64	0.000	0.000	0.000	0.000	0.000	0.040	0.164	0.203	0.232	0.253	0.285	0.306	0.338	0.357	0.376	0.386
2.65	0.000	0.000	0.000	0.000	0.000	0.037	0.157	0.195	0.223	0.244	0.274	0.295	0.327	0.346	0.365	0.375
2.66	0.000	0.000	0.000	0.000	0.000	0.034	0.149	0.186	0.213	0.234	0.265	0.285	0.316	0.335	0.353	0.363
2.67	0.000	0.000	0.000	0.000	0.000	0.031	0.143	0.179	0.205	0.225	0.255	0.275	0.305	0.324	0.342	0.352
2.68	0.000	0.000	0.000	0.000	0.000	0.028	0.136	0.171	0.197	0.217	0.246	0.266	0.296	0.314	0.332	0.342
2.69	0.000	0.000	0.000	0.000	0.000	0.025	0.129	0.164	0.190	0.209	0.238	0.257	0.286	0.304	0.321	0.331

TABLE A-30   Mil-Std-414 DEFINITIONS FOR RANGE METHOD

Symbol	Read	Definitions
n		Sample size for a single lot.
$\bar{X}$	X bar	Sample mean. Arithmetic mean of sample measurements from a single lot.
R		Range. The difference between the largest and smallest measurements in a subgroup. In this Standard, the subgroup size is 5 except for those plans in which n = 3, 4, or 7, in which case the subgroup is the same as the sample size.
R_1		Range of the first subgroup.
R_2		Range of the second subgroup.
$\bar{R}$	R bar	Average range. The arithmetic mean of the range values of the subgroups of the sample measurements from a single lot.
U		Upper specification limit.
L		Lower specification limit.
k		The acceptability constant given in Tables.
c		A factor used in determining the quality index when using the range method. The c values are given in Tables.
Q_U	Q sub U	Quality Index for use with Table.
Q_L	Q sub L	Quality Index for use with Table.
p_U	p sub U	Sample estimate of the lot percent defective above U from Table.
p_L	p sub L	Sample estimate of the lot percent defective below L from Table.
p		Total sample estimate of the lot percent defective $p = P_U + P_L$.
M		Maximum allowable percent defective for sample estimates given in Tables.
M_U	M sub U	Maximum allowable percent defective above U given in Tables. (For use when different AQL values for U and L are specified.)
M_L	M sub L	Maximum allowable percent defective below L given in Tables. (For use when different AQL values for U and L are specified.)
$\bar{p}$	p bar	Sample estimate of the process percent defective, that is, the estimated process average.
$\bar{p}_U$	p bar sub U	The estimated process average for an upper specification limit.
$\bar{p}_L$	p bar sub L	The estimated process average for a lower specification limit.
T		The maximum number of estimated process averages which may exceed the AQL given in Table. (For use in determining application to tightened inspection.)
f		A factor used in determining the Maximum Average Range (MAR). The f values are given in Table.
$>$	Greater than	Greater than.
$<$	Less than	Less than.
Σ	Sum of	Sum of.

TABLE A-31 Mil-Std-414 DEFINITIONS FOR STANDARD DEVIATION METHOD

Symbol	Read	Definitions
n		Sample size for a single lot.
$\overline{X}$	X bar	Sample mean. Arithmetic mean of sample measurements from a single lot.
s		Estimate of lot standard deviation. Standard deviation of sample measurements from a single lot.
U		Upper specification limit.
L		Lower specification limit.
k		The acceptability constant given in Tables.
Q_U	Q sub U	Quality index for use with Table.
Q_L	Q sub L	Quality index for use with Table.
p_U	p sub U	Sample estimate of the lot percent defective above U from Table.
p_L	p sub L	Sample estimate of the lot percent defective below L from Table.
p		Total sample estimate of the lot percent defective $p = p_U + p_L$.
M		Maximum allowable percent defective for sample estimates given in Tables.
M_U	M sub U	Maximum allowable percent defective above U given in Tables. (For use when different AQL values for U and L are specified.)
M_L	M sub L	Maximum allowable percent defective below L given in Tables. (For use when different AQL values for U and L are specified.)
$\overline{p}$	p bar	Sample estimate of the process percent defective, i.e., the estimated process average.
$\overline{p}_U$	p bar sub U	The estimated process average for an upper specification limit.
$\overline{p}_L$	p bar sub L	The estimated process average for a lower specification limit.
T		The maximum number of estimated process averages which may exceed the AQL given in Table. (For use in determining application of tightened inspection.)
F		A factor used in determining the Maximum Standard Deviation (MSD). The F values are given in Table.
$>$	Greater than	Greater than
$<$	Less than	Less than
Σ	Sum of	Sum of

TABLE A-32 Mil-Std-105D SAMPLE SIZE CODE LETTERS

Lot or batch size	Special inspection levels				General inspection levels		
	S-1	S-2	S-3	S-4	I	II	III
2 to 8	A	A	A	A	A	A	B
9 to 15	A	A	A	A	A	B	C
16 to 25	A	A	B	B	B	C	D
26 to 50	A	B	B	C	C	D	E
51 to 90	B	B	C	C	C	E	F
91 to 150	B	B	C	D	D	F	G
151 to 280	B	C	D	E	E	G	H
281 to 500	B	C	D	E	F	H	J
501 to 1200	C	C	E	F	G	J	K
1201 to 3200	C	D	E	G	H	K	L
3201 to 10000	C	D	F	G	J	L	M
10001 to 35000	C	D	F	H	K	M	N
35001 to 150000	D	E	G	J	L	N	P
150001 to 500000	D	E	G	J	M	P	Q
500001 and over	D	E	H	K	N	Q	R

TABLE A-33 Mil-Std-105D SINGLE SAMPLING PLANS FOR NORMAL INSPECTION (MASTER TABLE)

Acceptable quality levels (normal inspection)

Each cell shows the acceptance number (Ac) and rejection number (Re). ↓ = use first sampling plan below arrow. ↑ = use first sampling plan above arrow. Ac = acceptance number. Re = rejection number.

Code	Sample size	0.010	0.015	0.025	0.040	0.065	0.10	0.15	0.25	0.40	0.65	1.0	1.5	2.5	4.0	6.5	10	15	25	40	65	100	150	250	400	650	1,000
A	2	↓	↓	↓	↓	↓	↓	↓	↓	↓	↓	↓	↓	↓	↓	↓	↓	0 1	1 2	2 3	3 4	5 6	7 8	10 11	14 15	21 22	30 31
B	3	↓	↓	↓	↓	↓	↓	↓	↓	↓	↓	↓	↓	↓	↓	↓	0 1	1 2	2 3	3 4	5 6	7 8	10 11	14 15	21 22	30 31	44 45
C	5	↓	↓	↓	↓	↓	↓	↓	↓	↓	↓	↓	↓	↓	↓	0 1	1 2	2 3	3 4	5 6	7 8	10 11	14 15	21 22	30 31	44 45	↑
D	8	↓	↓	↓	↓	↓	↓	↓	↓	↓	↓	↓	↓	↓	0 1	1 2	2 3	3 4	5 6	7 8	10 11	14 15	21 22	30 31	44 45	↑	↑
E	13	↓	↓	↓	↓	↓	↓	↓	↓	↓	↓	↓	↓	0 1	1 2	2 3	3 4	5 6	7 8	10 11	14 15	21 22	30 31	44 45	↑	↑	↑
F	20	↓	↓	↓	↓	↓	↓	↓	↓	↓	↓	↓	0 1	1 2	2 3	3 4	5 6	7 8	10 11	14 15	21 22	30 31	44 45	↑	↑	↑	↑
G	32	↓	↓	↓	↓	↓	↓	↓	↓	↓	↓	0 1	1 2	2 3	3 4	5 6	7 8	10 11	14 15	21 22	30 31	44 45	↑	↑	↑	↑	↑
H	50	↓	↓	↓	↓	↓	↓	↓	↓	↓	0 1	1 2	2 3	3 4	5 6	7 8	10 11	14 15	21 22	30 31	44 45	↑	↑	↑	↑	↑	↑
J	80	↓	↓	↓	↓	↓	↓	↓	↓	0 1	1 2	2 3	3 4	5 6	7 8	10 11	14 15	21 22	30 31	44 45	↑	↑	↑	↑	↑	↑	↑
K	125	↓	↓	↓	↓	↓	↓	↓	0 1	1 2	2 3	3 4	5 6	7 8	10 11	14 15	21 22	30 31	44 45	↑	↑	↑	↑	↑	↑	↑	↑
L	200	↓	↓	↓	↓	↓	↓	0 1	1 2	2 3	3 4	5 6	7 8	10 11	14 15	21 22	30 31	44 45	↑	↑	↑	↑	↑	↑	↑	↑	↑
M	315	↓	↓	↓	↓	↓	0 1	1 2	2 3	3 4	5 6	7 8	10 11	14 15	21 22	30 31	44 45	↑	↑	↑	↑	↑	↑	↑	↑	↑	↑
N	500	↓	↓	↓	↓	0 1	1 2	2 3	3 4	5 6	7 8	10 11	14 15	21 22	30 31	44 45	↑	↑	↑	↑	↑	↑	↑	↑	↑	↑	↑
P	800	↓	↓	↓	0 1	1 2	2 3	3 4	5 6	7 8	10 11	14 15	21 22	30 31	44 45	↑	↑	↑	↑	↑	↑	↑	↑	↑	↑	↑	↑
Q	1,250	↓	↓	0 1	1 2	2 3	3 4	5 6	7 8	10 11	14 15	21 22	30 31	44 45	↑	↑	↑	↑	↑	↑	↑	↑	↑	↑	↑	↑	↑
R	2,000	↓	0 1	1 2	2 3	3 4	5 6	7 8	10 11	14 15	21 22	30 31	44 45	↑	↑	↑	↑	↑	↑	↑	↑	↑	↑	↑	↑	↑	↑

↓ = use first sampling plan below arrow. If sample size equals, or exceeds, lot or batch size, do 100 % inspection.
↑ = use first sampling plan above arrow.
Ac = acceptance number.
Re = rejection number.

502

TABLE A-34 Mil-Std-105D DOUBLE SAMPLING PLANS FOR NORMAL INSPECTION
(MASTER TABLE)

Acceptable quality levels (normal inspection)

Note: In the value cells below, each entry is given as "Ac Re" (Ac = acceptance number, Re = rejection number). For each code letter the "First" row gives the first-sample plan and the "Second" row gives the cumulative second-sample plan. ↓ = use first sampling plan below arrow. ↑ = use first sampling plan above arrow. † = use corresponding single sampling plan (or alternatively, use double sampling plan below, where available).

Code letter	Sample	Sample size	Cumulative sample size	0.010	0.015	0.025	0.040	0.065	0.10	0.15	0.25	0.40	0.65	1.0	1.5	2.5	4.0	6.5	10	15	25	40	65	100	150	250	400	650	1,000
A																												
B	First	2	2	↓	↓	↓	↓	↓	↓	↓	↓	↓	↓	↓	↓	↓	↓	↓	†	0 2	†	†	†	†	†	†	†	†	†
B	Second	2	4																	1 2									
C	First	3	3	↓	↓	↓	↓	↓	↓	↓	↓	↓	↓	↓	↓	↓	↓	†	0 2	0 3	†	†	†	†	†	†	†	†	†
C	Second	3	6																1 2	3 4									
D	First	5	5	↓	↓	↓	↓	↓	↓	↓	↓	↓	↓	↓	↓	↓	†	0 2	0 3	1 4	2 5	†	†	†	†	†	†	†	†
D	Second	5	10															1 2	3 4	4 5	6 7								
E	First	8	8	↓	↓	↓	↓	↓	↓	↓	↓	↓	↓	↓	↓	†	0 2	0 3	1 4	2 5	3 7	†	†	†	†	†	†	†	†
E	Second	8	16														1 2	3 4	4 5	6 7	8 9								
F	First	13	13	↓	↓	↓	↓	↓	↓	↓	↓	↓	↓	↓	†	0 2	0 3	1 4	2 5	3 7	5 9	7 11	†	†	†	†	†	†	†
F	Second	13	26													1 2	3 4	4 5	6 7	8 9	12 13	18 19							
G	First	20	20	↓	↓	↓	↓	↓	↓	↓	↓	↓	↓	†	0 2	0 3	1 4	2 5	3 7	5 9	7 11	11 16	†	†	†	†	†	†	†
G	Second	20	40												1 2	3 4	4 5	6 7	8 9	12 13	18 19	26 27							
H	First	32	32	↓	↓	↓	↓	↓	↓	↓	↓	↓	†	0 2	0 3	1 4	2 5	3 7	5 9	7 11	11 16	17 22	25 31	↑	↑	↑	↑	↑	↑
H	Second	32	64											1 2	3 4	4 5	6 7	8 9	12 13	18 19	26 27	37 38	56 57						
J	First	50	50	↓	↓	↓	↓	↓	↓	↓	↓	†	0 2	0 3	1 4	2 5	3 7	5 9	7 11	11 16	17 22	25 31	↑	↑	↑	↑	↑	↑	↑
J	Second	50	100										1 2	3 4	4 5	6 7	8 9	12 13	18 19	26 27	37 38	56 57							
K	First	80	80	↓	↓	↓	↓	↓	↓	↓	†	0 2	0 3	1 4	2 5	3 7	5 9	7 11	11 16	17 22	25 31	↑	↑	↑	↑	↑	↑	↑	↑
K	Second	80	160									1 2	3 4	4 5	6 7	8 9	12 13	18 19	26 27	37 38	56 57								
L	First	125	125	↓	↓	↓	↓	↓	↓	†	0 2	0 3	1 4	2 5	3 7	5 9	7 11	11 16	17 22	25 31	↑	↑	↑	↑	↑	↑	↑	↑	↑
L	Second	125	250								1 2	3 4	4 5	6 7	8 9	12 13	18 19	26 27	37 38	56 57									
M	First	200	200	↓	↓	↓	↓	↓	†	0 2	0 3	1 4	2 5	3 7	5 9	7 11	11 16	17 22	25 31	↑	↑	↑	↑	↑	↑	↑	↑	↑	↑
M	Second	200	400							1 2	3 4	4 5	6 7	8 9	12 13	18 19	26 27	37 38	56 57										
N	First	315	315	↓	↓	↓	↓	†	0 2	0 3	1 4	2 5	3 7	5 9	7 11	11 16	17 22	25 31	↑	↑	↑	↑	↑	↑	↑	↑	↑	↑	↑
N	Second	315	630						1 2	3 4	4 5	6 7	8 9	12 13	18 19	26 27	37 38	56 57											
P	First	500	500	↓	↓	↓	†	0 2	0 3	1 4	2 5	3 7	5 9	7 11	11 16	17 22	25 31	↑	↑	↑	↑	↑	↑	↑	↑	↑	↑	↑	↑
P	Second	500	1,000					1 2	3 4	4 5	6 7	8 9	12 13	18 19	26 27	37 38	56 57												
Q	First	800	800	↓	↓	†	0 2	0 3	1 4	2 5	3 7	5 9	7 11	11 16	17 22	25 31	↑	↑	↑	↑	↑	↑	↑	↑	↑	↑	↑	↑	↑
Q	Second	800	1,600				1 2	3 4	4 5	6 7	8 9	12 13	18 19	26 27	37 38	56 57													
R	First	1,250	1,250	†	†	0 2	0 3	1 4	2 5	3 7	5 9	7 11	11 16	17 22	25 31	↑	↑	↑	↑	↑	↑	↑	↑	↑	↑	↑	↑	↑	↑
R	Second	1,250	2,500			1 2	3 4	4 5	6 7	8 9	12 13	18 19	26 27	37 38	56 57														

↓ = use first sampling plan below arrow. If sample size equals or exceeds lot or batch size, do 100 % inspection.
↑ = use first sampling plan above arrow.
Ac = acceptance number.
Re = rejection number.
† Use corresponding single sampling plan (or alternatively, use double sampling plan below, where available).

TABLE A-35 DODGE–ROMIG SINGLE SAMPLING TABLES FOR LOT TOLERANCE PERCENT DEFECTIVE (LTPD)

SINGLE
SAMPLING

0.5%

LTPD

Single Sampling Table for
Lot Tolerance Per Cent Defective (LTPD) = 0.5%

Lot Size	Process Average 0 to 0.005%			Process Average 0.006 to 0.050%			Process Average 0.051 to 0.100%			Process Average 0.101 to 0.150%			Process Average 0.151 to 0.200%			Process Average 0.201 to 0.250%		
	n	c	AOQL %	n	c	AOQL %	n	c	AOQL %	n	c	AOQL %	n	c	AOQL %	n	c	AOQL %
1–180	All	0	0	All	0	0	All	0	0	All	0	0	All	0	0	All	0	0
181–210	180	0	0.02	180	0	0.02	180	0	0.02	180	0	0.02	180	0	0.02	180	0	0.02
211–250	210	0	0.03	210	0	0.03	210	0	0.03	210	0	0.03	210	0	0.03	210	0	0.03
251–300	240	0	0.03	240	0	0.03	240	0	0.03	240	0	0.03	240	0	0.03	240	0	0.03
301–400	275	0	0.04	275	0	0.04	275	0	0.04	275	0	0.04	275	0	0.04	275	0	0.04
401–500	300	0	0.05	300	0	0.05	300	0	0.05	300	0	0.05	300	0	0.05	300	0	0.05
501–600	320	0	0.05	320	0	0.05	320	0	0.05	320	0	0.05	320	0	0.05	320	0	0.05
601–800	350	0	0.06	350	0	0.06	350	0	0.06	350	0	0.06	350	0	0.06	350	0	0.06
801–1000	365	0	0.06	365	0	0.06	365	0	0.06	365	0	0.06	365	0	0.06	365	0	0.06
1001–2000	410	0	0.07	410	0	0.07	410	0	0.07	670	1	0.08	670	1	0.08	670	1	0.08
2001–3000	430	0	0.07	430	0	0.07	705	1	0.09	705	1	0.09	955	2	0.10	955	2	0.10
3001–4000	440	0	0.07	440	0	0.07	730	1	0.09	985	2	0.10	1230	3	0.11	1230	3	0.11
4001–5000	445	0	0.08	740	1	0.10	1000	2	0.11	1000	2	0.11	1250	3	0.12	1480	4	0.12
5001–7000	450	0	0.08	750	1	0.10	1020	2	0.12	1280	3	0.12	1510	4	0.13	1760	5	0.14
7001–10,000	455	0	0.08	760	1	0.10	1040	2	0.12	1530	4	0.14	1790	5	0.14	2240	7	0.16
10,001–20,000	460	0	0.08	775	1	0.10	1330	3	0.14	1820	5	0.16	2300	7	0.17	2780	9	0.18
20,001–50,000	775	1	0.11	1050	2	0.13	1600	4	0.15	2080	5	0.18	3060	10	0.20	4200	15	0.22
50,001–100,000	780	1	0.11	1060	2	0.13	1840	5	0.17	2590	8	0.19	3780	13	0.22	5140	19	0.24

1.0%

Single Sampling Table for
Lot Tolerance Per Cent Defective (LTPD) = 1.0%

Lot Size	Process Average 0 to 0.010%			Process Average 0.011 to 0.10%			Process Average 0.11 to 0.20%			Process Average 0.21 to 0.30%			Process Average 0.31 to 0.40%			Process Average 0.41 to 0.50%		
	n	c	AOQL %	n	c	AOQL %	n	c	AOQL %	n	c	AOQL %	n	c	AOQL %	n	c	AOQL %
1–120	All	0	0	All	0	0	All	0	0	All	0	0	All	0	0	All	0	0
121–150	120	0	0.06	120	0	0.06	120	0	0.06	120	0	0.06	120	0	0.06	120	0	0.06
151–200	140	0	0.08	140	0	0.08	140	0	0.08	140	0	0.08	140	0	0.08	140	0	0.08
201–300	165	0	0.10	165	0	0.10	165	0	0.10	165	0	0.10	165	0	0.10	165	0	0.10
301–400	175	0	0.12	175	0	0.12	175	0	0.12	175	0	0.12	175	0	0.12	175	0	0.12
401–500	180	0	0.13	180	0	0.13	180	0	0.13	180	0	0.13	180	0	0.13	180	0	0.13
501–600	190	0	0.13	190	0	0.13	190	0	0.13	190	0	0.13	190	0	0.13	305	1	0.14
601–800	200	0	0.14	200	0	0.14	200	0	0.14	330	1	0.15	330	1	0.15	330	1	0.15
801–1000	205	0	0.14	205	0	0.14	205	0	0.14	335	1	0.17	335	1	0.17	335	1	0.17
1001–2000	220	0	0.15	220	0	0.15	360	1	0.19	490	2	0.21	490	2	0.21	610	3	0.22
2001–3000	220	0	0.15	375	1	0.20	505	2	0.23	630	3	0.24	745	4	0.26	870	5	0.26
3001–4000	225	0	0.15	380	1	0.20	510	2	0.24	645	3	0.25	880	5	0.28	1000	6	0.29
4001–5000	225	0	0.16	380	1	0.20	520	2	0.24	770	4	0.28	895	5	0.29	1120	7	0.31
5001–7000	230	0	0.15	385	1	0.21	655	3	0.27	780	4	0.29	1020	6	0.32	1260	8	0.34
7001–10,000	230	0	0.16	520	2	0.25	660	3	0.28	910	5	0.32	1150	7	0.34	1500	10	0.37
10,001–20,000	390	1	0.21	525	2	0.26	785	4	0.31	1040	6	0.35	1400	9	0.39	1980	14	0.43
20,001–50,000	390	1	0.21	530	2	0.26	920	5	0.34	1300	8	0.39	1890	13	0.44	2570	19	0.48
50,001–100,000	390	1	0.21	670	3	0.29	1040	6	0.36	1420	9	0.41	2120	15	0.47	3150	23	0.50

n = sample size; c = acceptance number
"All" indicates that each piece in the lot is to be inspected
AOQL = Average Outgoing Quality Limit

TABLE A-35 DODGE–ROMIG SINGLE SAMPLING TABLES FOR LOT TOLERANCE PERCENT DEFECTIVE (LTPD)

DOUBLE SAMPLING

0.5% LTPD

Double Sampling Table for
Lot Tolerance Per Cent Defective (LTPD) = 0.5%

Lot Size	Process Average 0 to 0.005%						Process Average 0.006 to 0.050%						Process Average 0.051 to 0.100%					
	Trial 1		Trial 2			AOQL in %	Trial 1		Trial 2			AOQL in %	Trial 1		Trial 2			AOQL in %
	n_1	c_1	n_2	n_1+n_2	c_2		n_1	c_1	n_2	n_1+n_2	c_2		n_1	c_1	n_2	n_1+n_2	c_2	
1–180	All	0	–	–	–	0	All	0	–	–	–	0	All	0	–	–	–	0
181–210	180	0	–	–	–	0.02	180	0	–	–	–	0.02	180	0	–	–	–	0.02
211–250	210	0	–	–	–	0.03	210	0	–	–	–	0.03	210	0	–	–	–	0.03
251–300	240	0	–	–	–	0.03	240	0	–	–	–	0.03	240	0	–	–	–	0.03
301–400	275	0	–	–	–	0.04	275	0	–	–	–	0.04	275	0	–	–	–	0.04
401–450	290	0	–	–	–	0.04	290	0	–	–	–	0.04	290	0	–	–	–	0.04
451–500	340	0	110	450	1	0.04	340	0	110	450	1	0.04	340	0	110	450	1	0.04
501–550	350	0	130	480	1	0.05	350	0	130	480	1	0.05	350	0	130	480	1	0.05
551–600	360	0	150	510	1	0.05	360	0	150	510	1	0.05	360	0	150	510	1	0.05
601–800	400	0	185	585	1	0.06	400	0	185	585	1	0.06	400	0	185	585	1	0.06
801–1000	430	0	200	630	1	0.07	430	0	200	630	1	0.07	430	0	200	630	1	0.07
1001–2000	490	0	265	755	1	0.08	490	0	265	755	1	0.08	490	0	265	755	1	0.08
2001–3000	520	0	290	810	1	0:09	520	0	290	810	1	0.09	520	0	530	1050	2	0.10
3001–4000	530	0	310	840	1	0.09	530	0	570	1100	2	0.11	530	0	570	1100	2	0.11
4001–5000	540	0	305	845	1	0.09	540	0	580	1120	2	0.11	540	0	830	1370	3	0.12
5001–7000	545	0	315	860	1	0.10	545	0	615	1160	2	0.11	545	0	865	1410	3	0.12
7001–10,000	550	0	330	880	1	0.10	550	0	620	1170	2	0.12	550	0	1130	1680	4	0.14
10,001–20,000	555	0	345	900	1	0.10	555	0	925	1480	3	0.13	555	0	1185	1740	4	0.15
20,001–50,000	560	0	650	1210	2	0.12	560	0	940	1500	3	0.14	900	1	1400	2300	6	0.16
50,001–100,000	560	0	650	1210	2	0.12	560	0	1210	1770	4	0.15	905	1	1655	2560	7	0.17

Lot Size	Process Average 0.101 to 0.150%						Process Average 0.151 to 0.200%						Process Average 0.201 to 0.250%					
	Trial 1		Trial 2			AOQL in %	Trial 1		Trial 2			AOQL in %	Trial 1		Trial 2			AOQL in %
	n_1	c_1	n_2	n_1+n_2	c_2		n_1	c_1	n_2	n_1+n_2	c_2		n_1	c_1	n_2	n_1+n_2	c_2	
1–180	All	0	–	–	–	0	All	0	–	–	–	0	All	0	–	–	–	0
181–210	180	0	–	–	–	0.02	180	0	–	–	–	0.02	180	0	–	–	–	0.02
211–250	210	0	–	–	–	0.03	210	0	–	–	–	0.03	210	0	–	–	–	0.03
251–300	240	0	–	–	–	0.03	240	0	–	–	–	0.03	240	0	–	–	–	0.03
301–400	275	0	–	–	–	0.04	275	0	–	–	–	0.04	275	0	–	–	–	0.04
401–450	290	0	–	–	–	0.04	290	0	–	–	–	0.04	290	0	–	–	–	0.04
451–500	340	0	110	450	1	0.04	340	0	110	450	1	0.04	340	0	110	450	1	0.04
501–550	350	0	130	480	1	0.05	350	0	130	480	1	0.05	350	0	130	480	1	0.05
551–600	360	0	150	510	1	0.05	360	0	150	510	1	0.05	360	0	150	510	1	0.05
601–800	400	0	185	585	1	0.06	400	0	185	585	1	0.06	400	0	185	585	1	0.06
801–1000	430	0	200	630	1	0.07	430	0	200	630	1	0.07	430	0	200	630	1	0.07
1001–2000	490	0	500	990	2	0.09	490	0	500	990	2	0.09	490	0	500	990	2	0.09
2001–3000	520	0	530	1050	2	0.10	520	0	760	1280	3	0.11	520	0	980	1500	4	0.11
3001–4000	530	0	810	1340	3	0.11	530	0	1030	1560	4	0.12	840	1	1160	2000	6	0.13
4001–5000	540	0	1060	1600	4	0.13	845	1	1205	2050	6	0.14	845	1	1425	2270	7	0.14
5001–7000	545	0	1105	1650	4	0.13	860	1	1490	2350	7	0.15	860	1	1700	2560	8	0.16
7001–10,000	880	1	1300	2180	6	0.15	880	1	1770	2650	8	0.16	1170	2	2160	3330	11	0.17
10,001–20,000	900	1	1840	2740	8	0.18	1200	2	2250	3450	11	0.19	1740	4	2620	4360	15	0.21
20,001–50,000	1210	2	2330	3540	11	0.20	1500	3	2980	4480	15	0.22	2300	6	4240	6540	24	0.24
50,001–100,000	1210	2	2590	3800	12	0.21	1770	4	3690	5460	19	0.23	2560	7	5420	7980	30	0.26

Trial 1: n_1 = first sample size; c_1 = acceptance number for first sample
"All" indicates that each piece in the lot is to be inspected
Trial 2: n_2 = second sample size; c_2 = acceptance number for first and second samples combined
AOQL = Average Outgoing Quality Limit

SINGLE
SAMPLING

0.1%

AOQL

Single Sampling Table for Average Outgoing Quality Limit (AOQL) = 0.1%

Lot Size	Process Average 0 to 0.002%			Process Average 0.003 to 0.020%			Process Average 0.021 to 0.040%			Process Average 0.041 to 0.060%			Process Average 0.061 to 0.080%			Process Average 0.081 to 0.100%		
	n	c	p_t %	n	c	p_t %	n	c	p_t %	n	c	p_t %	n	c	p_t %	n	c	p_t %
1–75	All	0	–	All	0	–	All	0	–	All	0	–	All	0	–	All	0	–
76–95	75	0	1.5	75	0	1.5	75	0	1.5	75	0	1.5	75	0	1.5	75	0	1.5
96–130	95	0	1.4	95	0	1.4	95	0	1.4	95	0	1.4	95	0	1.4	95	0	1.4
131–200	130	0	1.2	130	0	1.2	130	0	1.2	130	0	1.2	130	0	1.2	130	0	1.2
201–300	165	0	1.1	165	0	1.1	165	0	1.1	165	0	1.1	165	0	1.1	165	0	1.1
301–400	190	0	0.96	190	0	0.96	190	0	0.96	190	0	0.96	190	0	0.96	190	0	0.96
401–500	210	0	0.91	210	0	0.91	210	0	0.91	210	0	0.91	210	0	0.91	210	0	0.91
501–600	230	0	0.86	230	0	0.86	230	0	0.86	230	0	0.86	230	0	0.86	230	0	0.86
601–800	250	0	0.81	250	0	0.81	250	0	0.81	250	0	0.81	250	0	0.81	250	0	0.81
801–1000	270	0	0.76	270	0	0.76	270	0	0.76	270	0	0.76	270	0	0.76	270	0	0.76
1001–2000	310	0	0.71	310	0	0.71	310	0	0.71	310	0	0.71	310	0	0.71	310	0	0.71
2001–3000	330	0	0.67	330	0	0.67	330	0	0.67	330	0	0.67	330	0	0.67	655	1	0.64
3001–4000	340	0	0.64	340	0	0.64	340	0	0.64	695	1	0.59	695	1	0.59	695	1	0.59
4001–5000	345	0	0.62	345	0	0.62	345	0	0.62	720	1	0.54	720	1	0.54	720	1	0.54
5001–7000	350	0	0.61	350	0	0.61	750	1	0.51	750	1	0.51	750	1	0.51	750	1	0.51
7001–10,000	355	0	0.60	355	0	0.60	775	1	0.49	775	1	0.49	775	1	0.49	1210	2	0.44
10,001–20,000	360	0	0.59	810	1	0.48	810	1	0.48	1280	2	0.42	1280	2	0.42	1770	3	0.38
20,001–50,000	365	0	0.58	830	1	0.47	1330	2	0.41	1870	3	0.37	2420	4	0.34	2980	5	0.33
50,001–100,000	365	0	0.58	835	1	0.46	1350	2	0.40	2480	4	0.33	3070	5	0.32	4270	7	0.30

0.25%

Single Sampling Table for Average Outgoing Quality Limit (AQQL) = 0.25%

Lot Size	Process Average 0 to 0.005%			Process Average 0.006 to 0.050%			Process Average 0.051 to 0.100%			Process Average 0.101 to 0.150%			Process Average 0.151 to 0.200%			Process Average 0.201 to 0.250%		
	n	c	p_t %	n	c	p_t %	n	c	p_t %	n	c	p_t %	n	c	p_t %	n	c	p_t %
1–60	All	0	–	All	0	–	All	0	–	All	0	–	All	0	–	All	0	–
61–100	60	0	2.5	60	0	2.5	60	0	2.5	60	0	2.5	60	0	2.5	60	0	2.5
101–200	85	0	2.1	85	0	2.1	85	0	2.1	85	0	2.1	85	0	2.1	85	0	2.1
201–300	100	0	1.9	100	0	1.9	100	0	1.9	100	0	1.9	100	0	1.9	100	0	1.9
301–400	110	0	1.8	110	0	1.8	110	0	1.8	110	0	1.8	110	0	1.8	110	0	1.8
401–500	115	0	1.8	115	0	1.8	115	0	1.8	115	0	1.8	115	0	1.8	115	0	1.8
501–600	120	0	1.7	120	0	1.7	120	0	1.7	120	0	1.7	120	0	1.7	120	0	1.7
601–800	125	0	1.7	125	0	1.7	125	0	1.7	125	0	1.7	125	0	1.7	125	0	1.7
801–1000	130	0	1.7	130	0	1.7	130	0	1.7	130	0	1.7	130	0	1.7	250	1	1.4
1001–2000	135	0	1.6	135	0	1.6	135	0	1.6	290	1	1.3	290	1	1.3	290	1	1.3
2001–3000	140	0	1.6	140	0	1.6	300	1	1.3	300	1	1.3	300	1	1.3	300	1	1.3
3001–4000	140	0	1.6	140	0	1.6	310	1	1.3	310	1	1.3	310	1	1.3	485	2	1.1
4001–5000	145	0	1.6	145	0	1.6	315	1	1.2	315	1	1.2	495	2	1.1	495	2	1.1
5001–7000	145	0	1.6	320	1	1.2	320	1	1.2	510	2	1.0	510	2	1.0	700	3	0.94
7001–10,000	145	0	1.6	325	1	1.2	325	1	1.2	520	2	1.0	720	3	0.91	720	3	0.91
10,001–20,000	145	0	1.6	330	1	1.2	535	2	1.0	750	3	0.89	970	4	0.81	1190	5	0.75
20,001–50,000	145	0	1.6	335	1	1.2	545	2	1.0	995	4	0.80	1240	5	0.74	1980	8	0.66
50,001–100,000	335	1	1.2	545	2	1.0	775	3	0.87	1250	5	0.73	1750	7	0.67	2810	11	0.62

n = sample size; c = acceptance number
"All" indicates that each piece in the lot is to be inspected
p_t = lot tolerance per cent defective with a Consumer's Risk (P_C) of 0.10

SINGLE SAMPLING

2.0%

AOQL

Single Sampling Table for
Average Outgoing Quality Limit (AOQL) = 2.0%

Lot Size	Process Average 0 to 0.04%			Process Average 0.05 to 0.40%			Process Average 0.41 to 0.80%			Process Average 0.81 to 1.20%			Process Average 1.21 to 1.60%			Process Average 1.61 to 2.00%		
	n	c	p_t %	n	c	p_t %	n	c	p_t %	n	c	p_t %	n	c	p_t %	n	c	p_t %
1–15	All	0	–	All	0	–	All	0	–	All	0	–	All	0	–	All	0	–
16–50	14	0	13.6	14	0	13.6	14	0	13.6	14	0	13.6	14	0	13.6	14	0	13.6
51–100	16	0	12.4	16	0	12.4	16	0	12.4	16	0	12.4	16	0	12.4	16	0	12.4
101–200	17	0	12.2	17	0	12.2	17	0	12.2	17	0	12.2	35	1	10.5	35	1	10.5
201–300	17	0	12.3	17	0	12.3	17	0	12.3	37	1	10.2	37	1	10.2	37	1	10.2
301–400	18	0	11.8	18	0	11.8	38	1	10.0	38	1	10.0	38	1	10.0	60	2	8.5
401–500	18	0	11.9	18	0	11.9	39	1	9.8	39	1	9.8	60	2	8.6	60	2	8.6
501–600	18	0	11.9	18	0	11.9	39	1	9.8	39	1	9.8	60	2	8.6	60	2	8.6
601–800	18	0	11.9	40	1	9.6	40	1	9.6	65	2	8.0	65	2	8.0	85	3	7.5
801–1000	18	0	12.0	40	1	9.6	40	1	9.6	65	2	8.1	65	2	8.1	90	3	7.4
1001–2000	18	0	12.0	41	1	9.4	65	2	8.2	65	2	8.2	95	3	7.0	120	4	6.5
2001–3000	18	0	12.0	41	1	9.4	65	2	8.2	95	3	7.0	120	4	6.5	180	6	5.8
3001–4000	18	0	12.0	42	1	9.3	65	2	8.2	95	3	7.0	155	5	6.0	210	7	5.5
4001–5000	18	0	12.0	42	1	9.3	70	2	7.5	125	4	6.4	155	5	6.0	245	8	5.3
5001–7000	18	0	12.0	42	1	9.3	95	3	7.0	125	4	6.4	185	6	5.6	280	9	5.1
7001–10,000	42	1	9.3	70	2	7.5	95	3	7.0	155	5	6.0	220	7	5.4	350	11	4.8
10,001–20,000	42	1	9.3	70	2	7.6	95	3	7.0	190	6	5.6	290	9	4.9	460	14	4.4
20,001–50,000	42	1	9.3	70	2	7.6	125	4	6.4	220	7	5.4	395	12	4.5	720	21	3.9
50,001–100,000	42	1	9.3	95	3	7.0	160	5	5.9	290	9	4.9	505	15	4.2	955	27	3.7

Single Sampling Table for
Average Outgoing Quality Limit (AOQL) = 2.5%

2.5%

Lot Size	Process Average 0 to 0.05%			Process Average 0.06 to 0.50%			Process Average 0.51 to 1.00%			Process Average 1.01 to 1.50%			Process Average 1.51 to 2.00%			Process Average 2.01 to 2.50%		
	n	c	p_t %	n	c	p_t %	n	c	p_t %	n	c	p_t %	n	c	p_t %	n	c	p_t %
1–10	All	0	–	All	0	–	All	0	–	All	0	–	All	0	–	All	0	–
11–50	11	0	17.6	11	0	17.6	11	0	17.6	11	0	17.6	11	0	17.6	11	0	17.6
51–100	13	0	15.3	13	0	15.3	13	0	15.3	13	0	15.3	13	0	15.3	13	0	15.3
101–200	14	0	14.7	14	0	14.7	14	0	14.7	29	1	12.9	29	1	12.9	29	1	12.9
201–300	14	0	14.9	14	0	14.9	30	1	12.7	30	1	12.7	30	1	12.7	30	1	12.7
301–400	14	0	15.0	14	0	15.0	31	1	12.3	31	1	12.3	31	1	12.3	48	2	10.7
401–500	14	0	15.0	14	0	15.0	32	1	12.0	32	1	12.0	49	2	10.6	49	2	10.6
501–600	14	0	15.1	32	1	12.0	32	1	12.0	50	2	10.4	50	2	10.4	70	3	9.3
601–800	14	0	15.1	32	1	12.0	32	1	12.0	50	2	10.5	50	2	10.5	70	3	9.4
801–1000	15	0	14.2	33	1	11.7	33	1	11.7	50	2	10.6	70	3	9.4	90	4	8.5
1001–2000	15	0	14.2	33	1	11.7	55	2	9.3	75	3	8.8	95	4	8.0	120	5	7.6
2001–3000	15	0	14.2	33	1	11.8	55	2	9.4	75	3	8.8	120	5	7.6	145	6	7.2
3001–4000	15	0	14.3	33	1	11.8	55	2	9.5	100	4	7.9	125	5	7.4	195	8	6.6
4001–5000	15	0	14.3	33	1	11.8	75	3	8.9	100	4	7.9	150	6	7.0	225	9	6.3
5001–7000	33	1	11.8	55	2	9.7	75	3	8.9	125	5	7.4	175	7	6.7	250	10	6.1
7001–10,000	34	1	11.4	55	2	9.7	75	3	8.9	125	5	7.4	200	8	6.4	310	12	5.8
10,001–20,000	34	1	11.4	55	2	9.7	100	4	8.0	150	6	7.0	260	10	6.0	425	16	5.3
20,001–50,000	34	1	11.4	55	2	9.7	100	4	8.0	180	7	6.7	345	13	5.5	640	23	4.8
50,001–100,000	34	1	11.4	80	3	8.4	125	5	7.4	235	9	6.1	435	16	5.2	800	28	4.5

n = sample size; c = acceptance number
"All" indicates that each piece in the lot is to be inspected
p_t = lot tolerance per cent defective with a Consumer's Risk (P_C) of 0.10

TABLE A-38 DODGE–ROMIG DOUBLE SAMPLING TABLES FOR AVERAGE OUTGOING QUALITY LIMIT (AOQL)

DOUBLE SAMPLING

1.0%
AOQL

Double Sampling Table for
Average Outgoing Quality Limit (AOQL) = 1.0%

Lot Size	Process Average 0 to 0.02% Trial 1 n_1	c_1	Trial 2 n_2	n_1+n_2	c_2	p_t %	Process Average 0.03 to 0.20% Trial 1 n_1	c_1	Trial 2 n_2	n_1+n_2	c_2	p_t %	Process Average 0.21 to 0.40% Trial 1 n_1	c_1	Trial 2 n_2	n_1+n_2	c_2	p_t %
1–25	All	0	–	–	–	–	All	0	–	–	–	–	All	0	–	–	–	–
26–50	22	0	–	–	–	7.7	22	0	–	–	–	7.7	22	0	–	–	–	7.7
51–100	33	0	17	50	1	6.9	33	0	17	50	1	6.9	33	0	17	50	1	6.9
101–200	43	0	22	65	1	5.8	43	0	22	65	1	5.8	43	0	22	65	1	5.8
201–300	47	0	28	75	1	5.5	47	0	28	75	1	5.5	47	0	28	75	1	5.5
301–400	49	0	31	80	1	5.4	49	0	31	80	1	5.4	55	0	60	115	2	4.8
401–500	50	0	30	80	1	5.4	50	0	30	80	1	5.4	55	0	65	120	2	4.7
501–600	50	0	30	80	1	5.4	50	0	30	80	1	5.4	60	0	65	125	2	4.6
601–800	50	0	35	85	1	5.3	60	0	70	130	2	4.5	60	0	70	130	2	4.5
801–1000	55	0	30	85	1	5.2	60	0	75	135	2	4.4	60	0	75	135	2	4.4
1001–2000	55	0	35	90	1	5.1	65	0	75	140	2	4.3	75	0	120	195	3	3.8
2001–3000	65	0	80	145	2	4.2	65	0	80	145	2	4.2	75	0	125	200	3	3.7
3001–4000	70	0	80	150	2	4.1	70	0	80	150	2	4.1	80	0	175	255	4	3.5
4001–5000	70	0	80	150	2	4.1	70	0	80	150	2	4.1	80	0	180	260	4	3.4
5001–7000	70	0	80	150	2	4.1	75	0	125	200	3	3.7	80	0	180	260	4	3.4
7001–10,000	70	0	80	150	2	4.1	80	0	125	205	3	3.6	85	0	180	265	4	3.3
10,001–20,000	70	0	80	150	2	4.1	80	0	130	210	3	3.6	90	0	230	320	5	3.2
20,001–50,000	75	0	80	155	2	4.0	80	0	135	215	3	3.6	95	0	300	395	6	2.9
50,001–100,000	75	0	80	155	2	4.0	85	0	180	265	4	3.3	170	1	380	550	8	2.6

Lot Size	Process Average 0.41 to 0.60% Trial 1 n_1	c_1	Trial 2 n_2	n_1+n_2	c_2	p_t %	Process Average 0.61 to 0.80% Trial 1 n_1	c_1	Trial 2 n_2	n_1+n_2	c_2	p_t %	Process Average 0.81 to 1.00% Trial 1 n_1	c_1	Trial 2 n_2	n_1+n_2	c_2	p_t %
1–25	All	0	–	–	–	–	All	0	–	–	–	–	All	0	–	–	–	–
26–50	22	0	–	–	–	7.7	22	0	–	–	–	7.7	22	0	–	–	–	7.7
51–100	33	0	17	50	1	6.9	33	0	17	50	1	6.9	33	0	17	50	1	6.9
101–200	43	0	22	65	1	5.8	43	0	22	65	1	5.8	47	0	43	90	2	5.4
201–300	55	0	50	105	2	4.9	55	0	50	105	2	4.9	55	0	50	105	2	4.9
301–400	55	0	60	115	2	4.8	55	0	60	115	2	4.8	60	0	80	140	3	4.5
401–500	55	0	65	120	2	4.7	60	0	95	155	3	4.3	60	0	95	155	3	4.3
501–600	60	0	65	125	2	4.6	65	0	100	165	3	4.2	65	0	100	165	3	4.2
601–800	65	0	105	170	3	4.1	65	0	105	170	3	4.1	70	0	140	210	4	3.9
801–1000	65	0	110	175	3	4.0	70	0	150	220	4	3.8	125	1	180	305	6	3.5
1001–2000	80	0	165	245	4	3.7	135	1	200	335	6	3.3	140	1	245	385	7	3.2
2001–3000	80	0	170	250	4	3.6	150	1	265	415	7	3.0	215	2	355	570	10	2.8
3001–4000	85	0	220	305	5	3.3	160	1	330	490	8	2.8	225	2	455	680	12	2.7
4001–5000	145	1	225	370	6	3.1	225	2	375	600	10	2.7	240	2	595	835	14	2.5
5001–7000	155	1	285	440	7	2.9	235	2	440	675	11	2.6	310	3	665	975	16	2.4
7001–10,000	165	1	355	520	8	2.7	250	2	585	835	13	2.4	385	4	735	1170	19	2.3
10,001–20,000	175	1	415	590	9	2.6	325	3	655	980	15	2.3	520	6	980	1500	24	2.2
20,001–50,000	250	2	490	740	11	2.4	340	3	910	1250	19	2.2	610	7	1410	2020	32	2.1
50,001–100,000	275	2	700	975	14	2.2	420	4	1050	1470	22	2.1	770	9	1850	2620	41	2.0

Trial 1: n_1 = first sample size; c_1 = acceptance number for first sample
"All" indicates that each piece in the lot is to be inspected
Trial 2: n_2 = second sample size, c_2 = acceptance number for first and second samples combined
p_t = lot tolerance per cent defective with a Consumer's Risk (P_C) of 0.10

Index

Index of Formulas